Insulin & Related Proteins –
Structure to Function and Pharmacology

Insulin & Related Proteins – Structure to Function and Pharmacology

Edited by

Markus Leyck Dieken
Novo Nordisk Pharma GmbH,
Mainz, Germany

Matthias Federwisch
Institut für Biochemie,
RWTH Aachen, Germany

and

Pierre De Meyts
Hagedorn Research Laboratory,
Gentofte, Denmark

SPRINGER-SCIENCE+BUSINESS MEDIA, B.V.

A C.I.P. Catalogue record for this book is available from the Library of Congress.

ISBN 978-94-017-3956-6 ISBN 978-0-306-47582-5 (eBook)
DOI 10.1007/978-0-306-47582-5

Printed on acid-free paper

CONTENTS

FOREWORD

This book contains contributions presented at the last of the Alcuin Symposia, held in April 2000, as a Festschrift to honour Prof. Axel Wollmer on the occasion of his retirement from the Rheinisch-Westfälische Technische Hochschule (RWTH) Aachen. The Alcuin Symposia were initially held alternatively in York and Aachen, in recent years in Aachen only, as joint workshops on insulin between the groups of Prof. Guy Dodson in York and the groups of Profs Dietrich Brandenburg and Axel Wollmer in Aachen. The Symposium was named after Alcuin, an Anglo-Latin poet, educator and cleric from York, who was invited to join the court of Charlemagne at Aachen. Alcuin's first 50 years were spent in Yorkshire, were he was first a pupil, later headmaster of the cathedral school of York, the most renowned of its day. Charlemagne was gathering at Aachen the leading Irish, English and Italian scholars of the age. Alcuin was appointed head of the Palatine school, where Charlemagne himself, his family, his friends, and his friends's sons were taught. Alcuin introduced the traditions of Anglo-Saxon humanism into Western Europe and was the foremost scholar of the revival of learning known as the Carolingian Renaissance. He also promoted the use of the beautiful Carolingian minuscule script, the ancestor of modern Roman typefaces.

The final Alcuin Symposium was not limited to the topic of insulin, as it gathered friends and colleagues from Universities other than York and Aachen and from industrial research laboratories with whom Axel Wollmer has collaborated and enjoyed long-standing exchanges, some of whom have since moved from insulin research to new pastures.

Thanks are due to everybody who supported the symposium and contributed to its programme or to this book or both. The most valuable help of Renate Kehren and Renate Urlichs with all Alcuin Symposia held in Aachen is gratefully appreciated.

Prof. Helmut Zahn introduced insulin as a second research topic when he moved to the German Wool Research Institute (Deutsches Wollforschungsinstitut or DWI) in Aachen in 1958. His work culminated in 1963 with the total synthesis of insulin (at the same time as Prof. Panayotis Katsoyannis in New York, another co-author in this book, and shortly afterwards in China). Prof. Zahn relates with humour his (unexpectedly long) transition from Heidelberg to Aachen in the first chapter of this book. The total synthesis work not only was a landmark in peptide chemistry, it created the chemical basis for the design of insulin analogues, which turned out to be an invaluable tool for the study of the structure-function relationships of insulin, especially after direct studies of insulin receptor binding became possible in the early seventies. Later taken over by genetic engineering and site-directed mutagenesis, this work ushered analogues into the area of designer insulins which in recent years have markedly improved the prospects for tighter glycemic control in diabetes therapy.

The work of the Aachen groups was from the beginning marked by a strong emphasis on uniting peptide chemistry, solution structure, crystallography and biology in order to understand the structure-function relationships of insulin. This led to the early and longstanding collaboration between the DWI group (Zahn, Brandenburg and colleagues) and RWTH group of Axel Wollmer who used biophysical and theoretical approaches, especially circular dichroism. Four chapters in this book are contributed by Axel Wollmer and his team, dealing with the self-association of insulin and what is probably Axel's major scientific contribution, the T-R transition in the insulin molecule. The theoretical aspects of the T-R transition are further discussed by Peter Krüger.

The solving of the crystal structure of insulin by Dorothy Hodgkin's group in 1969 marked another milestone in insulin chemistry, and led to further longstanding collaborations. These are recounted in detail in this book in the chapters by Dietrich Brandenburg, who provides an encompassing account of the scientific contributions from Aachen over more than 40 years as well as all their international collaborations, and on the 3-D structure by Guy Dodson and Jean Whittingham. Other former Hodgkin alumni such as Dan Mercola and Tom L. Blundell and colleagues have provided contributions on some of their newer ventures. Michael Weiss, Panayotis Katsoyannis and colleagues provide more discussion on structure-activity of insulin analogues. Molecular dynamics (on myoglobin, not insulin) is discussed by Robert Woody and colleagues. Another longstanding collaborator of the Aachen groups, Richard Jones, discusses with Fariba Shojaee-Moradie the biology of potentially clinically useful insulin analogues. More biology, now more focused on mechanisms of insulin resistance, is found in the chapter of another former DWI collaborator, Hans-Georg Joost, and colleagues.

As mentioned above, the work on insulin receptors has from the start relied heavily on the collaboration with the Aachen groups, first in the early seventies with Pierre Freychet and Jesse Roth at the NIH (later continued by Pierre De Meyts and colleagues) and Steen Gammeltoft and Jørgen Gliemann in Copenhagen. Brandenburg's group with Marlies Fabry have made their own primary contributions to the field with photoaffinity labelling. Recent advances in the structure-function of the insulin and insulin-like growth factor receptors are discussed in chapters from the two Novo Nordisk groups (De Meyts and Whittaker, and Kristensen and Andersen), as well as the first crystallographic data from Colin Ward's group in Australia. The chapters by L. Pellegrini et al. and Joachim Grötzinger (another Aachen alumnus) provide glimpses into other cell surface receptors.

It was only suitable that this small sampling of contributions from the vast network of friends and collaborators of Axel Wollmer and colleagues be put together on the occasion of his retirement. Axel, always a scientist and a gentleman, has put his stamp on nearly four decades of protein structure research, and he has made invaluable contributions to our understanding of structural transitions in insulin. Now that the academic rules have forced him to undergo his own T-R transition, all his friends hope to be able to continue to interact with him and pick on his wisdom for a long time.

Pierre De Meyts
Matthias Federwisch
Markus Leyck Dieken

CONTRIBUTORS

Prof. Dr. Dietrich Brandenburg
Sudetenstraße 63
64385 Reichelsheim
Germany
e-mail: Dietrich.Brandenburg@t-online.de
Tel. 06164-54530

Prof. Sir Tom Blundell, FRS
Department of Biochemistry
University of Cambridge
80 Tennis Court Road
Cambridge CB2 1GA
U.K.
e-mail: tom@cryst.bioc.cam.ac.uk

Pierre De Meyts, MD, PhD, F.A.C.E.
Hagedorn Research Laboratory
Niels Steensens-Vej 6
DK-2820 Gentofte
Denmark
e-mail: pdm@novonordisk.com

Prof. Dr. Guy G. Dodson, FRS
Protein Structure Research Group
Department of Chemistry
University of York
Heslington
York Y01 5DD
U.K.
e-mail: myers@ysbl.york.ac.uk

Dr. Matthias Federwisch
Institut für Biochemie
RWTH Aachen
Pauwelsstrasse 30
52057 Aachen
Germany

Priv.-Doz. Dr. Joachim Grötzinger
Institut für Biochemie
Christian-Albrechts-Universität Kiel
Olshausenstr. 40
24098 Kiel
Germany
e-mail: jgroetzinger@biochem.uni-kiel.de

Dr. Richard Jones
Department of Diabetes and Endocrinology
St. Thomas' Hospital
London SE1 7EH
U.K.
e-mail: richard.h.jones@kcl.ac.uk

Prof. Dr. Dr. Hans Georg Joost
Institut für Phamakologie und Toxikologie
RWTH Aachen
Wendlingweg 2
52057 Aachen
Germany
e-mail: joost@rwth-aachen.de

Dr. Claus Kristensen
Department of Insulin Research
Health Care Discovery
Novo Nordisk A/S
2880 Bagsvaerd
Denmark
e-mail: CLAK@novonordisk.com

Priv.-Doz. Dr. Peter Krüger
Laboratory of Biomolecular Dynamics
Department of Biochemistry
Katholieke Universiteit Leuven
Celestinenlaan 200D
B-3001 Leuven
Belgium
e-mail: Krueger@syncity.de

Prof. Daniel A. Mercola PhD, MD
Sidney Kimmel Cancer Center
10835 Altman Row
San Diego, California 92121
USA
e-mail: danmercola@skcc.org

Fariba Shojaee-Moradie
Department of Diabetes, Endocrinology and Internal Medicine
King's College London
St Thomas' Hospital Campus
Lambeth Palace Road
London SE1 7EH
UK
e-mail: fariba.shojaee-moradie@kcl.ac.uk

Dr. Thomas Mülders
Institut für Biochemie
RWTH Aachen
Pauwelsstrasse 30
52057 Aachen
Germany

Prof. Dr. Colin Ward
Biomolecular Research Institute
343 Royal Parade
Parkville
Victoria, 3052
Australia
e-mail: Colin.Ward@hsn.csiro.au

Prof. Michael A. Weiss MD, PhD
Department of Biochemistry
Case Western Reserve University
School of Medicine
Cleveland, Ohio 44016
USA
e-mail: weiss@biocserver.BIOC.CWRU.Edu

Dr. Jonathan Whittaker
Hagedorn Research Laboratory
Niels Steensens-Vej 6
DK-2820 Gentofte
Denmark
e-mail: JOW@novonordisk.com

Dr. Jean Whittingham
Protein Structure Research Group
Department of Chemistry
University of York
Heslington
York Y01 5DD
U.K.
e-mail: jean@yorvic.york.ac.uk

Prof. Dr. Axel Wollmer
Institut für Biochemie
RWTH Aachen
Pauwelsstrasse 30
52057 Aachen
Germany
e-mail: ABWollmer@aol.com

Prof. Dr. Robert W. Woody
Department of Biochemistry
Colorado State University
Fort Collins, Colorado 80523
USA
e-mail: rww@lamar.colostate.edu

Prof. Dr.-Ing. Drs. h.c. mult. Helmut Zahn
Deutsches Wollforschungsinstitut
Veltmanplatz 8
52062 Aachen
Germany
e-mail: zahn@dwi.rwth-aachen.de

H. ZAHN

HOW INSULIN CAME FROM HEIDELBERG TO AACHEN AND WHY IT TOOK FIVE YEARS.

Abstract. The author gives an autobiographic sketch on events that led to the initiation of wool research and insulin chemistry at the University of Technology in Aachen.

1. INTRODUCTION

First of all I want to express my sincere thanks to the organizers of the Alcuin Symposium on Insulin and Related Proteins for having invited me to take part with a short historical contribution on my move from Heidelberg to Aachen in the spring of 1957.

This is worth noting only because I was appointed director of the German Wool Research Institute Aachen already in 1952 – five years before I and my coworkers moved to Aachen.

So I was "wearing two hats" for 5 years: scientific assistant of Professor Karl Freudenberg in the Chemical Institute of the University of Heidelberg and Director of the German Wool Research Institute at the University of Technology in Aachen.

I will try to give evidence for the statement that my Heidelberg "interregnum" paid off in the end for the insulin synthesis in Aachen and the story of how protein science came into being in Aachen especially the extremely successful cooperation of Axel Wollmer and his coworkers with the Wool Research Institute.

My presentation is divided into three parts:
Insulin chemistry in Heidelberg
Insulin synthesis in Aachen
The benefits of the postponed relocation from Heidelberg to Aachen.

2. INSULIN CHEMISTRY IN HEIDELBERG

In the years 1949 to 1957 I had the privilege of working in Karl Freudenberg's Chemical Institute in Heidelberg as scientific assistant and lecturer. After the habilitation (1948) the first students reported to me for scientific work as diploma or PhD students. The subjects of the 25 coworkers who received their doctoral degree in chemistry from Heidelberg University were chemical crosslinking of wool, silk, collagen, and insulin as well as synthesis of nylon oligomers and silk and insulin peptides. Recently I have described my journey from wool research in Egon Elöd's laboratory in the years 1939 to 1949 to analytic and synthetic chemistry in Heidelberg (Zahn, 2000).

M. Federwisch, M. Leyck Dieken and P. De Meyts (eds.), Insulin & Related Proteins, 1–5.

Heidelberg has seen an older and larger period of insulin chemistry already in the years 1926 to 1940, when Karl Freudenberg initiated a long series of studies on the chemistry of insulin.

I quote Sanger (1960): "Before much was known about the structure of insulin many experiments were done, especially by Freudenberg and his collaborators (Freudenberg and Wegman, 1935), in which insulin was treated with various reagents that were specific for one type of group and the changes in biological activity were observed. It has long been realized that the disulphide bridges are important and must remain intact for activity. When they are broken by reduction (Stern and White, 1960) or oxidation (Freudenberg and Wegman, 1935) there is immediate inactivation".

Freudenberg played an important part in the recognition of the purity and biological activity of Abel's crystalline insulin (Murnaghan and Talalay, 1967).

3. INSULIN SYNTHESIS IN AACHEN

The progress of peptide synthesis, the discovery of new methods of protection, new procedures for the activation of the carboxyl group and novel methods of coupling encouraged several teams after the elucidation of insulin's structure by Sanger (Ryle et al., 1955) to synthesize corresponding peptide sequences.

Under the first approaches has been the synthesis of some cystine peptides related to sequences in insulin by Maclaren, Savige and Swan (1956).

Klostermeyer and Humbel (1966) have outlined strategy and tactics of the first synthesis of sheep insulin in Aachen and Pittsburgh: "The main obstacle to a total synthesis of the hormone seemed to be the combination of the chains. As soon as this problem proved to be solvable - however small the yield has been - the road to a synthesis of insulin was traced out: It should be possible to synthesize separately the A and B chains following du Vigneaud's work (1956) by using S-benzylated cysteine residues and permanent protection of functional side chain groups, then to remove all protecting groups with sodium in liquid ammonia, and finally to combine the SH- chains thus obtained by oxidation."

This however would not be a structure confirming total synthesis in the classical organic definition - for that in addition the stepwise assembly of the three disulfide bridges should occur by clearly arranged reactions - but it would be the easiest route to synthetic insulin.

The strategy was the right way. At the end of 1963 Zahn and coworkers (1963) and about the same time also Katsoyannis et al. (1963, 1964) in cooperation with Dixon (1964) succeeded in preparation of sheep insulin.

4. THE BENEFITS OF THE POSTPONED RELOCATION FROM HEIDELBERG TO AACHEN.

Today, many years after moving from Heidelberg to Aachen the question is whether it has been a benefit for the insulin synthesis in Aachen that I stayed in Heidelberg for 5 full years although I was director of the Wool Research Institute in Aachen already from April 1, 1952.

Four prerequisites are to be met for successful experimental work in chemistry: enthusiastic coworkers, sufficient laboratory space, inspiring atmosphere, and uncomplicated funding.

As far as the setup of a synthesis team is concerned, my lecture courses in Heidelberg on Polymer Chemistry proved to be attractive for graduate students.

By contrast I had no chair at the University of Technology in Aachen. I was offered a part-time lectureship only. Complicated negotiations between the board of trustees of the Wool Research Institute, the University of Technology, the State government and myself were going on five long years.

Finally a good chance to be offered a chair of Textile Chemistry at the Aachen University of Technology appeared in the spring of 1957.

5. LABORATORY SPACE

My coworkers were equipped with sufficient laboratory space in Freudenberg's Institute in Heidelberg. My desk was set up in one of the smaller so-called private laboratories. Generous laboratory facilities become at my team's disposal after the relocation of the Chemical Institute in new buildings in 1954. By contrast laboratories for my coworkers were allocated in Aachen not until spring of 1957.

The third prerequisite for successful research work, an inspiring atmosphere, was present in Heidelberg in the fifties. This has been aptly described by Rolf Appel (1999) with whom I shared several years in Freudenberg's Institute:

"The Heidelberg University had a glorious time between 1950 and 1960: Personalities such as Jaspers in the Philosophical Faculty, Bauer in the Medical Faculty, the Nobel laureates Bothe and Jensen in Physics, the chemist Kuhn at the Max Planck Institute and the mineralogist Ramdohr had worldwide fame and attracted scientists and students from all over the world to the Neckar river."

In the Chemistry Department it was Freudenberg who held together the special fields under the same roof, each working group was supported and could develop freely.

I should add to this text some remarks on the stimulation from the peptide and protein research work which had a great time in the fifties of the last century. Fortunately some of the pioneers in these fields happened to work at the University of Heidelberg and the Max Planck Institute for Medical Research or at the Mainz University.

A direct influence on our research activities had visitors and lecturers at the meetings of the Heidelberg Chemical Society such as Erwin Brand (New York) and F. Sanger (Cambridge). The former taught us how to apply the azide method for coupling peptides without racemisation, the latter had a great influence on our insulin research subjects.

Funding has been stated as the fourth important prerequisite for success. I was lucky enough to enjoy financial help even during my first years in Heidelberg before my nomination as director of the Wool Research Institute in Aachen opened new funding sources.

Dr. J. R. McPhee (1977), then Director, Planning and Services, International Wool Secretariat, London described the early funding in Heidelberg :

"The association between IWS and the German Wool Research Institute goes back many years; in fact, this association is almost as old as the institute itself, and rather older than the Institute's laboratories in Veltmanplatz. At the time when IWS and the Wool Research Institute first came together, Professor Zahn was working with a few colleagues on wool research projects, in temporary accommodation in the Chemical Institute of the University of Heidelberg. Two of his students in those days were Dr. Wilhelm and Dr. Würz. These two gentlemen were the very first doctoral graduates of the German Wool Research Institute. I am both pleased and proud to be able to say that both of these pioneers were supported from IWS funds."

I should now refer to the statement that my Heidelberg time paid off for the German Wool Research Institute in Aachen. There are, among other things, four arguments:

1. The Institute's research projects went ahead right away in Heidelberg. The enthusiasm of the Institute's founders was kept alive.
2. President and committees provided funds for increasing my Heidelberg staff. 30 staff members agreed to move to Aachen and settled in the new building during spring of 1957.
3. This great moving in, the influx of students and technical staff, the realization of the first workshops for the wool industry attracted attention in university and industry.
4. With consent of the board of trustees I could reach an agreement with Richard Wegler of Farbenfabriken Bayer to accommodate temporarily my former coworkers Eugen Schnabel and Johannes Meienhofer. The two together with the staff members Toru Okuda, Rudolf Zabel and Dietrich Brandenburg and the PhD students Hellmut Bremer, Otto Brinkhoff, Werner Sroka and Henning Klostermeyer completed the synthesis of insulin in December 1963 only 2 ½ years after the assumption of the insulin project by Johannes Meienhofer.

A detailed documentation on the early years of the Wool Research Institute has been compiled with Hans-Joachim Henning (Zahn and Henning, 1977). I quote from the address given by Professor Dr. H. Hammer, at that time deputy chairman of the Faculty of Chemistry and Biology:

"It was to the Faculty's greatest benefit that the Wool Research Institute with its scientific coworkers and far-reaching research program has brought to Aachen new research and teaching fields such as fibre and textile research and protein chemistry and anchored them firmly in lively cooperation with friendly institutes."

This applied especially for Axel Wollmer and his team in Aachen.

I close with my heartfelt wishes for Axel Wollmer's new activities after his retirement.

6. REFERENCES

Appel R. "Ein Lebensweg in und durch die Chemie." Chemie erlebt 50 Jahre GDCh, Frankfurt am Main 1999, 16-32.

Dixon G.H., *Excerpta Medica int. Congr. Ser.* 83 (1964): 1207.

du Vigneaud V. "Les Prix Nobel en 1955." *Kungl. Boktr. P. A. Norstedt and Söner*, (1956) Stockholm.

Freudenberg K. and Wegmann T. *Hoppe-Seylers Z. Physiol. Chem.* 233 (1935): 159-171.

Hammer, H. In: *Festschrift* l.c. (1977): 12.

Katasoyannis P.G., Fukuda K., Tometsko A., Suzuki K. and Tilak M. "Insulin Peptides. X. The Synthesis of the B-Chain of Insulin and Its Combination with Natural or Synthetic A-Chain to Generate Insulin Activity." *J. American Chemical Society* 86 (1964): 930-932.

Katasoyannis P.G., Tometsko A. and Fukuda K. "Insulin Peptides. IX. The Synthesis of the A-Chain of Insulin and ist Combination with Natural B-Chain to Generate Insulin Activity." *J. Amer. Chem. Soc.* 85 (1963): 2863-2865.

Klostermeyer H. and Humbel R.E. "Chemie und Biochemie des Insulins." *Angewandte Chemie* 78 Nr.18/19 (1966):871-886.

Maclaren J.A., Savige W.E. and Swan J.M. "Some Observations on the Synthesis of Cystine Peptides." *Proceedings of the International Wool Textile Research Conference Australia 1955.* Volume C (1956): C164 – C167.

McPhee J. R. In: *Festschrift* l.c. (1977): 15.

Meienhofer J., Schnabel E., Bremer H., Brinkhoff O., Zabel R., Sroka W., Klostermeyer H., Brandenburg D., Okuda T. and Zahn H. "Synthese der Insulinketten und ihre Kombination zu insulinaktiven Präparaten." *Z. Naturforsch.* 18b (1963):1120-1121.

Murnaghan J.H. and Talaly P. "John Jacob Abel and the Crystallization of Insulin." *Perspectives in Biology and Medicine* 10 (1967): 363-364.

Ryle A.P., Sanger F., Smith L.F. and Kitai R. "The Disulphide Bonds of Insulin." *Biochemical Journal* 60 (1955): 541- 556.

Sanger F. "Chemistry of Insulin." *British Medical Bulletin* 16 (1960): 183-188.

Stern K.G. and White A. *J. Biol. Chem.* 117 (1960): 95.

Zahn H. "My journey from wool research to insulin." *Journal of Peptide Science* 6 (2000): 1-10.

Zahn H. and Henning H.-J. "Festschrift zum 25jährigen Jubiläum am 1.April 1977." Deutsches Wollforschungsinstitut an der Technischen Hochschule Aachen.

7. ACKNOWLEDGEMENTS

Thanks are due to Professor Don Steiner, Chicago, for providing me with publications on the involvement of Karl Freudenberg in the corroboration of Abel's work. I thank Professor Hartwig Höcker, Director of the German Wool Research Institute, for encouragement and support. Finally I thank Katja Kettenhofen for text processing.

D. BRANDENBURG

INSULIN AT THE GERMAN WOOL RESEARCH INSTITUTE – RETROSPECT AND OUTLOOK

Abstract. This paper summarizes the research carried out between 1977 and 1999 in the German Wool Research Institute (Deutsches Wollforschungsinstitut, DWI) and in collaboration with other laboratories. Chemical synthesis and chemo-enzymatic semisynthesis yielded a large number of insulin analogues and special insulin probes, which allowed detailed structure-function studies and investigations based on photoaffinity labelling of insulin receptors. The outlook refers to some recent areas which are of potential interest for diabetes treatment and require further research.

1. INTRODUCTION

The successful synthesis of insulin – more precisely: the generation of insulin activity through the combination of two fully synthetic chains – was not the end, but rather the beginning of intensive studies on this fascinating protein hormone in Aachen. While the immediate goals of the years after 1963 were the optimisation of the synthesis of A- and B-chains with native sequence and the development of efficient procedures for chain combination, other areas developed. The discovery of proinsulin and the natural pathway of biosynthesis by Don Steiner was an immediate challenge to synthesise the prohormone. The availability of native bovine or porcine insulin in sufficient amounts offered the possibility to explore modifications of the molecule on a preparative scale. Besides being a source for A- and B-chains, semisynthetic operations were developed as shortcuts to analogues.

Structure-function studies began, which received a solid basis through the elucidation of the three-dimensional structure of 2Zn porcine insulin by Dorothy Hodgkin and her associates in 1969. Once there were first analogues to study, most fruitful collaborations started to bloom. In our early attempts to correlate biological activity to the spatial structure of insulin in solution, only circular dichroism offered information. This led to the most rewarding acquaintance and friendship with Axel Wollmer. Quickly, his contributions became an indispensable part of our work. Since then, insulin research in Aachen rested on two columns – chemistry and biological studies at DWI, and biophysical as well as theoretical studies at the Institute of Biochemistry.

At the Wool Research Institute, insulin, after wool, became the second research field. An annex "für Insulinforschung" was erected, and its inauguration celebrated with a Symposium "Chemie des Insulins und Proinsulins" in 1973. Four years later, DWI celebrated its 25th anniversary. In the Festschrift, Günter Weitzel (1977) has given a thorough review on the insulin work carried out since 1958, which had led to over 160 publications.

M. Federwisch, M. Leyck Dieken and P. De Meyts (eds.), Insulin & Related Proteins, 7–28.

In 1979, the Second International Insulin Symposium summoned all the leading insulin researchers. This Symposium and the Proceedings (Brandenburg and Wollmer, 1980), gave a vivid picture of the state of the art and were a fine platform to present our own and our collaborative studies. By then, our work had reached new dimensions by the beginning of enzymatic semisynthesis, receptor studies via photoaffinity labelling, and in-house bioassays. Thus, there was a continuous development from organic and peptide chemistry towards bio-organic chemistry, biochemistry and cell biology.

Other highlights were the Annual Congress of the German Diabetes Association with an International Satellite Symposium "Insulin and Insulin Action" in 1987, the Symposium "25 Years Insulin Synthesis" in 1989 and, of course, the Alcuin Symposia. Looking back, I should like to summarise the progress since 1977.

2. SYNTHESIS OF INSULIN, ANALOGUES AND PROINSULIN

The extensive work continued, both with respect to insulin and proinsulin, with overlaps. Continuous refinements of strategy and tactics were rewarded by improved yields, larger segments and higher purity. This led, for instance, to crystals of half-synthetic sheep insulin and their X-ray analysis (Cutfield et al., 1979).

2.1. Chain Combination / Sulfur Chemistry

An improved method ("buffer exchange method") for combination of insulin chains, especially for modified chains, had been developed (Gattner et al.,1981).

An equimolar mixture of the chains was reduced at pH 8.6, and then transferred to the oxidation buffer (pH 10.6) by gel filtration. During this step, approx. 50% of A(SH)4 were oxidised to $A(SS)(SH)_2$ which presumably contained the 6-11 ring. As found with 10 modified A- and 4 modified B-chains, combination yields were directly correlated to the biopotency of the analogue generated. Insulin was obtained in 15%, LysA1-insulin (activity 4%) in 1% yield. Recycling (1-2x) of the non-insulin fractions increased the yields markedly to 33-34%. Preparative RP-HPLC was successfully applied for fine purification (Schartmann et al., 1983).

Much work was invested into elucidation of the use of the S-*tert.*-butyl-mercapto group for sulfur protection as, for instance, in the synthesis of various A-chain segments of sheep and chicken insulin. It allowed the preparation of a [Lys13]A-chain (bovine), designed for cross-linking to B-chain (Wolf et al., 1979).

2.2. Insulin Analogues

Several analogues were prepared from a synthetic A-chain and its natural counter chain (see Table 1). All syntheses were carried out by segment condensation in solution, using global protection with acid-labile groups for the side-chains and terminal carboxyl groups, and trityl groups for cysteine. Final couplings (DCC/HOBt) were 1-12 to 13-21 for A-chains, and 1-16 to 17-30 for B-chains. The purified S-sulfonate chains were either co-reduced and oxidised ("buffer exchange method"), or $A(SH)_4$ was combined with $B(SSO_3H)_2$. A fully synthetic insulin/IGF-I A+D-chain was similarly constructed and combined with semisynthetic human B-chain.

Almost 20 years after our first, at that time disappointing experiments with Merrifield's ingenious method, solid phase peptide synthesis proved successful. Application of the "Fmoc procedure" allowed the assembly of two B-chains, which gave $[Ala^{B12}]$- and $[Ile^{B12}]$-insulin (R. Casaretto, 1986). $[Tyr(NO_2)^{A14}]$A-chain (Lenz et al., 1996) and, for the construction of mini-proinsulins, 3 $[X^{A3}]$A-chains and 3 $[X^{B15}]$B-chains (Weber, 1998) could also be successfully synthesised .

An elaborate combination of available techniques was necessary to synthesise an analogue with a non-cleavable bond between A- and B-chains (Videnov et al., 1990). The key intermediate, B1-protected $[Dsu^7]$B(1-8) was assembled on a solid support according to the Fmoc-procedure. It was acylated at the A7 amino group with Boc-(A1-6). The resulting "U-peptide" was then joined to A(8-21), which had been synthesised in solution. The C-terminal B-chain segment, (B9-B25)-NH_2, also obtained via solid phase synthesis, was then coupled to B8 through the azide.

Subsequently, the S-tert.butyl thio groups at cysteines 6 and 11 were cleaved, and finally the Acm groups at cysteines A20 and B19. The synthesis comprised a considerable number of problems, as isolation of the pure dicarba-DPPI-amide. However, the low bioactivity and receptor binding are an indication, that the A7-B7 disulfide bond plays a structural, but not a functional role in the signal-generating process.

2.3. *Proinsulin*

Proinsulin was the greatest challenge for our peptide chemists. After initial work on the porcine prohormone, human proinsulin became the goal in 1971. The plan was to construct the N-terminal "half" from 1-23 + 24–45 and the C-terminal "half" from 46–70 + 71–86. Global protection with acid-labile groups, and trityl protection for all 6 cysteines corresponded to the tactics tested with insulin. Model experiments, in which bovine proinsulin was transformed into the hexa-S-trityl derivative, which in turn was re-converted into proinsulin, confirmed the validity of this approach (Büllesbach et al., 1980).

The segment 71-86 could be obtained by coupling of two readily soluble cystine peptides, but only in low yield (Berndt, 1979). Coupling of S-tritylated segments was superior, position 77-78 being more favourable than 74-75 or 78-79 (Berndt, 1979; Sasaki, 1979). Peptide 46-70 was synthesised either from 46–64 + 65–70, or 46–59 + 60–70. The yields from both routes were the same, but the latter synthesis was more convenient (Naithani and Schwertner, 1983).

The assembly of the sulfur-free peptide 24-45 was, contrary to expectation, very difficult and afforded several approaches. Finally, the joining of Bpoc-(24-39) and (40-45)-OBzl was successful. Segment 1-23 was synthesised in the S-trityl form from three peptides, and final coupling of 1-16 to 17-23 (Danho and Föhles, 1980).

Both protected large segments 1-45 and 46-86 could be obtained by coupling of their N- and C-terminal components, their good solubility allowed efficient purification by counter-current distribution. A small-scale coupling, followed by deblocking, sulfitolysis and gel chromatography, gave impure proinsulin hexa-S-sulfonate. However, the small amount, and lack of appropriate refined purification – RPHPLC was not yet available – precluded conversion into proinsulin (Danho et al., 1980). Then, progress in genetic engineering brought this work to an end. For a detailed review, see Naithani and Zahn (1984).

3. SEMISYNTHESIS

The three major goals of semisynthesis were to obtain analogues for structure-function studies, insulins with reporter groups (photo-activatable, fluorescent, spin-labelled) or to simplify the access to proinsulin and proinsulin models.

3.1. N-Terminal Semisynthesis

Access for systematic alterations at the N-termini had been developed over 10 years, with increasing refinement. New partially protected insulins became essential intermediates for further selective modifications. Two acid-labile groups, Boc and Cit, and two alkali-labile groups, Msc and Tfa, have been studied in detail. A1,B29- and A1,B1-Msc_2-insulins (Schüttler and Brandenburg, 1979) as well as A1-Cit-insulin (Naithani and Gattner, 1982) became the most important intermediates. A series of crystallizable analogues with replacement of Phe^{B1}, inter alia by Tyr, D-Phe, Met, desamino-Tyr (Weimann, 1977) or Trp (Brandenburg et al., 1980) was synthesized via A1,B29- Msc_2 –insulin, as well as [B1-Abz]insulin (Lenz et al., 1995). In contrast, substitutions at, and replacements of Gly^{A1}, required multistep strategies (Friesen, 1978). Examples are replacements by acetyl, propyl, t.-butoxyacetyl groups, by ß-Ala, γ-Abu, D-Pro (Trindler and Brandenburg, 1982) or by (4-azidophenylacetyl)-D-α-, γ-diaminobutyryl (Saunders and Brandenburg, 1981). Recently, the fluorogenic A1-Abz, B29-Tyr(NO_2)-insulin could be obtained (Lenz et al., 1994). Sequential four-fold Edman degradation and quick coupling, which minimised pyrGlu formation, with Boc-tetrapeptides gave several analogues with A4 replacements (Saunders et al., 1983) . To study the effect of simultaneous modifications at A1 and the C-terminus of the B-chain, des-pentapeptide(B26-B30)-insulin analogues were synthesised by acylation of B1-Msc-des-Gly-DPPI with nitrophenyl esters of Boc_2-Lys, Msc-L-Ala, and Msc-D-Ala (Chu et al., 1981).

3.2. C-terminal Semisynthesis

In the early phase of chemical semisyntheses, protection of COOH groups was mandatory. However, saponification of insulin hexamethyl ester yielded mainly [A21-asparaginimide]insulin, showing that this access was not feasible (Gattner and Schmitt, 1977). At last, enzymatic semisynthesis opened the way for the conversion of porcine to human insulin and the preparation of many analogues. Two shortened insulins were the key intermediates, des-(B23-B30)octapeptide insulin (DOPI) and des-Ala^{B30}-insulin (DAI).

3.2.1. *At B29 – with Des-Ala^{B30}-insulin*

While DOPI was used for the first semisynthesis of human insulin (Inouye et al., 1979), DAI, for which an improved access had been found (Schmitt and Gattner, 1978) allowed the most rational pathway by necessitating the exchange of one amino acid only. This was subsequently exemplified independently with Thr-OMe (Gattner et al., 1980a) and with Thr-OBut (Morihara et al.,1979). Later, Hans Gattner and colleagues synthesized a considerable number of analogues from DAAI and amino acid esters or amides.

A large step forward was the trypsin-catalyzed transpeptidation of porcine insulin with L-Thr-t.-butylester (Jonczyk and Gattner, 1981) as impressively demonstrated by a one-pot semisynthesis of human insulin. High yields were obtained even without optimization, the Arg-Gly bond was not affected.

3.2.2. *At B22 – with Des-(B23-B30)octapeptide insulin*

The first application for the semisynthesis of analogues was accomplished in Aachen. A mutant forms of human insulin, containing Leu in position B24 or B25 had been found in a diabetic patient. To identify the mutant, the 2 isomers were obtained by solution synthesis of octapeptides, amino protection of DOPI with $(Boc)_2O$, and trypsin-catalysed coupling with the peptides (Gattner et al., 1980b). [Leu^{B24}, Leu^{B25}] human insulin, several analogues, as well as des-(B27-B30)tetrapeptide-insulin methylester (DTPI-Ome), were synthesised shortly afterwards.

The method was adopted to prepare DPPI-NH_2 (Fischer et al., 1985) and a series of shortened [Leu^{B25}] analogues (Fischer et al., 1986). The finding, that des-(B26-B30)pentapeptide-insulin amide (DPPI-NH_2) had full insulin activity, triggered extensive, detailed studies. In Aachen, these yielded seven analogues, *inter alia* with D-Phe, Trp, Ty or His in B25, and/or D-Phe in B24 (Casaretto et al., 1987).

The semisynthesis of seven DTPI analogues, in which the side-chain and/or the carboxy group of B26-tyrosine had been modified, was performed without protection of the amino groups of DOPI. Transpeptidation of insulin with Gly-Phe-Phe-Tyr-OMe was also successful (Lenz et al., 1991).

The role of the invariant Tyr in position B26 was further studied in detail (Sievert, 1991, Sievert et al., 1993). Trypsin-catalyzed semisynthesis at B22, or chymotrypsin-catalyzed synthesis (see below), also with free DPPI in the thermodynamically-controlled synthesis, led to more than 30 analogues in yields between 55 and 65% (on DPPI or DPPI-OMe).

Finally, a series of several full length analogues and about 40 shortened insulins with varied chain length, side chain and/or amide function, as well as incorporated ß-amino acids, or a depsi-peptide bond, was similarly prepared. (Spoden et al., 1995, Leyer et al., 1995, Siedentop, 1997, Wollmer et al., 1994). The use of unprotected DPOI was advantageous with respect to simplicity and overall yields.

Synthetic peptides were increasingly obtained via solid phase synthesis using Fmoc stategy. Extending enzyme application further, the phenylacetyl residue served for lysine protection, and was later removed by immobilized penicillin amidohydrolase (Svoboda et al., 1994).

3.2.3. At B25 – with Des-(B26-B30)pentapeptide insulin

A new route for the semisynthesis of insulin analogues was successfully explored: The key intermediate was the methyl ester of DPPI, obtained via papain-catalyzed esterification of DPPI. The kinetically controlled chymotrypsin-catalyzed synthesis with various amino acid esters or amides in 50% aqueous DMF gave the corresponding DTPI analogues in representative yields of 60-90% (Sievert, 1991).

3.3. Crosslinked Insulins

Our previous work with A1,B29-cross-linked insulins, which had given interesting structure-function information and were fine proinsulin models, was carried on. A novel cross-linked monomeric analogue, A1-B29 Dsu-[D-Ala^{A1,B1}]insulin, could be obtained through a multistep procedure, combining replacement and crosslinking techniques (Saunders and Freude, 1982).

A successful test of *intramolecular* enzymatic peptide synthesis has been performed with A1,B29-(carbonyl-bis-methionyl)-insulin. It was first cleaved at Arg^{B22}with trypsin at pH 9.6. Then, the B22-B23 bond was successfully reformed by trypsin in organic-aqueous buffer at pH 6.5, and crystallisable insulin was obtained upon deblocking. Although the split insulin does not contain structural information, the high yield of 60% in the 1:1 coupling was remarkable (Chu et al., 1981).

Extending the range of cross-linking further, Achim Schüttler synthesised all the 6 possible isomers with bridged amino acids: symmetrical dimers by linking Msc-protected monomers with suberic acid bis-p-nitrophenyl esters, asymmetrical dimers via selective activation of diprotected insulins at B29 or B1 with an excess of reagent. Several dimers were also prepared with oxalyl or dodecane-dioyl bridges (Schüttler and Brandenburg, 1982).

C-Terminal crosslinking was achieved through the combination of chemical and enzymatic methods. First, an N-protected B30-hydrazide was synthesised via lysylendopeptidase-catalyzed transpeptidation. Chemical coupling by the azide method with ethylene or hexamethylene diamine gave amides, which were then dimerized with succinic acid bis nitrophenyl ester. This approach proved also successful for immobilisation of insulin (Gattner and Naithani, 1989).

3.4. Proinsulin and Mini-proinsulins

While all semisynthetic operations with insulin were main-chain modifications on the intact disulfide structure, proinsulin studies started either with the native molecule, or separated insulin chains. Lysyl-endopeptidase played an important role for B29-specific operations. For example, acylation of bovine proinsulin with Boc-Ala-Gln-Ala gave des-(1-21)-preproinsulin (Naithani et al., 1979). The side-chain protected (Lys^{29}, Lys^{59})-bis-Msc derivative of bovine proinsulin could be obtained via temporary protection of the α-amino group (Büllesbach and Naithani, 1980).

The semisynthesis of a shortened open-chain proinsulin is a fine illustration of the elaborate chemo-enzymatic strategies developed already in 1982. Arg-A-$(SS)_2$ was elongated sequentially with Boc-Gly-Lys(Msc)OH and then Boc-Thr-Arg-Arg-Ala. The carboxyl component, B(1-29)(SS), was first fully esterified, and then selectively saponified at its C-terminus with trypsin. Chemical coupling to the A-chain and deblocking gave **B**-Arg-Arg-Ala-Gly-Lys-Arg-**A** (Büllesbach, 1982)

Such strategies were developed further, and several mini-proinsulins were synthesized for studies of the folding process. A unique model has been designed (Lenz et al., 1996). Semisynthetic Boc-[(AbzB1](B1-B29) and synthetic [$Tyr(NO_2)^{A14}$](A1-A21) were linked between B29 and A1 using lysyl endopeptidase catalysis. After deblocking, reduction and oxidation gave the mini-proinsulin in 93% purity (HPLC). The course of SS formation / folding was followed by fluorescence emission spectroscopy which showed an increasing energy transfer as a consequence of tertiary structure formation.

Several single-chain insulins **B**(1-29)-NHNH-Sub-Lys-**A** possessed a better potential for chain combination than native proinsulin, but helix-breaking residues in A3 or B14 disturbed the folding process (Weber, 1998). Apparently, during refolding of reduced proinsulin a considerable amount of incorrectly paired tri-disulfide is formed and later re-shuffled (Bota, 1999). In contrast, there seems to be a direct pathway towards correct pairing in a reduced mini proinsulin.

4. STRUCTURE-FUNCTION RELATIONSHIPS

The analogues have been subjected to receptor binding and biological assays, predominantly in vitro, at DWI and in many collaborative studies. Structural information was provided mainly by CD spectroscopy, and thus Axel Wollmer's contribution were ubiquitous. Within the context and limitations of this report, it is only possible to summarize the results by listing, and to add a few comments. For a recent review, see Brandenburg (1999).

4.1. Synthetic Insulin Analogues

At A19, bioactivity can be modulated by changes in the 4-position of the ring, and by increasing bulkiness. 1) The mitogenic activity of [D-Arg^{B22}]insulin was clearly higher (16%). An antiparallel isomer was isolated with very low metabolic, but relatively high mitogenic activity (15%). The mitogenic activity was also high for the B17 analogues, 35% (D-Leu) and even 100% (Nle). 2) A hybrid in which the D-chain of IGF-I was linked to A21 exhibited reduced insulin activity, but was 2.8x more potent than insulin with respect to mitogenesis.

4.2. Semisynthetic Insulin Analogues

1) The semisynthesis was performed in order to determine the Leu position in a mutant found by Howard Tager and colleagues in 1979. [Leu^{B24}]insulin (the mutant insulin Chicago) was claimed to be an antagonist of insulin, thus causing diabetes. A careful study showed only additive effects of this analogue (also [Leu^{B25}]insulin) and insulin, but no antagonism (Diaconescu et al., 1982).

2) The effect of simultaneous modifications at A1 and the C-terminus of the B-chain had effects similar to those in insulin, but the role of A1 in the shortened analogue maybe less pronounced.

3) The full potency of this analogue, as compared to only ≈ 20% for DPPI-OH, was a striking surprise. This finding became fundamental to future research. CD spectra revealed characteristic differences compared to insulin. The molecule is monomeric.

4) The inactivation by Phe ⇒ Leu replacement was smaller than with insulin. The phenyl side chain in B25 seems to be essential.

5) Polar aromatic residues in B25 raised the activity even higher. Activity differences were, however, not reflected in CD spectra.

6) C-Terminal and ring modifications modulate potency, the D-Tyr analogue was highly active. CD spectra indicated that free DTPI has lost its ability to associate, while carboxyamidation or methylation restore the typical self-association of insulin. The results thus underline the importance of position B26 in the fine-tuning of activity and solution structure of insulin.

Table 1. Properties of Synthetic Insulin Analogues

Analogue	% Bioact	% Bind.	Author	Com.
[Phe^{A14}]	96		Danho et al., 1980c	
[Phe^{A19}]	23		Danho et al., 1980b	
[Tyr(I)^{A19}]	24		Wieneke et al., 1983	
[Tyr(I)$_2$^{A19}]	3		Wieneke et al., 1983	
[Phe(F)^{A19}]	60		Stoev et al.,1988	
[D-Arg^{B22}]	2	1-2	Knorr et al., 1982	1
[D-Arg^{B22}] antipar.	0.2	0.1	Knorr et al., 1982	1
[D-Leu^{B17}]	6	3	Knorr et al., 1983	1
[Nle^{B17}]	16	26	Knorr et al., 1983	1
Insulinyl-D^{A21}	20	28	King et al., 1982	2
[Ala^{B12}] / [Nle^{B12}]	30/10		Casaretto, 1986	

Table 2. Semisynthetic Insulins

Analogue	% Bioact.	% Bind.	Author	Comment
[Guanidinoacetyl^{A1}]	88		Friesen (1978)	
A1/B1/B29-Cit	15/100/15		Naithani & Gattner (1982)	
[(Apa)-D-Dab^{A1}]	66		Saunders & Brandenburg (1981)	
[D-Pro^{A1}]	100		Trindler & Brandenburg (1982)	
[Asp-imide^{A21}]	40		Gattner & Schmitt (1977)	
[Leu^{B24}] / [Leu^{B25}]	12/3.4		Gattner et. al. (1980b)	1
[LeuB^{24}, LeuB^{25}]	0.6	0.4	Jonczyk et al. (1981)	
[X^{A1}]DPPI			Chu et al. (1981)	2
D-Ala/L-Ala	6.5/2			
Lys^{A0}-DPPI	8		Chu et al. (1981)	2
DPPI-NH_2	105		Fischer et al. (1985)	3
[Leu^{B25}]DPPI-NH_2	18	9	Fischer (1986)	4
[X^{B25}]DPPI-NH_2	313/228		Casaretto et al. (1987)	5
His/Tyr/Trp/D-Phe	54/0.5			
[D-Phe^{B24}]DPPI-NH_2	8		Casaretto et al. (1987)	5
DTPI-OH/-NH_2	116/110	95/110	Lenz et al. (1991)	6
[D-Tyr^{B26}]DTPI-NH_2	293	203	Lenz et al. (1991)	6
[Tyr(R)]DTPI-NH_2	126/128/5	79/182/15	Lenz et al. (1991)	6
R = NH_2/NO_2/I_2	0	1		
[X^{B26}]DTPI-OH/	13-251	15-449	Sievert (1991)	7
-OMe/-NH_2			Sievert et. al. (1993)	
[D-Phe^{B24,B25}]insulin	< 1	0.8	Svoboda et al. (1994)	
[B24-B25CO-O]	3-4		Wollmer et al. (1994)	8
insulin/DPPI- NH_2			Kurapkat et al. (1997)	
DPPI-NH-R	51-197	28-216	Spoden et al. (1995)	9
DHxPI-NH-R	1-11	1–7	Spoden et al. (1995)	9
DD ⇒ DHpPI-NH_2			Spoden et al. (1995)	9
[Asn^{B25}]DPPI-NH_2	292	227	Spoden et al. (1995)	9
[B26–B30] analogues		67–149	Leyer et al. (1995)	10
[Ala^{B26}]DTPI-NH_2		273	Leyer et al. (1995)	10
[Sar^{B26}]DTPI-NH_2		1100	Leyer et al. (1995)	11
[ß-X^{B23-26}]DPPI-NH_2		2-22	Siedentop (1997)	12
[ß-X^{B23-26}]insulins		4-15	Siedentop (1997)	12
[D-Ala^{B26}]		18	Siedentop (1997)	12
[D-Ala^{B26}]DTPI-NH_2		1252	Siedentop (1997), Kurapkat et al. (1999)	13

7) The role of the invariant Tyr in position B26 was further studied in detail with > 30 analogues. In vitro receptor binding and biopotency determinations showed, that hydrophilic or D-amino acids were 2-4-fold as active as insulin. Aliphatic/hydrophobic amino acids led to a marked decrease. Extremes were Glu (>400%) and Leu (ca 15%). CD spectra and gel filtration indicated that all analogues except those with Tyr + protected COOH group are monomeric.

With the exception of the basic residues of K and R, increasing hydrophilicity induced significant changes of secondary structure (relative to DPPI) concomitant with enhanced affinity for the receptor. Thus, residue B26 modulates structural and functional properties of des(B27-B30)-insulin – via induction of favourable main-chain adjustments rather than via contribution to direct hormone-receptor interactions.

8) Contrary to expectation, the depsipeptide bond caused pronounced inactivation. Instead of enhanced flexibility, a stabilisation of the monomer by interaction between Phe^{B25} and the core was found.

9) Altogether 10 analogues of DPPI, and 5 analogues of DHxPI have been studied. The highest value was found for DPPI-hydroxypropylamide, the lowest for the dimethylamide. The surprising jump in activity upon COOH ⇒ $CONH_2$ transition was only found with DPPI, not with longer or shorter truncated insulins. An unexpected finding was that Asn in position B25 of DPPI-NH_2 could replace the aromatic residue.

10) When testing 9 analogues, there was only a small reduction in receptor affinity upon Gly-replacements in positions 27-30, or Tyr^{B26} ⇒ Ala replacements in des-B30- to des-(B28-B30)-NH_2 insulins. Pro ⇒ D-Pro in des-(B29-B30)-NH_2 enhanced binding to 143%.

11) A striking observation was that sarcosine in position B26 brings about a dramatic increase in receptor affinity.

12) ß-Ala in position B23 and homo-ß-Phe in position B24 reduced receptor binding of insulin or DPPI-NH_2 greatly (typically <5%), while homo-ß-PheB25 was less detrimental (15 or 22%, respectively). Association to dimers was impaired in the same sense.

13) Like sarcosine, D-Ala in position B26 gave rise to extreme superpotency. Surprisingly, the NMR structure gave no structural explanation for this phenomenon.

4.3. Crosslinked Insulins

As previously found, A1,B29-crosslinked insulins have high structural resemblance to insulin, but low bioactivity. From the unaltered bioactivity of A1-B29-crosslinked insulin upon Gly ⇒ D-Ala replacement it was concluded that the low activity is not a consequence of local distortion around A1, but rather due to impaired structural adjustment necessary for binding to the insulin receptor (Saunders and Freude, 1982).

Three each of symmetrical and asymmetrical insulin dimers, in which amino groups were bridged with dicarboxylic acids, present a new class of insulin derivatives. Fat cell biopotencies (on a molar basis) fell from B1-Sub-B29' = 110% > A1-Sub- B1' =36% > A1-Sub-A1' = 26% to B29-B29'-dimers = 4%. Strikingly, receptor binding to fat cells was 2 – 10 fold higher, with maximal values for the B29-B29'-dimer (Schüttler and Brandenburg, 1982). The activity of dimers crosslinked between B30 carboxy groups with diamines was also low and decreased markedly with the length of the bridge (Gattner and Naithani, 1989).

Ten years later, the B29-B29'-suberoyl-insulin dimer gained fame. It was the first insulin analogue with impaired maximal stimulation of glucose oxidation, and turned out to be an antagonist of insulin in 3T3L1 cells (Weiland et al., 1990).

5. PHOTO-REACTIVE INSULIN DERIVATIVES AND RECEPTOR STUDIES

In 1976, we entered an exciting new field, photoaffinity labelling. P. Thamm synthesized [4-azido-Phe^{B1}]insulin as the first homogeneous photo-insulin, and then a whole series of well-defined derivatives. They allowed specific light-induced cross-linking of insulin to receptors, and thus three major lines of research, studies on the bioactivity of cells containing insulin-receptor conjugates, structural analyses of receptors, and investigations on the fate of insulin/receptor complexes in cells. On this basis, our receptor group was born.

5.1. Synthesis of Photo-activatable Insulins and IGF-I

The first generation comprised photoinsulins with azido- or nitro-azido residues attached to B1, B2 (in des-Phe-insulin) or B29 (Thamm et al., 1980). B2-(2-nitro,4-azidophenylacetyl)-des-Phe-insulin (Napa-DP-insulin), was the key for a large number of studies.

Later, various new photo-insulins were designed and synthesized in order to meet the requirements of refined studies. For example, (4-azidophenylacetyl)-[D-α-,γ-diaminobutyryl^{A1}]insulin was a highly active analogue substituted at the A-chain (Saunders and Brandenburg, 1981). To increase the hydrophilicity, 4-azido-2-sulfobenzoic acid was synthesized and incorporated into positions B1 and B29. It also gave the bifunctional B1+B29 P-insulin. Since the the perfluorophenyl ring was reported to be advantageous, we prepared B1- and B29- azido-perfluorobenzoyl (Atf) insulins. The five photo insulins possessed receptor affinities of 40-70% (Brandenburg et al., 1993).

The advantage of the 4-azidosaliciloyl (Asa) residue is its propensity for iodination. This group proved to be very valuable for labelling of insulin and also IGF-I (Fabry, 1999). While all photo-insulins were prepared via unequivocal protecting group tactics, mixtures of mono- to tetra-substituted recombinant Asa-IGF-I derivatives were separated by RP-HPLC. Tryptic digestion and MS analysis of the fragments showed that the Asa-group was attached to D68, B1, or B27. The latter Asa-IGF-I was of particular interest due to its close relationship to B29-Asa-insulin.

Three C-terminally labelled photo-insulins could be obtained by trypsin-catalyzed coupling of Gly-Phe-Phe-ED-Nap to DPOI, or incorporation of 4-benzoylphenylalanine (Bpa) into position B30 of insulin, and B26 of DTPI. The shortened analogues are related to DPPI-NH_2 and thus are "active-site" probes for the receptor. All photo insulins labelled the α-subunit of the intact insulin receptor or its soluble ectodomain.

For our novel concept PAMAC (photoaffinity-mediated avidin complexing) altogether three insulins were designed and synthesized. The first was B29-(2-nitro-4-azidophenyl)-biocytinyl-insulin (Wedekind et al., 1989). Later, a ten-step semisynthesis gave B1-Asa-[Bct^{B1}, Lys^{B2}]insulin, which contains a new site for tryptic cleavage at B2 (Fabry and Brandenburg, 1992). Thirdly, (Atf-Bct)-OSu was developed as a novel, now commercially available label, from which B29-(Atf-Bct)-insulin was synthesized.

5.2. Irreversible Activation of Cells

It was a striking observation to see already in our first experiments, that covalent linking of photoreactive insulin to viable adipocytes produced a prolonged signal (Brandenburg et al., 1980). Subsequently, this phenomenon was studied in considerable detail (Klotz et al., 1984, Saunders et al., 1984). When permanent activation was quantitatively compared with the potency of reversible complexes, irreversibly bound ^{125}I-Napa-DP-insulin gave 25% of the potency, while ^{125}I-B29-Napa-insulin reached even 50%. Thus, the site of cross-linking had a marked effect, and it appeared that attachment through the C-terminus of B-chain leads to a better stabilisation of the active complex. The relative potency of the covalent complex with ^{125}I-Napa-DP-insulin rose from 25 to 75%, that of the complex with ^{125}I-B29-Napa-insulin from 30 to nearly 100% upon raising the pH from 7 to 8. This indicated that the bound insulin is able to flex about the cross-link, moving from an inactive to an active complex as a function of pH. Effective interaction obviously required the deprotonation of an amino group.

A careful study on the time-dependence of biological activity induced by irreversible cross-linking to adipocytes resulted in a half-life of 80 minutes for the active complex at 37°C. In contrast, at 20°C the signal persisted for > 180 minutes, similar to insulin under reversible conditions. We concluded that internalisation is the causative event for deactivation. Function of the insulin-receptor complex may, however, be maintained for some time after internalization (Schüttler et al., 1985).

Finally, it was found that the signal produced by N-terminally bound Napa-DP insulin, but not by bound B29-Napa-insulin, could be suppressed by an anti-insulin antibody directed against the C-terminus of insulin. This indicated competition of receptor and antibody for a common C-terminal binding site and its importance for activity (Mucke et al., 1986).

5.3. Receptor Analysis

Our photo-insulins have been used in a large number of studies in several laboratories. We concentrated on the biological studies (above), carried out an initial series of analytical experiments (Neffe, 1985), and then started towards identifying insulin binding sites at the receptor. To this end, we developed PAMAC (Wedekind et al., 1989).

A trifunctional reagent was designed for PAL and subsequent selective isolation of covalent conjugates or their fragments. B29-Napa-Bct-insulin was radio-iodinated and cross-linked to purified insulin receptor (0.9 mg). After extensive tryptic digestion, affinity chromatography on streptavidin gave a fragment of M_r 14,000. Microsequencing showed the N-terminus to be Leu20-His21 of the α-subunit. Thus, a first binding site was detected at the N-terminus (sequence 20 – 120) in a remarkably stable domain.

The availability of soluble insulin receptor ectodomain in mg amounts facilitated further work enormously. By comparison of the labelling pattern obtained upon PAL of highly purified insulin receptor from human placenta and receptor ectodomain with ^{125}I-Asa-[Bct^{B1},Lys^{B2}]insulin and tryptic fragmentation it could be established that insulin is bound to identical contact sites in both cases (Fabry and Brandenburg, 1992).

Then, ^{125}I-Asa-[Bct^{B1},Lys^{B2}]insulin made it possible to identify a second hormone contact site (Fabry et al., 1992). The ectodomain was labelled, the conjugate digested, and the resulting tryptic fragment (≈ 18 KDa) purified, either by HPLC or streptavidin-affinity chromatography. Its amino terminus was identified as Gly 390, the fragment corresponding to the sequence 390 ≈ 488. Thus, a site carboxyl-terminal to the cystine-rich region contributes to the insulin binding site. Based on these findings, a model of the receptor with a corresponding binding pocket was proposed.

Other studies followed: Photoaffinity labelling of receptor ectodomain with the bivalent 4-azido-2-sulfobenzoyl-insulin gave no indication of dimeric α-subunit. We concluded that N- and C-terminal insulin binding domains are located in the same α-subunit. Enzymatic fragmentation and microsequencing of the conjugate gave the N-terminal region 1- 190 and thus confirmed our earlier findings with the intact holoreceptor (Brandenburg et al., 1993). The bivalency of the tetrameric αß-αß-receptor could unequivocally be demonstrated by sequential covalent labelling of receptor ectodomain with B29-(^{125}I-Asa)-insulin and B29-(Atf-Bct)-insulin, followed by affinity chromatography on a streptavidin column. The presence of radioactivity and retention on the column were proof that these receptors contained both insulin labels.

In order to compare the homologous receptors for insulin and IGF-I, ^{125}I-Asa-IGF-I and ^{125}I-Asa-insulin were used for photoaffinity labelling of both receptors. Two insulin receptor chimeras, which contained either only the cystine-rich region of the IGF-I receptor, or also its C-terminal domain (Schumacher et al., 1993), were also included. Preliminary results confirmed the importance of N- and C-terminal domains for insulin, and of the cystine-rich domain for IGF binding. Surprisingly, ^{125}I-Asa-insulin was a very good probe for the IGF-R. On the other hand, both ^{125}I-B28-Asa-IGF-I and ^{125}I-B29-Asa-insulin bound covalently to the same region of the insulin receptor ectodomain. The sequence 704 – 717 is further downstream than our second site (see above), but agrees well with results from Don Steiner's laboratory (Fabry, 1999).

5.4. Other Receptor Studies

Promising tools for analyzing the reversible interaction between insulin and its receptor were spin-labelled insulins (SL-insulins). 2 SL-insulins carry a 2,2,5,5-tetramethyl-3-carboxy-pyrrolidine-1-oxyl residue as a stable nitroxide radical in positions B1 or B29. C-terminal labelling via trypsin-catalyzed semisynthesis gave DPPI-tempamide (2,2,6,6-tetramethyl-4-amino-piperidine-1-oxyl residue).
Preliminary studies with B1 or B29-SL-insulins with receptor ectodomain gave no indication of specific immobilization of the spin label. In contrast, with the SL-DPPI the first direct demonstration of site-specific interaction between insulin and receptor in a reversible system was possible (Brandenburg et al., 1993). Furthermore, the stoichiometry of insulin:receptor (soluble ectodomain) was found to be 1.8:1, thus proving independently the bivalency of the receptor. Spin labelling thus proved to be a very valuable approach for receptor studies under reversible conditions.

6. CO-OPERATION

Peptide and protein chemistry at DWI were the solid grounds on which further studies could be founded. They had provided a large number of insulin analogues as well as a series of special insulin derivatives, which were available for the elucidation of relationships between primary structure and structures of higher order on one hand, and receptor binding, biological activity in vitro and in vivo, and also immunological properties, on the other. Furthermore, they proved to be valuable tools for studying receptor structure, hormone binding sites as well as activity and fate of the insulin-receptor complex.

These were the questions – the answers could only be found by intensive co-operation with friends and colleagues from all the fields involved. The outcome speaks for itself. Of the 260 publications between 1977 and 1999, 50% are joint publications. About 90 papers originate from the co-operating laboratories, dealing with extended and refined studies with analogues or photo-insulins made in Aachen. Except for some hints, the results could not be included in the text above and await valuation (Brandenburg, in preparation). At present, a few statements have to suffice:

1. In order to evaluate the effects brought about by chemical alterations on biological properties, it was deemed necessary at a very early stage to test how the 3D structure in solution was affected. Thus, analysis by CD began already in 1970 to

give information on monomer structure and dynamics of assembly. The most fruitful work with Axel Wollmer and his group is reflected in more than 30 joint papers. In addition to CD, fluorescence spectroscopy and recently NMR gained increasing importance, and these approaches were well supplemented by X-ray crystal analysis (Guy Dodson) and recently ESR spectroscopy (Heinz-Jürgen Steinhoff).

2. In the earlier phase (1968 – 74), all bioassays in vitro were performed by Henri Ooms in Belgium and Jørgen Gliemann in Denmark. Then, in order to have a quick feedback, we established fat cell assays at DWI with his help. After the initiation of receptor binding assays by Pierre Freychet in France, our analogues were subjected to extended studies by Pierre DeMeyts, pinpointing down, *inter alia*, the phenomenon of negative cooperativity. He helped us to establish a receptor binding assay at DWI. Many others, as Don Steiner, George King, Ron Kahn, Peter Rösen, or Richard A. Roth made valuable contributions.

The cooperation with Hans-Georg Joost intensified upon his moving to Aachen, which added now biology as a third focal point to insulin chemistry and physics. Very exciting was his finding of the first case of antagonism to insulin, and most useful his help to establish cell culture and bioassays with 3T3 L1 cells at DWI.

3. Altogether, not too many in-vivo studies have been carried out – quite a bit of valuable information may still be hidden. After previous evaluations by Walter Puls at Bayer AG, St. Thomas's Hospital in London took the lead. In vivo experiments in greyhounds established already in 1976 correlations between in-vivo and in-vitro efficacy of insulin analogues, especially cross-linked ones. Such findings were much later re-discovered by others. The stimulating friendship with Peter Sönksen and Richard Jones involved other biological and immunological studies as well, in particular the first experiments on photoaffinity labelling. It persists (see section "outlook"), and has led to about 20 papers.

4. Various aspects of immunology could be investigated with insulin analogues made in Aachen, and yielded about 15 papers by K.G. Petersen, Klaus Keck, Franz Jansen and others. The extensive work on C-peptide by Kumar Naithani with the Novo group deserves special mentioning.

5. A domain of co-operation were the receptor studies – particularly by means of photoaffinity labelling. As soon as our first photo-insulins had been synthesised, they were in high demand. About 40 papers presented detailed investigations on receptor structure and intracellular fate. Martin Wisher and Peter Sönksen were the first, then Jerry Olefsky with Kim Heidenreich and Paulos Berhanu in the US, Max Fehlmann and Yannick Marchand-Brustel in France joined in, and finally Falk Fahrenholz and Elzbieta Kojro with microsequencing of our isolated receptor fragments. In these analyses, the ectodomain of Erik Schaefer as well as the mutated receptors obtained by Ralf Schumacher in Axel Ullrich's laboratory were extremely helpful.

7. OUTLOOK

The last three doctoral theses have been completed, and the habilitation thesis of Marlies Fabry has come to its experimental end. Why, and how, outlook? Well, there are several aspects which point into the future – questions which have been raised, but not yet answered. All may be of relevance for diabetes therapy.

7.1. Phosphorylated Insulins

It has been claimed that phospho-insulins which, however, were ill defined, lower the blood sugar in diabetic animals without causing hypoglycemia (Albisser, 1993). We have now prepared, by unequivocal semisynthesis, several phospho-insulins like $Tyr(PO(OH)_2)^{B0}$- or $Thr(PO(OH)_2)^{B0}$-insulins and their B29-substituted isomers. Receptor affinities were from 30 to 70%. These are the first homogeneous phospho-insulins – expecting now in vivo testing and further feedback work (Lauter, 1999, Höcker et al., 1997).

7.2. Hepatospecific Insulins

Hepatospecificity would be a highly desirable property. Based on ideas of Richard Jones and Achim Schüttler, T4-insulin was synthesised as a molecule which in the organism can be bound by specific endogeneous binding proteins (Jones et al., 2000). Due to their high molecular weight, such complexes would have free access to insulin receptors in the liver, but not in the periphery, and thus could mimick the physiological mode of action. The first experiments (Shojaee-Moradie et al., 1998) were encouraging, the principle seemed to work. As a further development, a series of such conjugates has now been synthesised (Sundermann, 1999). We eagerly wait for the results of biological testing.

7.3. Novel Insulin Dimers

Continuing our earlier work, novel, covalently cross-linked insulin dimers have been designed and obtained via chemo-enzymatic semisynthesis. We wanted to increase the discrepancy between binding and activity, aiming at insulin antagonists. When several superpotent shortened insulins, $[X^{B26}]DPPI\text{-}NH_2$ (X = Glu, D-Ala, Sar) were cross-linked between their B1 amino groups, bioassays of the resulting dimers gave unexpected results. Compared to insulin, receptor binding was 2-4-fold, but bioactivity at least 8-fold (Glu, D-Ala). The activity of the $[Sar^{B26}]$dimer was dramatically increased to 1700 +/- 600 %. To our knowledge, it is the most active insulin, at least in this in vitro system 3T3 L1 cells (Havenith, 1999; Havenith and Brandenburg, 2000). Again, there is a group of highly interesting insulins. How will their in vivo properties be, can they be novel leads?

After more than 40 years, insulin research at DWI has almost come to its end. Open aspects of receptor research persist and require continuation – in the skilful hands of Marlies Fabry. But with respect to studies based on preparative bioorganic chemistry, we have to hand over to our scientific successors, wherever they are.

8. REFERENCES

Albisser A.M. "Treatment of Diabetes Using Phosphorylated Insulin." US Patent 5242900 (1993).

Bota J. "Untersuchungen zum Faltungsweg monochenarer Insulinderivate." Thesis, RWTH Aachen (1999).

Brandenburg D. and Wollmer A. (Eds.) "Insulin, Chemistry, Structure and Function of Insulin and Related Hormones." Ed., Walter de Gruyter, Berlin, New York, (1980).

Brandenburg D. "Insulin – structure, function, design." *Exp. Clin. Endocrinol. Diabetes* 107, Suppl. 2 (1999): 6-12.

Berndt, H. "Synthese der Sequenz 71-86 des Humanproinsulins, III. Synthese über die Fragmente 71-78 und 79-86." *Hoppe-Seyler's Z. Physiol. Chem.* 360 (1979): 765-772

Bota J. "Untersuchungen zum Faltungsweg monochenarer Insulinderivate." Thesis, RWTH Aachen (1999).

Brandenburg D. "Insulin – Structure, Function, Design." *Exp. Clin. Endocrinol. Diabetes* 107, Suppl. 2 (1999): 6-12.

Brandenburg D. and Wollmer A. (Eds.) "Insulin, Chemistry, Structure and Function of Insulin and Related Hormones." Ed., Walter de Gruyter, Berlin, New York, (1980).

Brandenburg D., Fabry M., Pirrwitz J., Strack U., Thevis W., Wittkamp A., Schaefer E., Ellis L., Keana J.E. and Cai S.X.. "Novel Derivatives for Analyzing the Interaction between Insulin and its Receptor." In: *Peptides 1992* (Schneider C.H. and Eberle, A.N. eds.) Escom Science Publishers B.V., Leiden, (1993): pp 111-112.

Brandenburg, D., Diaconescu, C., Saunders, D. and Thamm, P. "Covalent Linking of Photoreactive Insulin to Adipocytes Produces a Prolonged Signal." *Nature* 286 (1980a): 821-822.

Brandenburg, D., Lei, K., Wang, Z., Dong, B., Binggen, R. and Zhu, S.: "Preparations and Properties of Crystalline Porcine and Bovine (Trp) B1 Insulins." *Scientia Sinica* 23, 1443-1452 (1980b).

Büllesbach E.E. "Semisynthesis of a Shortened Open-Chain Proinsulin." *Tetrahedron Letters* 23 (1982): 1877-1880.

Büllesbach, E.E. and Naithani, V.K. "Semisynthesen, ausgehend von Proinsulin, II. Darstellung und Anwendung von $N\varepsilon^{29}$, $N\varepsilon^{59}$-Bis(methylsulfonylethoxy-carbonyl)-proinsulin." *Hoppe-Seyler's Z. Physiol. Chem.* 361 (1980): 723-734.

Büllesbach, E.E., Danho, W., Helbig, H.-J. and Zahn, H. "Human Proinsulin, VIII. Studies on the S-Tritylation of Reduced Proinsulin, Insulin A and B Chains and their Detritylation." *Hoppe-Seyler's Z.Physiol.Chem.* 361 (1980): 865-873.

Casaretto M., Spoden M., Diaconescu C., Gattner H.-G., Zahn H., Brandenburg D. and Wollmer A. "Shortened Insulins with Enhanced in vitro Potency." *Biol. Chem. Hoppe-Seyler* 368 (1987): 709-716.

Casaretto R. "Merrifieldsynthese von [Ala^{12}]- und [Ile^{12}]Insulin-B-Ketten und Kombination mit natürlicher A-Kette." Thesis, RWTH Aachen (1986).

Chu, S.-C., Wang, C.-C. and Brandenburg, D. "Preparation and Properties of Des-pentapeptide(B26-30)-Insulin Analogues." In Brunfeldt, K. (Ed.) Peptides 1980, Proceedings. of the Sixteenth European Peptide Symposium. The Danish Institute of Protein Chemistry, Copenhagen, 1981a, pp. 359-364.

Chu S., Wang, C. and Brandenburg D. "Intramolecular Enzymatic Peptide Synthesis: Trypsin-mediated Coupling of the Peptide Bond between B22-Arginine and B23-Glycine in a Split Crosslinked Insulin." *Hoppe-Seyler's Z. Physiol. Chem.* 362 (1981b): 647-654.

Cutfield S.M., Dodson G.G., Schwertner E. and Zahn H. "X-Ray Diffraction of Crystals of Half-Synthetic Sheep Insulin." *Hoppe-Seyler's Z. Physiol Chem.* 360 (1979): 783-785.

Danho, W. and Föhles, J. "Human Proinsulin, IV. Synthesis of a Protected Peptide Fragment Corresponding to the Sequence 1-23 of the Prohormone." *Hoppe-Seyler's Z.Physiol.Chem.* 361

(1980): 839-847

Danho, W., Naithani, V.K., Sasaki, A.N., Föhles, J., Berndt, H., Büllesbach, E.E. and Zahn, H. "Human Proinsulin, VII. Synthesis of Two Protected Peptides Corresponding to the Sequences 1-45 and 46-86 of the Prohormone." *Hoppe-Seyler's Z.Physiol.Chem.* 361 (1980a): 857-863.

Danho, W., Sasaki, A., Büllesbach, E., Föhles, J. and Gattner, H.-G. "[A19-Phenylalanine]Insulin: A New Synthetic Analogue." *Hoppe-Seyler's Z. Physiol. Chem.* 361 (1980b): 735-746.

Danho W., Sasaki A., Büllesbach E., Gattner H.-G. and Wollmer A. "[A1^{4}-Phenylalanine] Insulin: A New Synthetic Analogue." *Hoppe-Seyler's Z. Physiol. Chem.* 361 (1980c): 747-754.

Diaconescu C., Saunders D., Gattner H.-G. and Brandenburg D. "[LeuB^{24}]- and [LeuB^{25}]Insulins are not Antagonists of Lipogenesis in Adipocytes." *Hoppe Seyler's Z. Physiol. Chem.* 363 (1982): 187-192.

Fabry M. "Synthesis and Characterization of Photoreactive Insulin-like Growth Factor 1 Derivatives for Receptor Analysis." *Bioconjugate Chem.* 10 (1999): 200 – 205.

Fabry M., Schaefer E., Ellis L., Kojro E., Fahrenholz F.and Brandenburg D. "Detection of a New Hormone Contact Site within the Insulin Receptor Ectodomain using a Novel Photoreactive Insulin." *J. Biol. Chem.* 267 (1992): 8950-8956.

Fabry, M. and Brandenburg, D. "Design and Synthesis of a Novel Biotinylated Photoreactive Insulin for Receptor Analysis." *Biochemistry Hoppe-Seyler* 373 (1992a):143-150.

Fabry, M. and Brandenburg, D. "Analysis of the Human Placental Insulin Receptor." *Biochemistry Hoppe-Seyler* 373 (1992b): 915-923.

Fischer W.H., Saunders D., Brandenburg D., Wollmer A. and Zahn H. "A Shortened Insulin with Full in vitro Potency." *Biol. Chem. Hoppe-Seyler* 366 (1985): 521-525.

Fischer, W.H., Saunders, D., Brandenburg, D., Diaconescu, C., Wollmer, A., Dodson, G., De Meyts, P. and Zahn, H. "Structure-Function Relationships of Shortened (LeuB25) Insulins, Semisynthetic Analogues of a Mutant Human Insulin." Biol. Chem. Hoppe-Seyler 367(1986): 999-1006

Friesen, H.-J., Naithani, V.K. and Gattner, H.-G. "Preparation and Application of N^{α} -B1,$N^{\varepsilon l}$-B29-Bis(tert.-butyloxycarbonyl)insulin." *Hoppe-Seyler's Z. Physiol. Chem.* 359 (1978): 103-111.

Gattner H.-G., Danho W. und Naithani V.K. "Enzyme-catalyzed Semisynthesis with Insulin Derivatives." In: *Insulin, Chemistry, Structure and Function of Insulin and Related Hormones,* Ed. D. Brandenburg, A. Wollmer, Walter de Gruyter, Berlin, New York, (1980a): 117-123.

Gattner H.-G., Danho W., Behn C. and Zahn H.. "The Preparation of Two Mutant Forms of Human Insulin, Containing Leucine in Position B24 or 25, by Enzyme-Assisted Synthesis." *Hoppe-Seyler's Z. Physiol. Chem.* 361 (1980b): 1135-1138.

Gattner H.-G., Krail G., Danho W., Knorr R., Wieneke H.-J., Buellesbach E.E., Schartmann B., Brandenburg D. and Zahn H.. "Eine verbesserte Methode der Kombination von Insulinketten zur Darstellung von Insulinanalogen." *Hoppe-Seyler's Z. Physiol. Chem.* 362 (1981): 1043-1049.

Gattner, H.-G. and Naithani, V.K. "C-Terminal Carboxyl Function (Thr-B3O) Mediated Cross-Linking and Immobilization of Insulin." In Jung,G. and Bayer, E. (Eds.) Peptides. Walter de Gruyter, Berlin, New York , 1989, pp. 262-264.

Gattner, H.-G. and Schmitt, E.W. "(A21-Asparaginimide) Insulin. Saponification of Insulin Hexamethyl Ester, I." *Hoppe-Seyler's Z. Physiol. Chem.* 358 (1977) 105 – 113.

Havenith C. "Neue kovalente Insulindimere: Semisynthese und biologische Eigenschaften." Thesis RWTH Aachen (1999).

Havenith C. and Brandenburg D. "Kovalent verbrückte Insulindimere." DOS (31.8.2000) DE 19908041 A1 (2000).

Höcker H., Brandenburg D., Lenz V. and Kleinjung J. "Phosphorylierte Insuline und deren Derivate, Verfahren zu deren Herstellung, ihre Verwendung und eine sie enthaltende Zubereitung." DOS DE 195 35 701 A1 (1997).

Jonczyk A. and Gattner H.-G. "Eine neue Semisynthese des Humaninsulins -Tryptisch-katalysierte Transpeptidierung von Schweineinsulin mit L-Threonin-tert-butylester." *Hoppe-Seyler's Z. Physiol. Chem.* 362 (1981): 1591-1598.

Jonczyk, A., Keefer, L.M., Naithani, V.K., Gattner, H.-G., De Meyts, P. and Zahn, H. "Preparation and

Biological Properties of [LeuB[24], LeuB[25]] Human Insulin." *Hoppe-Seyler's Z. Physiol. Chem.* 362 (1981) 557-561.
Jones R.H., Shojaee-Moradi F., Sönksen P.H., Brandenburg D., Schüttler A. and Eckey H. "Hepatoselective Pharmaceutical Actives." USP 6063761 (2000).
King G.L., Kahn C.R., Samuels B., Danho W., Büllesbach E.E. and Gattner, H.G. "Synthesis and Characterization of Molecular Hybrids of Insulin and Insulin-like Growth Factor I. The Role of the A-chain Extension Peptide." *Biol. Chem.* 257 (1982): 10869-10873.
Klotz, G., Saunders, D.J. and Brandenburg, D. "The Biological Potency of Covalent Insulin - Receptor Complexes: Dependence on Site of Cross-linkage." *Hoppe-Seyler's Z. Physiol. Chem.* 365 (1984a) 493-498.
Klotz, G., Saunders, D. and Brandenburg, D. "Biological Potency of Covalently Linked Insulin-Receptor in Rat Adipocyte. Comparison with the Potency of Reversible Complexes." *Biochim. Biophys. Acta* 803 (1984b) 78-84
Knorr R., Danho W., Büllesbach E.E., Gattner H.-G., Zahn H., King G.L. and Kahn C.R. "[B22-D-Arginine]Insulin: Synthesis and Biological Properties." *Hoppe-Seyler's Z. Physiol. Chem.* 363 (1982): 1449-1460.
Knorr, R., Danho, W., Büllesbach, E.E., Gattner, H.G., Zahn, H., King, G.L. and Kahn, C.R. "[B17-D-Leucine]Insulin and [B17-Norleucine]Insulin: Synthesis and Biological Properties." *Hoppe Seyler's Z. Physiol. Chem.* 364 (1983): 1615-1626.
Kurapkat G., Siedentop M., Gattner H.-G., Hagelstein M., Brandenburg D., Grötzinger J. and Wollmer A. "The Solution Structure of a Superpotent B-chain-shortened Single-replacement Insulin Analogue." *Protein Sci.* 8 (1999): 499-508.
Kurapkat, G., De Wolf, E., Grötzinger, J. and Wollmer, A. "Inactive Conformation of an Insulin Despite its Wild-type Sequence." *Protein Sci* 6. (1997): 580-587.
Lauter S. "Phosphorylierte Insulin-Hybride mit Affinität zu Plasmaproteinen." Thesis RWTH Aachen (1999).
Lenz V., Gattner H.-G., Seivert D., Wollmer A., Engels M. and Höcker H. "Semisynthetic Des-(B27-B30)-insulins with Modified B26-Tyrosine." *Biol. Chem. Hoppe-Seyler* 372 (1991): 495-504.
Lenz V., Leithäuser M., Casaretto M., Gattner H.-G., Brandenburg D. and Höcker H. "An Enzymatic-chemical Synthesis of Insulins with Non-proteinogenic Amino Acid Residues via Chain Condensation and Subsequent Disulfide Folding." In: *Peptides: Chemistry, Structure and Biology.* Eds.: P.T.P. Kaumaya, R.S. Hodges, Mayflower Scientific Ltd., (1996): 637-638.
Lenz V.J., Federwisch M., Gattner H.-G., Brandenburg D., Höcker H., Hassiepen U. and Wollmer A. "Structure and Rotational Dynamics of Fluorescently Labeled Insulin in Aqueous Solution and at the Amphiphile-Water Interface of Reversed Micelles." *Biochemistry* 34 (1995): 6130-6141.
Lenz V.J., Gattner H.-G., Leithäuser M., Brandenburg D., Wollmer A. and Höcker H. "Proteolysis of a Fluorogenic Insulin Derivative and Native Insulin in Reversed Micelles Monitored by Fluorescence Emission, Reversed-phase High-performence Liquid Chromatography, and Capillary Zone Electrophoresis." *Anal. Biochem.* 221 (1994): 85-93.
Leyer S., Gattner H.-G., Leithäuser M., Brandenburg D., Wollmer A. and Höcker H. "The Role of the C-terminus of the Insulin B-chain in Modulating Structural and Functional Properties of the Hormone." *Int. J. Peptide Protein Res.* 46 (1995): 397-407.
Mucke, P., Diaconescu, C., Klotz, G., Jorgensen, P., Saunders, D. and Brandenburg, D. "Recognition of Covalent Insulin - Receptor Complexes on Viable Adipocytes by Anti-insulin Antibodies." *Biol. Chem. Hoppe-Seyler* 368 (1987) 85-92.
Naithani V.K. and Zahn H. "Synthesis of Proinsulin." *Peptide and Protein Reviews* 3 , Ed. Milton T.W. Hearn, Marcel Dekker, Inc., New York and Basel (1984).
Naithani, V.K. and Gattner, H.G. "Preparation and Properties of Citraconyl-insulins." *Hoppe-Seyler's Z. Physiol. Chem.* 363, 1443-1448 (1982)
Naithani, V.K. and Schwertner, E. "Human Proinsulin VI. Synthesis of Protected Segment 46-70 of Prohormone." *Hoppe Seyler's Z. Physiol. Chem.* 364(1983), 1603-1613
Naithani, V.K., Büllesbach, E.E. and Zahn, H. "Semisynthesis of a Des-(1-21)-Preproinsulin Derivative."

Hoppe-Seyler's Z. Physiol. Chem. 360 (1979) 1363-1366.
Neffe J. "Struktur und Internalisierung des Insulinrezeptors". Thesis, RWTH Aachen (1985).
Sasaki, N.A. "Synthesis of Human Proinsulin Sequence 71-86, II. Improved Synthesis of the Fragment." *Hoppe-Seyler's Z.Physiol.Chem.* 360 (1979) 761-764.
Saunders D. and Freude K. "Cross-Linked [DalaA1]Insulins. Evidence for a Change in the Conformation of the Insulin Monomer at its Receptor." *Hoppe Seyler's Z. Physiol. Chem.* 363 (1982): 655-659.
Saunders D., Klotz G. and Brandenburg D. "Biological Activity of Covalently-linked Insulin-receptor Complexes in Rat Adipocytes. Effect of pH." *Biochim. Biophys. Acta* 803 (1984): 85-89.
Saunders D.J. and Brandenburg D. "Iodinated A1-NY-(4-Azidophenylacetyl)-D-α-,γ-diaminobutyric acid Insulin: Its Preparation and Properties." *Hoppe-Seyler's Z. Physiol. Chem.* 362 (1981): 1237-1245.
Saunders, D., Freude, K., Naithani, V.K., Brandenburg, D. and Stoev, S. "Semisyntheses of Insulin Analogues by Replacement of the Sequence A 1-4." In Blaha, K. and Malon, P. (Eds.) Peptides 1982. Walter de Gruyter & Co., Berlin-New York, 1983, pp. 371-374.
Schartmann, B., Gattner, H.-G., Danho, W. and Zahn, H. "Erhöhte Insulinausbeuten durch „Recycling" der nicht-kombinierten Ketten." *Hoppe-Seyler's Z. Physiol. Chem.* 364 (1983), 179-186.
Schmitt, W.E. and Gattner, H.-G. "Verbesserte Darstellung von Des-alanyl^{B30}-insulin." *Hoppe-Seyler's Z. Physiol. Chem.* 359 (1978): 799-802.
Schumacher R., Soos M.A., Schlessinger J., Brandenburg D., Siddles K. and Ullrich A. "Signaling-competent Receptor Chimeras Allow Mapping of Major InsulinReceptor Binding Domain." *J. Biol. Chem.* 268 (1993): 1087-1094.
Schüttler A. and Brandenburg, D. "Preparation and Properties of Covalently Linked Insulin Dimers." *Hoppe-Seyler's Z. Physiol. Chem.* 363 (1982): 317-330.
Schüttler A., Diaconescu C., Saunders D.J. and Brandenburg D. "Time-dependence of Biological Activity Induced by Covalent Insulin- receptor Complexes in Rat Adipocytes." *Biochem. J.* 232 (1985): 49-53.
Schüttler, A. and Brandenburg, D. "Darstellung von N,N-Bis(methylsulfonyl-ethoxycarbonyl) insulinen." *Hoppe-Seyler's Z.Physiol.Chem.* 360 (1979) 1721-1725.
Shojaee-Moradie F., Eckey H., Jackson N.C., Schüttler A., Brandenburg D., Sönksen P.H. and Jones R.H. "Novel Hepatoselective Insulin Analogues: Studies with Covalently Linked Thyroxyl-insulin Complexes." *Diabet. Med.* 15 (1998): 928-936.
Siedentop M. "Erhöhung der Segmentflexibilität im Rezeptorbindungsbereich B23-B26 des Insulins." Thesis, RWTH Aachen (1997).
Sievert D. "Semisynthese und Struktur-Funktionsuntersuchung von Des-Tetrapeptid(B27-B30)-Insulinen mit Austausch von B26-Tyrosin." Thesis, RWTH Aachen (1991).
Sievert D., Lenz V., Gattner H.-G., Brandenburg D., Höcker H. and Wollmer A. "Residue B26 Modulates Structural and Functional Properties of Des(B27-B30)-insulin." In: *Chemistry of Peptides and Proteins* (D. Brandenburg, V. Ivanov, W. Voelter, eds.) Vol. 5/6 DWI Reports 112B Aachen (1993): pp 425-431.
Spoden M., Gattner H.-G., Zahn H. and Brandenburg D. "Structure-Function Relationship of Des-(B26-B30)-insulin." *Int. J. Peptide Protein Res.* 46 (1995): 221-227.
Stoev S., Zakhariev S., Golovinsky E., Gattner H.-G., Naithani V. K., Wollmer A. and Brandenburg D. "Synthesis and Properties of A19-(p-Fluorophenylalanine) Insulin." *Biol. Chem. Hoppe-Seyler* 369 (1988): 1307-1315.
Sundermann E. "Heteromere Insulin-Hybride mit Affinität zu Plasmaproteinen." Thesis RWTH Aachen (1999).
Svoboda, I., Brandenburg, D., Barth, T., Gattner, H.-G., Jiracek, J., Velek, J., Blaha, I., Ubik., K., Kasicka., V., Pospisek., J. and Hrbas, P. "Semisynthetic Insulin Analogues Modified in Position B24, B25 and B29." *Biochemistry Hoppe-Seyler* 375 (1994) 373-378
Thamm P., Saunders D. and Brandenburg D. "Photoreactive Insulin Derivatives: Preparation and Characterization." In: *Insulin, Chemistry, Structure and Function of Insulin and Related Hormones*, Ed. D. Brandenburg, A. Wollmer, Walter de Gruyter, Berlin, New York (1980): 309-316.

Trindler, P. and Brandenburg, D. "Semisynthetic Modification of the N-terminus of the Insulin-A-chain." In Chemistry of Peptides and Proteins 1, Walter de Gruyter & Co., Berlin-New York, 1982, pp. 308-314.

Videnov G., Buettner K., Casaretto M., Foehles J., Gattner H.-G., Stoev S. and Brandenburg D. "Studies on the Total Synthesis of an A7,B7-Dicarba-Insulin, III. Assembly of Segments and Generation of Biological Activity." *Biol. Chem. Hoppe-Seyler* 371 (1990): 1057-1066.

Weber J. "Synthese und Rückfaltungsstudien von verbrückten Insulinen mit Aminosäureaustausch in den Positionen A3 und B15." Thesis, RWTH Aachen (1998).

Wedekind F., Baer-Pontzen K., Bala-Mohan S., Choli D., Zahn H. and Brandenburg D. "Hormone Binding Site of the Insulin Receptor: Analysis Using Photoaffinity-mediated Avidin Complexing." *Biol. Chem. Hoppe-Seyler* 370 (1989): 251-258.

Weiland M., Brandenburg C., Brandenburg D. and Joost H.G. "Antagonistic Effects of a Covalently Dimerized Insulin Derivative on Insulin Receptors in 3T3-L1 Adipocytes." *Proc. Natl. Acad. Sci. USA* 87 (1990): 1154-1158.

Weimann H.-J. "Partialsynthese von Insulinanalogen mit modifiziertem N-Terminus der B-Kette". Thesis, RWTH Aachen (1977).

Weitzel G. "Insulinforschung im Deutschen Wollforschungsinstitut an der RWTH Aachen". In: *Festschrift zum 25jährigen Jubiläum.* Zahn, H. and Ziegler, K. (Eds.) Deutsches Wollforschungsinstitut Aachen , Aachen (1977): pp. 439–476.

Wieneke H.J., Danho W., Büllesbach E., Gattner H.G. and Zahn H. "The Synthesis of [A19-3-Iodotyrosine] and [A19-3,5-Diiodotyrosine] insulin (porcine)." *Hoppe-Seyler's Z. Physiol. Chem.* 364 (1983): 537-550.

Wolf G., Berndt H. and Brandenburg D. "Synthese der [Lys A13] Rinderinsulin-A-Kette in der Form [Lys(Tfa)A13]A(SO3H)4 und Nα-Msc-[LysA13]A(SO3H)4 unter Verwendung des S-tert-Butylmercapto-Restes als Thiolschutz." *Hoppe-Seyler's Z. Physiol. Chem.* 360 (1979): 1569-1578.

Wollmer A., Gilge G., Brandenburg D. and Gattner H.-G. "An Insulin with the Native Sequence but Virtually no Activity." *Biochemistry Hoppe-Seyler* 375 (1994): 219-222.

9. ACKNOWLEDGEMENTS

I wish to express my sincere thanks to Prof. Helmut Zahn, Prof. Hartwig Höcker, all colleagues and coworkers at DWI, and all friends and colleagues with whom we collaborated over the years. It is a great pleasure to see so many here at this Symposium. A very special "Thank You" goes to Axel Wollmer, in conjunction with my best wishes for the future.

10. ABBREVIATIONS

Abz	aminobenzoyl
Ac	acetyl
Acm	acetamidomethyl
Asa	4-azidosalicyloyl
Atf	4-azido-tetrafluorobenzoyl
Bct	biocytinyl
Boc	tert. butyloxycarbonyl
Bpa	4-benzoyl-phenylalanine
Bpoc	biphenylpropyloxycarbonyl
But	tert. butyl
Bzl	benzyl

Cit	citraconyl
DCC	dicyclohexyl carbodiimide
DDPI	des-(B29-B30)-insulin, des-dipeptide insulin
DhpPI	des-(B24-B30)-insulin, des-heptapeptide insulin
DhxPI	des-(B25-B30)-insulin, des-hexapeptide insulin
DOPI	des-(B23-B30)-insulin, des-octapeptide insulin
DP-insulin	des-Phe^{B1}-insulin
DPPI	des-(B26-B30)-insulin, des-pentapeptide insulin
Dsu	2,7-diaminosuberoyl
DTPI	des-(B27-B30)-insulin, des-tetrapeptide insulin
Fmoc	9-fluorenylmethoxycarbonyl
HOBt	1-hydroxybenzotriazole
Msc	methylsulfonylethoxycarbonyl
Napa	2-nitro-4-azidophenylacetyl
Np	4-nitrophenyl
OSu	succinimido-oxy
PA	photoaffinity
PAL	photoaffinity labelling
SL-insulin	spin-labelled insulin
Sub	suberoyl
Tfa	trifluoroacetyl

G.G. DODSON AND J.L. WHITTINGHAM

INSULIN: SEQUENCE, STRUCTURE AND FUNCTION - A STORY OF SURPRISES

Abstract. The determination of the structure of the insulin hexamer has made it possible to design amino acid substitutions which alter insulin's activity and association properties. This has given insight into insulin's biological function and has led to improvement in clinical preparations of insulin.

1. INTRODUCTION

The discovery of insulin in 1921 presented Chemistry, Biology and Medicine with an enormous challenge, namely the mechanism of the hormone's action, which in spite of immense progress, is still unresolved owing to the particular problems associated with the receptor structure and behaviour. The determination of the insulin sequence in 1954 (Brown et al., 1955), followed by the discovery of proinsulin (Steiner et al., 1967) and the successful analysis of insulin crystals (Adams et al., 1969), established the chemical and structural nature of the molecule. This explained its properties of self-assembly and provided a molecular basis through which the hormone's biosynthesis could be understood. Experiments in a number of laboratories have characterised insulin's aggregation from monomers to dimers, which in the presence of zinc assemble to hexamers. This pattern of self-association has been characterised by a range of biophysical and spectroscopic methods, and because of the knowledge of 3-dimensional structures of insulin monomers, dimers and hexamers provided by X-ray analysis, the roles of the individual residues in different stages of assembly are now generally understood. Nevertheless, interesting and fundamental puzzles still exist.

It was the advent of protein engineering which made it possible to change the hormone's self-association properties, by altering residues whose interactions in the dimer or hexamer were recognised as energetically or conformationally important for stability. There was an extremely important, practical reason for these studies. The insulin hexamer is too big for rapid absorption by the tissues, therefore the movement of insulin from the injection site to the serum is sluggish (Brange et al., 1988).

The obvious solution to the slow absorption of the insulin hexamer was to produce monomeric insulins by protein engineering. However, solutions of monomeric insulins are not stable enough for clinical preparations, hexamer formation being necessary to protect the molecule from chemical and physical degradation. The answer to this problem was to create stable hexamers that rapidly disassembled in the tissues and released active monomers into the blood. It is

M. Federwisch, M. Leyck Dieken and P. De Meyts (eds.), Insulin & Related Proteins, 29–39.

interesting that in the pancreatic B-cells where insulin is synthesised, Nature preserves the fragile insulin monomer by storing it as hexamers in specific membrane enclosed vesicles. The hexamers rapidly break down to monomers when they are released from the vesicles into the blood stream.

1.1. The Pattern of Assembly in Insulin

The structure of the insulin monomer, and its assembly to dimers and hexamers is illustrated in Figure 1. The surface that is buried during dimer formation, which consists of non-polar, mostly aromatic residues, has an essentially flat appearance. When the dimer is formed, the two monomers are arranged around a local two-fold axis. This arrangement presents the B-chain C terminal strands in an anti-parallel arrangement and allows the formation of four main chain - main chain hydrogen bonds that provide direction and favourable energies, and help to ensure compactness in the dimer. Inspection of the dimer shows that the main non-polar contacts at the interface involving the B-chain C terminal residues are made by B28 Pro packing against B23 Gly, B26 Tyr against B16 Tyr and B24 Phe, and B24 Phe against B24 Phe and B26 Tyr. There are other non-polar side chain contacts made by residues located on the main B-chain helix, including the packing of B12 Val against its B12 Val equivalent in the adjacent monomer, at the centre of the interface. Residue B13 Glu is adjacent to its two-fold related equivalent in the hexamer, however, the side chains in the dimer are turned away to reduce the effects of electrostatic repulsion. Comparisons of the insulin molecule as a dimer and as the monomer shows there are no significant structural changes, in particular the side chains alter little. The dimer-forming surfaces are essentially preformed.

The surfaces on the dimer buried by hexamer formation are entirely different. They consist of non-polar residues from the B-chain N terminus, including residues B1 Phe, B2 Val and B6 Leu. From the central part of the B-chain B9-B19 helix there are charged, polar and non-polar contacts made by residues B9 Ser, B10 His, B14 Val, B17 Leu and B18 Val. Residue B10 His has a particularly important role in hexamer formation since it is involved in zinc ion co-ordination. In contrast to the dimers, there are also important charged and non-polar contacts made by the A-chain, involving residues A13 Leu, A14 Tyr and A17 Glu.

1.2. Hexamer Conformations

Figure 1 also illustrates insulin's assembly to three kinds of hexamer. First there is the 2Zn insulin (solved in 1969), which is essentially symmetrical with all six monomers having the B1-B8 segment in an extended conformation (Baker et al., 1988). There are two zinc ions on the 3-fold axis, about 16Å apart. Each of these is co-ordinated in an octahedral arrangement to three B10 His side chains and three water molecules. The B1-B8 extended conformation is referred to as T, and thus in this notation the 2Zn hexamer can be described as T_6. Second is the 4Zn insulin hexamer, generated by high concentration of chloride. This crystal form was

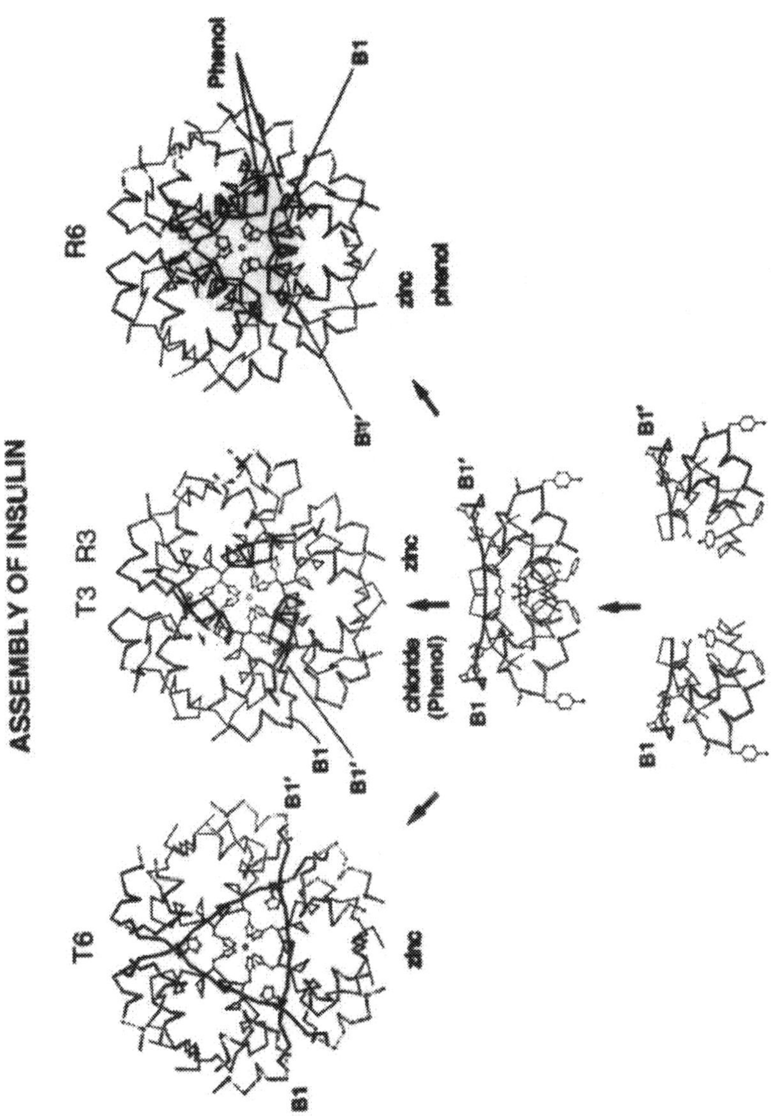

Figure 1. The assembly of insulin: monomer to dimer to hexamer. The main chain atoms of residues B1 to B8, which vary in conformation in the three hexamers, are highlighted in black.

discovered by Jørgen Schlichtkrull, and it had great value in diabetes therapy owing to its protracted action. The crystal structure was solved by Graham Bentley and Dan Mercola in 1976 (Bentley et al., 1976). To everyone's great surprise, the hexamer was no longer symmetrical. Three of the monomers, related by the hexamer's 3-fold axis, have B1-B8 extended as in T_6. In the other three monomers the B1-B8 segment forms a continuation of the B9-B19 helix seen in the T-state. These B1-B8 helical segments pack together about the central 3-fold axis. In this arrangement there are close contacts made by B6 Leu which closes off the channel of the central axis, greatly reducing the exchange of the zinc ion and consequently increasing the hexamer's stability. In this hexamer, the B10 His is sometimes turned away from the 3-fold axis and instead co-ordinates an off-axial zinc ion which is also co-ordinated to the B5 His side chain presented by the now helical B1-B8 segment. When this happens there are four zinc ion sites per hexamer – explaining Schlichtkrull's original analytical observation. The movement of residues B1-B8 in the T $\rightarrow$ R transition is about 25 Å. Remarkably, both the T $\rightarrow$ R and R $\rightarrow$ T transformations can be achieved in the rhombohedral crystals, indicating the insulin hexamer is a very plastic structure. It was the report that these crystal forms could be interconverted that led Axel Wollmer to investigate the transition in solution in a series of spectroscopic experiments that defined the influence of such factors as counter ions, additives and metal ions (Renscheidt et al., 1984).

Thirdly, there is the symmetrical insulin hexamer with all six monomers in the R state, the R_6 hexamer (Derewenda et al., 1989). Like 4Zn insulin, the structure of this hexamer caused surprise. This hexameric organisation depends on such reagents as phenol; so far it has not been generated by the chloride or thiocyanate ions. The phenol is buried between the dimers and lies under the B1-B8 helix making favourable van der Waals contacts. There are also well defined hydrogen bonds between the phenol hydroxyl group and the main chain CO of residue A6 Cys and NH of A11 Cys (Figure 2). As with the T_6 structure, the R_6 hexamer has two central zinc ions, co-ordinated by the B10 His side chains. In this case, however, the metal co-ordination is tetrahedral (usually involving chloride ion) owing to the bulk of the three B6 Leu side chains which prevent octahedral geometry.

The contacts made by B1-B8 are inevitably different in the T and R hexamers. In the T-state the α-amino group is salt-bridged to residue A17 Glu. The B1 Phe is buried nicely in a pocket between A14 Tyr and A13 Leu of the adjacent dimer: thus the dimers are cross-linked by this extended N-terminal segment. Finally, the B6 Leu contributes to a non-polar envelope between the dimers. In the R-state trimers the B1-B8 helix places B1 Phe and B2 Val on the surface where they are essentially hydrated. The favourable contacts between the three equivalent B6 Leu side chains serve to stabilise both trimer and to trap the zinc ions, and thus creating a much more robust hexamer than the T_3R_3 hexamer. The removal of B1 Phe from the non-polar pocket made by A14 Phe and A13 Leu leads to these side chains becoming much more mobile in the R state hexamer.

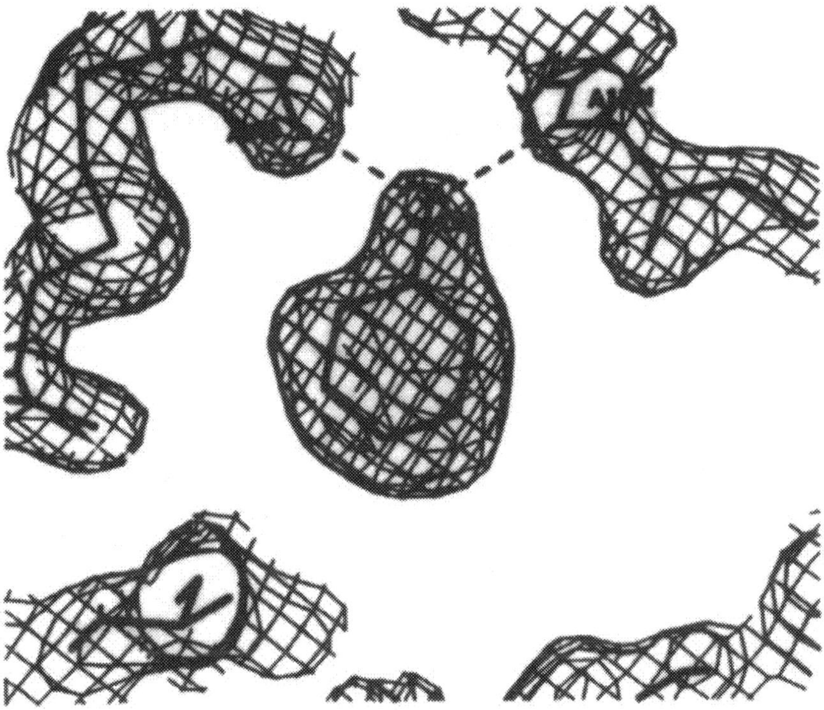

Figure 2. An electron density map, showing the binding of phenol in the R_6 insulin hexamer. Hydrogen bonds, securing the position of the phenol molecule, are shown by dashed lines.

2. MUTATIONS IN B9 SER, B10 HIS, B13 GLU AND B28 PRO

2.1. B9 Ser → Asp

Residue B9 Ser forms part of the hexamer's central channel in which axial zinc ions are located. The residue is close to B10 His and B13 Glu of adjacent dimers. As a result of the B9 Ser → Asp mutation, there are twelve carboxylic acids at the centre of the hexamer, and in the dimer a collection of four. Accordingly, at neutral pH B9 Asp insulin is a monomer in the presence and absence of zinc ions. This mutated insulin was used to try and crystallise a monomer. All attempts failed for a period of years.

In a systematic series of experiments carried out by Jens Brange, it was found that when zinc ions, calcium ions and phenol were added crystals appeared.

Previously, it had been assumed that because of the electrostatic repulsion between the six naturally occurring B13 glutamic acid side chains and the six mutated B9 aspartic acids in the centre of the hexamer, the crystals would contain monomeric insulin. But it was a hexamer that was formed – again a surprise. The Ca^{2+} ion was seen to sit between B9 Asp and B13 Glu, which had formed a new Ca^{2+} ion binding site (Figure 3). It acted across dimers, neutralising the repulsion, while the zinc co-ordinated the B10 His and the phenol added extra favourable dimer-dimer contacts (Dodson et al., 1993).

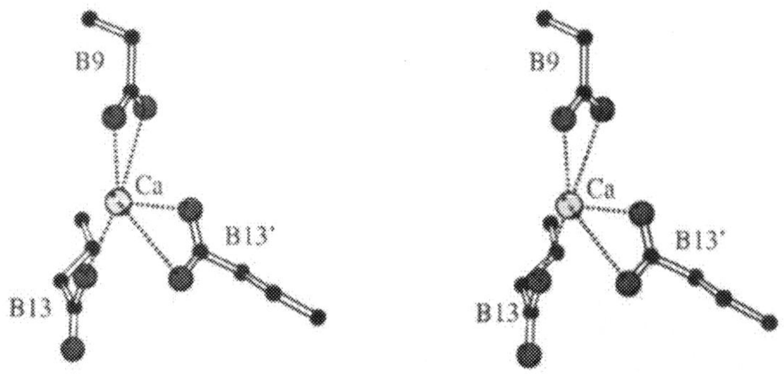

Figure 3. Stereoview representation showing the co-ordination geometry for one of the calcium ions bound in the centre of the B9Asp, B27Glu insulin hexamer. This figure was made using MOLSCRIPT (Kraulis, 1991).

2.2. B9 Ser → His

The mutation of B9 Ser to His was made to create further stability in the hexamer through extra zinc ion binding – to generate by this means a long acting low level insulin reservoir in order to try and create the natural pattern of continuous low-level insulin release from the pancreas. B9 His insulin crystallised as a T_3R_3 hexamer. The extra histidine side chains lead to a remarkable arrangement of zinc binding sites in the R_3 trimer involving B9 and B10 His, as seen in Figure 4. These extra zinc ion binding sites are located between the dimers and have stabilised the T_3R_3 conformation, which is inherently more stable than T_6. Hence, the B9 Ser → His mutation did indeed lead to a much more stable hexamer (M. Turkenburg, unpublished results).

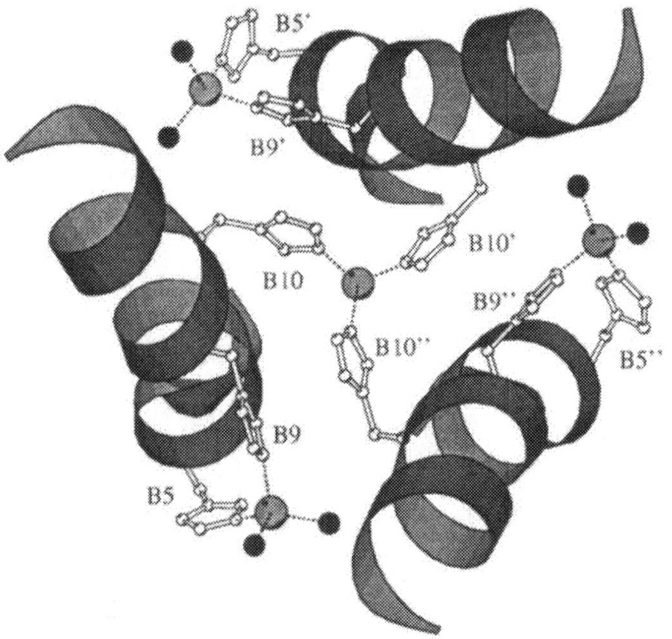

Figure 4. Zinc ion interactions at the centre of the R3 trimer of the T_3R_3 B9 His insulin hexamer. At each dimer-dimer interface, residues B9 His, B5 His and two water molecules (black spheres) form a binding site for a tetrahedrally co-ordinated zinc ion (grey sphere), off the hexameric 3-fold axis. Residues of different dimers in the hexamers are labelled as unprimed, single primed or double primed. This figure was made using MOLSCRIPT (Kraulis, 1991).

2.3. B10 His → Asp

The mutation of the B10 His to Asp prevents hexamer formation since the hexamer structure can no longer co-ordinate zinc along its 3-fold axis. However, in the presence of zinc ions very fragile crystals can be grown in which the B5 His acts to co-ordinate zinc ions. This was less of a surprise: the need for zinc suggested the B5 His might be involved in some way. The assembly generated by this B5 His zinc binding, however, was unexpected. It is a dodecamer made up of six T_2 dimers each presenting one of its two B5 His to the 3-fold axis where it binds a central zinc and creates a co-ordination geometry very similar to that seen in the hexameric insulin (Dodson et al., 1993), Figure 5.

Figure 5. Main chain atom trace of the dodecameric structure of B10 Asp, B28 Asp insulin. Six dimers co-ordinate around two zinc ions (white sphere) through the only remaining histidine side chain in the molecule, B5 His

2.4. B13 Glu → Gln

In the native insulin hexamer, the six essentially invariant B13 glutamate side chains are packed closely together. At neutral pH insulin does not form hexamers unless zinc is present. Clearly, the binding energy made available by the zinc co-ordination overcomes the electrostatic repulsion between the six B13 Glu side chains. When B13 Glu is mutated to Gln, there is no electrostatic repulsion and favourable hydrogen bonding interactions can be made (Figure 6). Hence, B13 Gln insulin forms hexamers in the absence of zinc, crystallising as a T_3R_3 structure (Bentley et al., 1992). It would be interesting to investigate the T → R kinetics of this mutant insulin since its behaviour suggests that the residues affect the relative stability of the two states. Axel Wollmer's laboratory would have been the best place to see this happen!

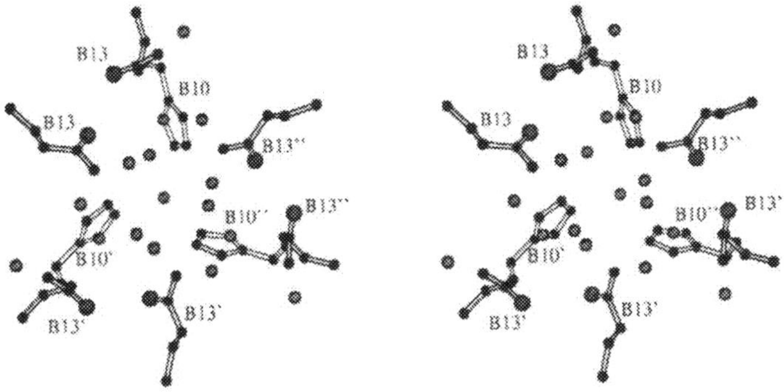

Figure 6. Stereoview representation of the centre of the zinc-free B13Gln insulin hexamer. In the absence of zinc ions the B10 His and B13 Gln side chains form hydrogen bonding to water molecules (grey spheres) clustered at the centre. This figure was made using MOLSCRIPT (Kraulis, 1991).

2.5. B28 Pro → Asp

In the 2Zn insulin hexamer residue B28 Pro sits nicely against the residue B23 Gly of an adjacent monomer. Its mutation to Asp would reduce the favourable van der Waals interactions (destabilising the dimer) and lead to an increase in electrostatic repulsion involving A4 Glu and the B21 Glu of adjacent dimers. Not surprisingly, experiments show that B28 Asp insulin is indeed monomeric at millimolar concentrations. The mutation of B28 Pro → Lys and B29 Lys → Pro prepared by the Lilly laboratories also produces a highly monomeric molecule – here, however, there is not an electrostatic repulsion, suggesting the van der Waals forces from the close contacts generated by this residue are a critical factor in hexamer stability. Crystals of B28 Asp have been grown with zinc ions, *m*-cresol and other phenols. The crystal structure shows how the B28 Pro → Asp mutation facilitates the loss of a crucial monomer-monomer van der Waals contact (Figure 7).

When crystals of the single-chain B28Asp-B29Lys-Ala-Ala-A1Gly insulin were grown in hexamer forming conditions using *m*-cresol, the R_6-state was expected, and was obtained. In additions to the six *m*-cresol molecules at the dimer-dimer interfaces, however, two extra *m*-cresols were inserted within one of the dimers, something that came as another complete surprise. The *m*-cresols were located between the two monomers, breaking the hydrogen bonds in the β-sheet that normally exists between them. Furthermore, this interaction breaks the 3-fold symmetry of the hexamer, something not previously seen in the studies on the hexamer. The insertion of the *m*-cresol into the dimer puts them into a non-polar environment but allows the hydroxyl groups to form hydrogen bonds. The breaking of the β-sheet hydrogen bonds between the monomers in the dimer is partly compensated by a water molecule bridging the two peptide backbones (Whittingham et al., 1998).

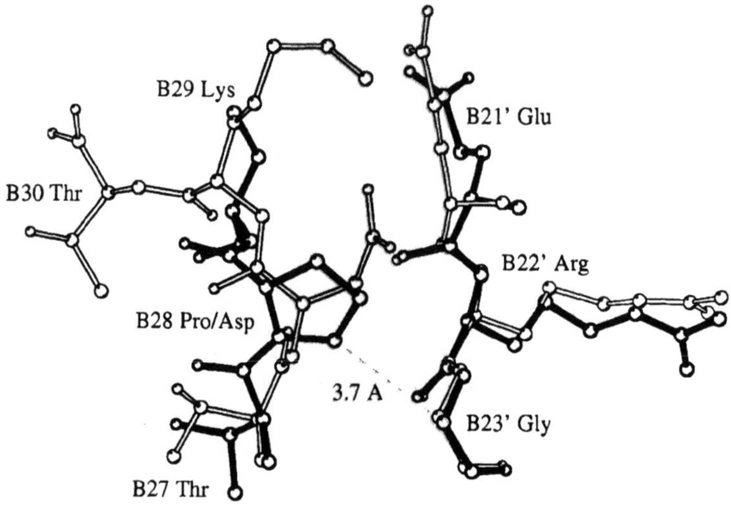

Figure 7. Comparison of B28 Asp insulin (white bonds) and the native R_6 insulin (black bonds), showing the effect of the B28 Pro → Asp mutation. The dashed line shows the close van der Waals contact between residues B28 Pro and B23 Gly in the native insulin dimer. This figure was made using MOLSCRIPT (Kraulis, 1991).

3. CONCLUSION

The knowledge of insulins hexameric structure and behaviour has made it possible to design mutants which give the hexamer new properties. The experiments on the hexamers and the protein engineering experiments particularly have continued to surprise, just as the earlier studies on the 4Zn insulin and on the R_6 phenol insulin did.

As a result of the research over the last 30 years we now have a wonderfully detailed picture of the insulin hexamer structures and a clear understanding of the kinetics of their transitions. Thus, the union of Crystallography, Spectroscopy and Chemistry has brought a fundamental and satisfying description to the pattern of insulin's assembly. Those studies, to which Axel Wollmer has made a consistent and important contribution over the years, have also revealed the adaptability of protein surfaces, the variety of interactions that they can make and their capacity for conformational alternatives.

4. REFERENCES

Adams M.J., Blundell T.L., Dodson E.J., Dodson G.G., Vijayan M., Baker E.N., Harding M.M., Hodgkin D.C., Rimmer B. and Sheat S. "Structure of rhombohedral 2 zinc insulin crystals." *Nature* 224 (1969): 491-495.

Baker E.N., Blundell T.L., Cutfield J.F., Cutfield S.M., Dodson E.J., Dodson G.G., Crowfoot D.M., Hodgkin D.C., Hubbard R.E., Isaacs N.W., Reynolds C.D., Sakabe K., Sakabe N. and Vijayan N.M. "The structure of 2Zn pig insulin crystals at 1.5 Å resolution." *Philosophical Transactions of the Royal Society of London* 319 (1988): 371-456.

Bentley G., Dodson E., Dodson G., Hodgkin D. and Mercola D. "Structure of insulin in 4-zinc insulin." *Nature* 261 (1976): 166-168.

Bentley G.A., Brange J., Derewenda Z., Dodson E.J., Dodson G.G., Markussen J., Wilkinson A.J., Wollmer A. and Xiao B. "Role of B13 in insulin assembly. The hexamer structure of recombinant mutant (B13 Glu to Gln) insulin." *Journal of. Molecular Biology* 228 (1992): 1163-1176.

Brange J., Ribel U., Hansen J.F., Dodson G., Hansen M.T., Havelund S., Melberg S.G., Norris F., Norris K., Snel L., Sørensen A.R. and Voigt H.O. "Monomeric insulins obtained by protein engineering and their medical implications." *Nature* 333 (1988): 679-682.

Brown H., Sanger F. and Kitai R. "The Structure of Pig and Sheep Insulin." *Biochemistry Journal* 60 (1955): 556-565.

Derewenda U., Derewenda Z., Dodson E.J., Dodson G.G., Reynolds C.D., Smith G.D., Sparks C. and Swenson D. "Phenol stabilizes more helix in a new symmetrical zinc insulin hexamer." Nature 338 (1989): 594-596.

Dodson E.J., Dodson G.G., Hubbard R.E., Moody P.C.E., Turkenburg J., Whittingham J., Xiao B., Brange J., Kaarsholm N. and Thogersen H. "Insulin assembly: its modification by protein engineering and ligand binding." *Philosophical Transactions of the Royal Society, London A* 345 (1993): 153-164.

Kraulis P.J. "MOLSCRIPT: a program to produce both detailed and schematic plots of protein structures." *Journal of Applied Crystallography* 24 (1991): 946-950.

Renscheidt H., Strassburger W., Glatter U., Wollmer A., Dodson G.G. and Mercola D.A. "A solution equivalent of the 2Zn $\rightarrow$ 4Zn transformation of insulin in the crystal." *European Journal of Biochemistry* 142 (1984): 7-14.

Steiner D.F. and Oyer P.E. "The biosynthesis of insulin and probable precursor of insulin by a human islet cell adenoma." *Proceedings of the National Academy of Sciences USA* 57 (1967): 473-480.

Whittingham J.L., Edwards D.J., Antson A.A., Clarkson J.M. and Dodson G.G. "Interactions of Phenol and *m*-Cresol in the Insulin Hexamer, and their Effect on the Association Properties of B28 Pro $\rightarrow$ Asp Insulin Analogues." *Biochemistry* 37 (1998): 11516-11523.

U. HASSIEPEN, J. STAHL, B. MOTZKA AND M. FEDERWISCH

THE ASSOCIATION/DISSOCIATION EQUILIBRIA OF INSULIN IN THE PRESENCE OF METAL IONS: A FLUORESCENCE ENERGY TRANSFER AND CIRCULAR DICHROISM STUDY

Abstract. The association/dissociation behaviour in the presence of co-ordinating metal ions and T-R transforming ligands is too complex for rigorous analysis. In the present study equimolar mixtures of insulin labelled with a fluorescence donor and with an acceptor group, respectively, were used to measure fluorescence resonance energy transfer as a function of concentration. These experiments were performed in the presence of 2 Zn or 2Co ions per hexamer and in the absence and presence of KSCN or 3-pentanol, respectively. The results were analysed on the basis of a monomer-dimer-hexamer model in terms of apparent equilibrium constants K_{12} and K_{26}. The experiments were parallelled by measurements of the near-UV circular dichroism on unlabelled insulin and analysed accordingly. Both sets of experiments show that K_{26} is not much more increased by Zn than by Co ions and much less than under R_6 conditions.

1. INTRODUCTION

The association of insulin has been the subject of several investigations (see references in Kadima et al., 1993). The techniques used so far are sedimentation equilibrium studies (Pekar and Frank, 1972; Goldman and Carpenter, 1974; Jeffrey et al., 1976; Holladay et al., 1977; Mark and Jeffrey, 1990; Brems et al., 1992), difference spectroscopy (Strazza et al., 1985), dynamic light scattering (Hvidt, 1991; Kadima et al., 1993), circular dichroism (CD) (Goldman and Carpenter, 1974; Wood et al., 1975; Pocker and Biswas, 1981; Melberg and Johnson, 1990) and fluorescence energy transfer (Hassiepen et al., 1998, 1999). Most of the investigations focussed on metal-free insulin. Metal bound insulin is of importance with respect to storage in the ß-cells as well as on the shelf and to pharmacokinetics. To the best of our knowledge neither the influence of different co-ordinating metals nor of the T_3R_3 and R_6 conformations of the insulin hexamer on its association/dissociation behaviour have been quantitatively evaluated, except in the publication of Hvidt (1991).

While the use of CD spectroscopy for the study of insulin association is well established (see above), fluorescence energy transfer was used only recently. Both techniques were applied in the present study and their respective results compared. The present approach evaluates the effect of dilution on the fluorescence of

M. Federwisch, M. Leyck Dieken and P. De Meyts (eds.), Insulin & Related Proteins, 41–51.

equimolar mixtures of donor- and acceptor-labelled insulin (I_d and I_a). 2-Aminobenzoyl (Abz) and 3-nitrotyrosyl residues Tyr(NO_2), respectively, attached to the N_ε-amino group of LysB29, served as fluorescence donor/acceptor pair. Their Förster distance (Förster, 1948) of 29 Å closely corresponds to the separation of the attachment sites in the insulin dimer and hexamer. Labelling of insulin at the ε-amino group of LysB29 neither affects the tertiary structure nor interferes with self-association, hence, the labelled insulins are representative of the native hormone (Hassiepen et al., 1998). Energy transfer from the donor group to the quenching acceptor group only occurs when I_d and I_a coexist as subunits in the very same oligomer. Therefore, the fluorescence intensity, normalised to the donor concentration, increases upon dilution reflecting the dissociation of the oligomers.

The same experiments were carried out using the near-UV CD spectrum to monitor the dissociation. In the case of insulin self-association, ligand induced quaternary structure formation and T→R transformation are reflected in its CD spectrum (Wood et al., 1975; Renscheidt et al., 1984; Wollmer et al., 1987). In the near ultraviolet the main band at 275 nm, the tyrosyl L_b band, increases with dimer, tetramer and hexamer formation. This increase originates in additional perturbation of the tyrosyl chromophore by chromophores in neighbouring subunits, occurring as the quaternary structure assembles (Goldman and Carpenter, 1974; Strickland and Mercola, 1976; Wollmer et al., 1977; Mercola and Wollmer, 1981) as well as in a loss of conformational mobility for chromophoric side chains buried in the interfaces between subunits. Changes around 250 nm are related to changes in the disulphide chromophores occurring upon T→R transition. There are concomitant changes also in the far ultraviolet.

To analyse the insulin association different models were used as reviewed in (Mark and Jeffrey, 1990). Basically there are two models: 1. The monomer-dimer-tetramer-hexamer model (Goldman and Carpenter, 1974; Jeffrey et al., 1976; Wollmer et al., 1980; Pocker and Biswas, 1981). The association not necessarily stops at the hexamer (Brems et al., 1992). 2. The monomer-dimer-hexamer model (Pekar and Frank, 1972; Holladay et al., 1977; Hvidt, 1991). This work uses the second model. The binding metal also promotes the association of metal binding oligomers providing further binding sites for the metal. Because of these difficulties only "apparent" association constants rather than true thermodynamic parameters are obtainable. Since the results were produced for internal comparison only, this may be acceptable.

2. MATERIALS AND METHODS

2.1. Reagents and Solvents

All reagents and solvents were of analytical grade and are commercially available. Porcine $N\varepsilon^{B29}$-2-aminobenzoyl-insulin ($N\varepsilon^{B29}$-Abz-insulin, I_d) carrying the fluorescence donor group and porcine $N\varepsilon^{B29}$-3-nitro-L-tyrosyl-insulin ($N\varepsilon^{B29}$-

$Tyr(NO_2)$-insulin, I_a) carrying the acceptor group were prepared as described earlier (Lenz et al., 1994; Hassiepen et al., 1998). Human insulin was a gift from Novo Nordisk A/S (Copenhagen, Denmark).

2.2. Solutions for Spectroscopy

The solvent for insulin used throughout the study was 25 mM Tris, pH 7.8, with 0.33 eq. divalent metal ions/monomer. At this insulin / metal ratio only the high affinity zinc binding site is occupied and the aggregation stops at the hexamer (Grant et al., 1972). In some cases the metal was left out or 700 mM KSCN or 365 mM 3-pentanol were added. At these concentrations the hexamer conformation is T_3R_3 or R_6, respectively (Renscheidt et al. 1984; Pittman and Tager, 1995). The near-UV transparency of 3-pentanol is better than that of aromatic transforming agents. Insulin concentrations were determined using a Pye Unicam PU 8800 UV/Vis spectrophotometer (Philips, Kassel, Germany). The molar extinction coefficients at 276 nm used are: ε_d = 7,090 $M^{-1}cm^{-1}$, ε_a = 11,000 $M^{-1}cm^{-1}$ (Hassiepen et al., 1998) and ε = 6,070 $M^{-1}cm^{-1}$ for insulin (Frank and Veros, 1968).

2.3. Fluorescence Spectroscopy

Fluorescence measurements were made on a Spex Fluorolog 211 photon counting spectrometer (Spex Industries, NY, USA). The system provides corrections for changes in the lamp intensity and for the spectral sensitivity of the emission monochromator and photomultiplier.

The measurements were carried out at 20 °C with a bandwidth of 1.8 nm on both the excitation and emission side. Corrections for inner filter effects were calculated as described both on the excitation and emission side (Lakowicz, 1983).

The excitation wavelength for the Abz fluorophore was 330 nm throughout. The fluorescence intensity was integrated in the range between 400 and 430 nm to reduce noise. To analyse the measurements the relative quantum yield normalised to the fluorescence intensity of the Abz labelled insulin was calculated. In the case of pure Abz insulin the quantum yield is independent of the concentration.

2.4. Circular Dichroism Spectroscopy

CD measurements were carried out on a Jasco J-600 spectropolarimeter (Japan Spectroscopic Co., Ltd., Tokyo, Japan), calibrated according to Chen and Yang (1977). The spectral bandwidth was 1 nm. The temperature was 27°C.

2.5. Analysis

Knowing the Förster distance of the donor/acceptor pair, the distances of the labels in the oligomers and assuming random distribution of the subunits within the oligomer, it is possible to calculate the quantum yield. The relative quantum yield of a monomer within a dimer or a hexamer is equal to 0.815 and 0.5, respectively (Hassiepen et al., 1998, 1999). Together with the law of mass action this allows to calculate the relative quantum yield of all fluorophores in the mixture at a given concentration. The following equations were used:

$$c_{tot} = c_{Mon} + 2 \cdot c_{Dim} + 6 \cdot c_{Hex} \quad (1)$$

$$Q_{rel} = v_l \cdot [c_{Mon} + 2 \cdot 0.815 \cdot c_{Dim} + 6 \cdot 0.5 \cdot c_{Hex}] / c_{tot} \quad (2)$$

$$K_{12} = c_{Dim} / (c_{Mon})^2 = 0.72 \cdot 105 [M^{-1}] \quad (3)$$

$$(K_{12})^3 \cdot (K_{26})^2 = c_{Hex} / (c_{Mon})^6 \quad (4)$$

All parameters influencing the association behaviour of insulin (co-ordinating metal, conformation of the hexamer, etc.) are included in K_{26} while K_{12} is assumed to be unaffected. This analysis assumes ideality and neglects charge effects so that "apparent" association constants are obtained rather than true thermodynamic parameters. Since the results were meant for internal comparison only, this may be acceptable. Furthermore, accounting for non-ideality or using more complex models would introduce additional parameters accompanied by greater uncertainties while little is gained (Mark and Jeffrey, 1990).

This system of equations cannot be solved analytically. To fit the measured data from a dilution series K_{26} was varied. For a given combination of K_{12} and K_{26} the corresponding concentration of oligomers was calculated using equations (3) and (4). Then the total concentration was calculated using equation (1). The monomer concentration was increased successively until the given total concentration of the measurement was reached. Having calculated the oligomer distribution, Q_{rel} was calculated using equation (2). Then v_l was varied until an optimal fit for each measured point of the curve was achieved. K_{26} was slightly changed and the next round of calculations started until the fit did not improve further.

To analyse the circular dichroism measurements the same set of equations was used except that (2) was replaced by equation (5):

$$[\Theta]^{MRW} = v_l \cdot \{- 83 [deg \cdot cm^2 \cdot dmol^{-1}] \cdot c_{Mon} - 215 [deg \cdot cm^2 \cdot dmol^{-1}] \cdot 2 \cdot c_{Dim} - 290 [deg \cdot cm^2 \cdot dmol^{-1}] \cdot 6 \cdot c_{Hex}\} / c_{tot} \quad (5)$$

The $[\Theta]^{MRW}$ values of the monomer and dimer were calculated on the basis of the signal ratios of the oligomers (Strickland and Mercola, 1976) and a signal of the insulin hexamer bound to Zn^{2+} of 290 [deg · cm² · dmol⁻¹].

3. RESULTS

3.1. Fluorescence

The 1:1 mixtures of $N\varepsilon^{B29}$-Abz-insulin and $N\varepsilon^{B29}$-Tyr(NO_2)-insulin were prepared as described in Materials and Methods. The solution was successively diluted adding buffer lacking only the insulin and the co-ordinating metal. Hence, the concentration of the transforming agent was constant. For each concentration the quantum yield was determined. A typical curve for each buffer condition is shown in Figure 1. The curve in the presence of Zn^{2+} was left out for the sake of clarity. The data for insulin alone were added for comparison. The other curves were fitted as described in Materials and Methods. Each condition was measured three times and the resulting mean value and standard deviation are given in Table 1.

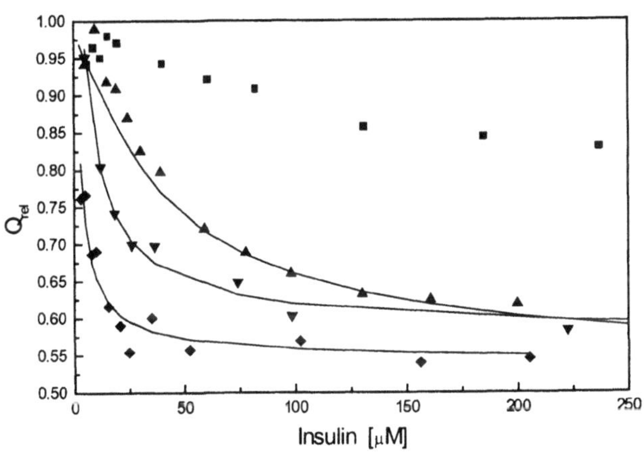

Figure 1. The decrease of the relative quantum yield, Q_{rel}, with increasing insulin concentration. Buffer conditions and the calculation of Q_{rel} are described in Materials and Methods. Insulin only: ■; insulin plus Co^{2+}: ▲; insulin plus Zn^{2+} and KSCN: ▼; insulin plus Zn^{2+} and 3-pentanol: ◆. Continuous line: Optimum fit of the measured values with equation 2. The fitted values are given in Table 1.

Table 1. Hexamerisation constant K_{26} determined by fluorescence energy transfer and dilution in three independent experiments. Given are the mean values and the standard deviations. Typical curves are shown in Figure 1.

metal	transforming agent	$K_{26} \times 10^{-3}$ [M^{-1}]
Co^{2+}	-	68.2 ± 0.5
Zn^{2+}	-	255.0 ± 2.1
Zn^{2+}	KSCN	1,240.0 ± 440.0
Zn^{2+}	3-pentanol	9,860.0 ± 2,600.0

3.2. *Circular Dichroism*

Solutions of human insulin were prepared as described in Materials and Methods. The solution was successively diluted adding buffer in which only the insulin and the co-ordinating metal were lacking. Hence, the concentration of the transforming agent was constant. For each concentration the $[\Theta]^{MRW}$ was determined. The measured points for each buffer condition are shown in Figure 2. The curves were fitted as described in Materials and Methods and the fitted curves are included in Figure 2. The resulting values are given in Table 2.

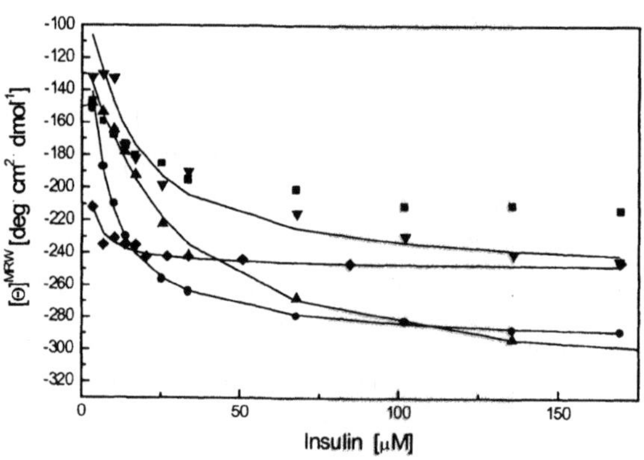

Figure 2. The increase of CD signal at 275 nm with increasing insulin concentration. Buffer conditions and the fitting procedure are described in Materials and Methods. Insulin only: ■; insulin plus Zn^{2+} and KSCN: ▼; insulin plus Zn^{2+}: ●; insulin plus Co^{2+}: ▲; insulin plus Zn^{2+} and 3-pentanol: ◆. Continuous line: Optimum fit of the measured values with equation 5. The fitted values are given in Table 2.

Table 2. Hexamerisation constant K_{26} determined by circular dichroism spectroscopy. The underlying curves and fits are shown in Figure 2.

metal	transforming agent	$K_{26} \times 10^{-3}$ [M^{-1}]
Co^{2+}	-	86.8
Zn^{2+}	-	1,210.0
Zn^{2+}	KSCN	250.0
Zn^{2+}	3-pentanol	109,800.0

It has to be taken into account that the CD signal is also dependent on the co-ordinating metal and transforming agent which explains for example the smaller signal amplitude in the presence of 3-pentanol although the equilibrium is shifted more on the hexamer side in this case. Diluting the 3-pentanol sample with pure buffer brought the signal back to that of the free insulin monomer.

It is now possible to calculate on the basis of the above assumed association scheme the fractional distribution of oligomers at any given total protein concentration (Figure 3).

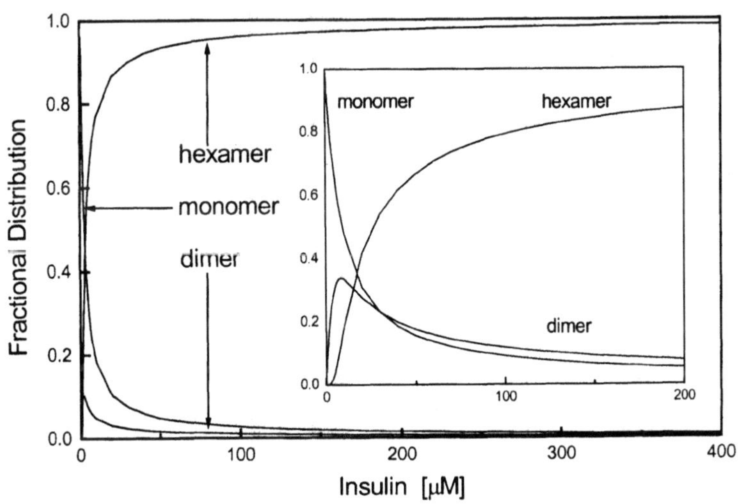

Figure 3: The fractional distribution (based on the number of monomers involved) of insulin monomer, dimers and hexamer as a function of the insulin concentration. The curves were calculated for insulin in the presence of Zn^{2+} and 3-pentanol or Zn^{2+} alone (inset), hence, $K_{26} = 9.86 \times 10^6\ M^{-1}$ and $K_{26} = 2.55 \times 10^5\ M^{-1}$, respectively (see Table 1).

4. DISCUSSION

The measurements were analysed with the monomer-dimer-hexamer model. The fits are reasonably good, hence, based on the present experiments there is no need to use one of the other association schemes discussed in the literature. Discrimination between alternative models might have required higher final concentrations (Jeffrey et al., 1976) or more measured points.

The advantage of the fluorescence experiments is threefold: 1. Donor fluorescence and acceptor absorption are far away from the absorption of the transforming agents. 2. The relative quantum yield of the oligomers involved can be calculated on basis of the Förster distance of the donor/acceptor pair, the distances in the oligomer known from the crystal structure and the proven statistical distribution of donor and acceptor in the oligomer (Hassiepen et al., 1998, 1999). This is an advantage because under experimental conditions insulin solutions always contain a mixture of oligomer species. 3. Monomer and dimer have a similar relative quantum yield which is remote from that of the hexamer. Since the determination of the hexamer equilibrium constant is the aim of this study their distribution compared to the monomer and dimer distribution can be determined with better accuracy.

The monomer-dimer-hexamer model can also be used to fit the circular dichroism series. In this case the signal amplitude of the oligomers has to be taken from the literature. Since under experimental conditions there are always dimers present, the monomer signal has to be calculated on the basis of a dimerization constant. The problem may well be circumvented by a global analysis of all CD measurements taking also the monomer and dimer signal as variables, but assuming them to be the same in all experimental set-ups. On the contrary, the hexamer signal is dependent on the co-ordinating metal (Goldman and Carpenter, 1974; Thomas and Wollmer, 1989) and the transforming agent. Because of these uncertainties the fluorescence measurements are more reliable. Therefore, the latter were carried out three times to allow the determination of the standard deviation.

Nevertheless, both series yield approximately the same equilibrium constants and, except for KSCN, also the same order of the equilibrium constants. It has to be taken into account that the KSCN series suffers most from strong absorption. The stability of the hexamer is dependent on the co-ordinating metal and on the conformation of the hexamer. The stability is significantly higher in the R_6 state. This has already been suggested by Derewenda and colleagues (1989) for phenol-like molecules because they also participate in the contacts between the subunits of the trimers in the hexamer. The decreased tendency of R_6 hexamers to dissociate is in accordance with the finding that the subunit interchange between them is dramatically slower than between T_6 hexamers (Hassiepen et al., 1999).

Table 3. Apparent association constants for insulin.
The data refer to metal-free conditions except for those provided by Hvidt ($Zn^{2+}R_6$).
The method used was equilibrium ultracentrifugation, only Pocker and Biswas and Hvidt used circular dichroism measurements and light scattering, respectively.
*+IIH model, ideal treatment. *Even aggregation model.*
S = species: p, porcine; b, bovine; h, human.

Reference	S	pH	$K_{12}\times10^{-5}$ $[M^{-1}]$	K_{24} $[M^{-1}]$		
Pekar and Frank (1972)	p	7.0	1.4		$K_6 = 4.0\times10^8$ M^{-2}	$K_{26} = K_6^{1/2}$ 20,000 M^{-1}
Goldman and Carpenter (1974)	b	8.0	2.22	40	$K_{46} = 220$ M^{-1}	$K_{26} = (K_{24} \cdot K_{46})^{1/2} =$ 94 M^{-1}
Jeffrey et al. (1976)	b	7.0	0.11	17,000*	$K_{46}^{*} = 17,000$ M^{-1}	$K_{26} = K_{24} = K_{46} =$ 17,000 M^{-1}
Holladay et al. (1977)	b	7.4	0.22		$K_6 = 8.6\times10^9$ M^{-2}	$K_{26} = K_6^{1/2} =$ 92,700 M^{-1}
Pocker and Biswas (1981)			7.5	5,000	$K_{46} = 45$ M^{-1}	$K_{26} = (K_{24} \cdot K_{46})^{1/2} =$ 474 M^{-1}
Mark and Jeffrey (1990)	b	7.0	0.85		$K_6^{+} = 5.17\times10^8$ M^{-2}	$K_{26} = K_6^{1/2} =$ 22,700 M^{-1}
Brems et al. (1992)	h	7.2	3	12,000*	$K_{46}^{*} = 12,000$ M^{-1}	$K_{26} = K_{24} = K_{46} =$ 12,000 M^{-1}
Hvidt (1991)	p	7.3	1.4		$K_6 = 1.5\times10^{11}$ M^{-2}	$K_{26} = K_6^{1/2} =$ 3.9×10^5 M^{-1}

There is only one value of K_{26} for metal bound insulin available in the literature (Hvidt, 1999) which refers to R_6 conditions (see Table 3). The value determined here is reasonably close to it. On the whole the values derived from the circular dichroism measurements are less consistent. A shortcoming of the CD analysis is that assumed signals for monomers and dimers had to be used. Improvements could be expected if the accurate values for these signals became accessible

5. REFERENCES

Brems D.N., Alter L.A., Beckage M.J., Chance R.E., DiMarchi R.D., Green L.K., Long H.B., Pekar A.H., Shields J.E. and Frank B.H. "Altering the association properties of insulin by amino acid replacement." *Protein Eng.* 5 (1992): 527-533.

Chen G.C. and Yang J.T. "Two-point calibration of circular dichrometer with d-10-camphersulfonic acid." *Anal. Lett.* 10 (1977): 1195-1207.

Derewenda U., Derewenda Z., Dodson E.J., Dodson G.G., Reynolds C.D., Smith G.D., Sparks C. and

Swenson D. "Phenol stabilizes more helix in a new symmetrical zinc insulin hexamer." *Nature* 338 (1989): 594-596.

Förster T. "Zwischenmolekulare Energiewanderung und Fluoreszenz." *Ann. Phys.* 2 (1948): 55-75.

Frank B.H. and Veros A.J. "Physical studies on proinsulin-association behavior and conformation in solution." *Biochem. Biophys. Res. Commun.* 32 (1968): 155-160.

Goldman J. and Carpenter F.H. "Zinc binding, circular dichroism, and equilibrium sedimentation studies on insulin (bovine) and several of its derivatives." *Biochemistry* 13 (1974): 4566-4574.

Grant P.T., Coombs T.L. and Frank B.H. "Differences in the nature of the interaction of insulin and proinsulin with zinc." *Biochem. J.* 126 (1972): 433-440.

Holladay L.A., Ascoli M. and Puett D. "Conformational stability and self-association of zinc-free bovine insulin at neutral pH." *Biochim. Biophys. Acta. 494* (1977): 245-254.

Hassiepen U., Federwisch M., Mülders T., Lenz V.J., Gattner H.-G., Krüger P. and Wollmer A. "Analysis of protein self-association at constant concentration by fluorescence-energy transfer." *Eur. J. Biochem.* 255 (1998): 580-587.

Hassiepen U., Federwisch M., Mülders T. and Wollmer A. "The lifetime of the insulin hexamer." *Biophys. J.* 77 (1999): 1638-1654.

Hvidt S. "Insulin association in neutral solutions studied by light scattering." *Biophys. Chem.* 39 (1991): 205-213.

Jeffrey P.D., Milthorpe B.K. and Nichol L.W. "Polymerization pattern of insulin at pH 7.0." *Biochemistry* 15 (1976): 4660-4665.

Kadima W., Ogendal L., Bauer R., Kaarsholm N., Brodersen K., Hansen J.F. and Porting P. "The influence of ionic strength and pH on the aggregation properties of zinc-free insulin studied by static and dynamic laser light scattering." *Biopolymers* 33 (1993): 1643-1657.

Lakowicz J.R. "Principles of fluorescence spectroscopy." Plenum Press, N.Y. (1983).

Lenz V.J., Gattner H.G., Leithäuser M., Brandenburg D., Wollmer A. and Höcker H. "Proteolyses of a fluorogenic insulin derivative and native insulin in reversed micelles monitored by fluorescence emission, reversed-phase high-performance liquid chromatography, and capillary zone electrophoresis." *Anal. Biochem.* 221 (1994): 85-93.

Mark A.E. and Jeffrey P.D. "The self-association of zinc-free bovine insulin. Four model patterns and their significance." *Biol. Chem. Hoppe Seyler* 371 (1990): 1165-1174.

Melberg S.G. and Johnson W.C. "Changes in secondary structure follow the dissociation of human insulin hexamers: a circular dichroism study." *Proteins* 8 (1990): 280-286.

Mercola D. and Wollmer A. "The crystal structure of insulin and solution phenomena: Use of the high-resolution structure in the calculation of the optical activity of the tyrosyl residues." in: Dodson, G., Glusker, J.P., and Sayre, D. (Eds.), *Structural studies on molecules of biological interest*, Clarendon Press, Oxford (1981), 557-582.

Pekar A.H. and Frank B.H. "Conformation of proinsulin. A comparison of insulin and proinsulin self-association at neutral pH." *Biochemistry* 11 (1972): 4013-4016.

Pittman I. and Tager H.S. "A spectroscopic investigation of the conformational dynamics of insulin in solution." *Biochemistry.* 34 (1995): 10578-10590.

Pocker Y. and Biswas S.B. "Self-association of insulin and the role of hydrophobic bonding: a thermodynamic model of insulin dimerization." *Biochemistry* 20 (1981): 4354-4361.

Renscheidt H., Strassburger W., Glatter U., Wollmer A., Dodson G.G. and Mercola D.A. "A solution equivalent of the 2Zn->4Zn transformation of insulin in the crystal." *Eur. J. Biochem.* 142 (1984): 7-14.

Strazza S., Hunter R., Walker E. and Darnall D.W. "The thermodynamics of bovine and porcine insulin and proinsulin association determined by concentration difference spectroscopy." *Arch. Biochem. Biophys.* 238 (1985): 30-42.

Strickland E.H. and Mercola D. "Near-ultraviolet tyrosyl circular dichroism of pig insulin monomers, dimers, and hexamers. Dipole-dipole coupling calculations in the monopole approximation." *Biochemistry* 15 (1976): 3875-3884.

Thomas B. and Wollmer A. "Cobalt probing of structural alternatives for insulin in solution." *Biol. Chem. Hoppe Seyler* 370 (1989): 1235-1244.

Wollmer A., Fleischhauer J., Strassburger W. and Thiele H. "Side-chain mobility and the calculation of tyrosyl circular dichroism of proteins. Implications of a test with insulin and des-B1-phenylalanine insulin." *Biophys. J.* 20 (1977): 233-243.

Wollmer A., Straßburger W., Hoenjet E., Glatter U., Fleischhauer J., Mercola D.A., de Graaf R.A.G., Dodson E.J., Dodson G.G., Smith D.G., Brandenburg D. and Danho W. "Correlation of structural details of insulin in the crystal and in solution." in: Brandenburg, D., Wollmer, A. (Eds.), *Insulin: Chemistry, structure and function of insulin and related hormones*, Walter de Gruyter, Berlin (1980), 27-35.

Wollmer A., Rannefeld B., Johansen B.R., Hejnaes K.R., Balschmidt P. and Hansen F.B. "Phenol-promoted structural transformation of insulin in solution." *Biol. Chem. Hoppe Seyler* 368 (1987): 903-911.

Wollmer A., Rannefeld B., Stahl J. and Melberg S.G. "Structural transition in the metal-free hexamer of protein-engineered [B13 Gln]insulin." *Biol. Chem. Hoppe Seyler* 370 (1989): 1045-1053.

Wood S.P., Blundell T.L., Wollmer A., Lazarus N.R. and Neville R.W. "The relation of conformation and association of insulin to receptor binding; x-ray and circular-dichroism studies on bovine and hystricomorph insulins." *Eur. J. Biochem.* 55 (1975): 531-542.

6. ACKNOWLEDGEMENTS

The authors thank Axel Wollmer for encouragement and helpful discussions.

A. WOLLMER AND T. MÜLDERS

SELF-ASSOCIATION OF INSULIN RECONSIDERED

I. A Scheme Based on Contact Surfaces

Abstract. The self-association of insulin has been a challenge for half a century. According to the crystal structure there are only two types of contacts between subunits in the hexamer, a strong and a weak one. This suggests that two corresponding equilibrium constants should suffice to describe the association/dissociation behaviour of insulin in solution.
Twenty years ago we put forward a scheme based on this notion. However, it turned out to be at variance with certain experimental observations and to leave open questions. A recent publication on a strategy applied in drug design to increase the binding potency of ligands to target proteins made us realize that the same principle is effective also in the self-association of insulin, which induced us to reconsider the old model.

1. INTRODUCTION

The structure analysis of the insulin hexamer has revealed two types of contacts between neighbour subunits, differing in strength. The stronger contact about the approximate two-fold axis 0P ($K_P \approx 10^5$ M^{-1}) is largeley composed of nonpolar side chains and a stretch of antiparallel ß-structure between residues 24-28 of the B-chains. Though involving more residues in nonpolar and hydrogen-bonding interactions in all, the contact about the approximate 2-fold axis 0Q is the weaker one ($K_Q \approx 10^3$ M^{-1}) (Blundell et al., 1972). Monomers held together by the strong contact are designated "classical" dimers D, those by the weak contact "alternative" dimers D'. Because of the existence of these two types of contacts only, it should be possible to describe the association/dissociation of insulin as a function of concentration in terms of the two corresponding contact constants K_P and K_Q.

2. ASSOCIATION SCHEME BASED ON TWO CONTACTS

Twenty years ago we put forward an association scheme based on this notion (Dodson et al., 1980; Wollmer et al., 1980; Hoenjet, 1981). Its reproduction in Figure 1 illustrates that the self-assembly of insulin, in striking contrast to, e.g., isodesmic head-to-tail association, is finite. This implies that there should be an intrinsic strive for completing the hexamer, as any final step involves two contacts at the same time. The completion of the hexamer thus very much resembles the completion of a vault by insertion of the keystone. Calculated on the basis of this scheme, the solution population as a function of the total insulin concentration more or less reduces to monomers, dimers and hexamers. As a consequence of the

M. Federwisch, M. Leyck Dieken and P. De Meyts (eds.), Insulin & Related Proteins, 53–65.

disparity of K_P and K_Q and of the keystone effect, association largely overrides the other species. However, for metal-free neutral conditions this prediction is at variance with experimental experience which in itself is highly contradictory.

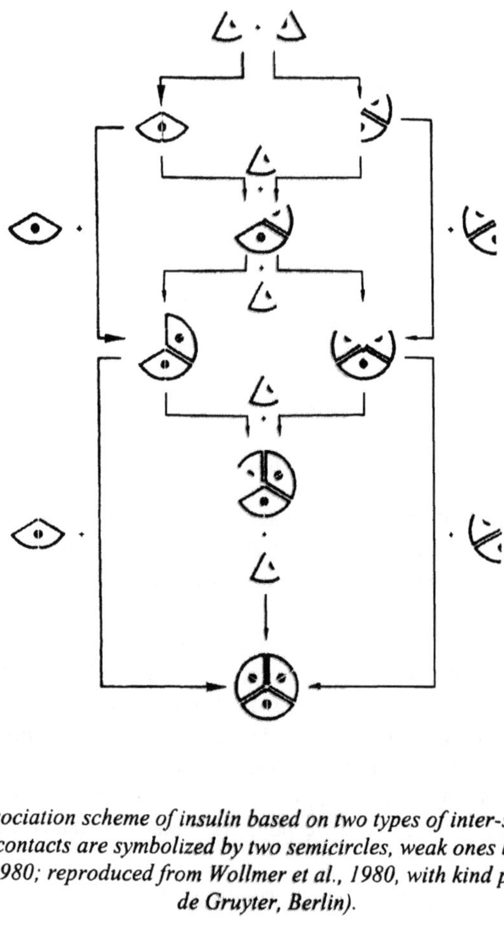

Figure 1. Association scheme of insulin based on two types of inter-subunit contacts. Strong contacts are symbolized by two semicircles, weak ones by two bars (Dodson et al., 1980; reproduced from Wollmer et al., 1980, with kind permission of Walter de Gruyter, Berlin).

Whereas hexamer formation at concentrations ≤ 12 g/l was derived primarily from molecular mass measurements (Pekar and Frank, 1972; Mark et al., 1987; Hansen, 1991; Kadima et al., 1993), spectroscopic methods capable of specifying the nature of the species present have provided convincing evidence that tetramers do form to a substantial extent (Kadima et al., 1992; Lin and Larive, 1995) and hexamers do not. The concentration dependence of the NMR spectrum at pH 9 in the range 8 – 28 g/l and presence of 100 mM NaCl was interpretable only in the sense of a dimer/tetramer equilibrium, since the persistence of two resonances for the C2 HisB10 protons of equal intensity was incompatible with hexamer formation (Kadima et al., 1992). This is in accordance with the observation that the near-UV

CD spectrum of metal-free insulin under these very conditions is markedly weaker than that of insulin at 1 g/l, hexameric owing to presence of 0.33 M Zn^{2+}/ M monomer (Figure 2). Essentially the same difference is observed at pH 7.5 with 10.4 g/l and 33 mM NaCl (Figure 2), i.e., under conditions for which the very detailed light-scattering study of Kadima et al. (1993) reported an average molecular weight also corresponding to the tetramer. Hence, the spatial arrangement of the chromophores generating the tyrosyl CD by inter-subunit interaction in the hexamer (Strickland and Mercola, 1976; Wollmer et al., 1977; Mercola and Wollmer, 1981) is not established. It should however be noted that the hexamer spectrum was not matched either at pH 7.5 and 100 mM NaCl, neither at 12 g/l for which conditions the reported average molecular weight was ≈36,000 Dalton (Kadima et al., 1993), nor even at 30 g/l (Figure 2). Contrary to wild-type insulin the spectra of metal-free [GlnB13]insulin at elevated concentration (12 g/l) and hexameric 2Zn insulin coincide (Figure 2). Therefore, it seems obvious that the insertion of the keystone dimer to complete the hexamer is essentially impeded by electrostatic repulsion of the GluB13 carboxylate groups in its centre (Wollmer et al., 1989; Kadima et al., 1992). In the tetramer the GluB13 side chains may still be sufficiently mobile that their repulsion does not force it apart. When the GluB13 charge is neutralized by protonation at pH 2, the tyrosyl band again does not match that of hexameric 2Zn insulin (Figure 3). At pH 2, of course, there is new repulsion because of protonation of the histidines B10 (and B5), and bovine insulin at 1 g/l has been reported to be monomeric in this milieu (Jeffrey and Coates, 1965). According to more recent results insulin at 5 g/l insulin is dimeric at pH2 established with HCl or H_2SO_4, whereas 20% acetic acid is required to make it monomeric (Whittingham and Fink, personal communication). Interestingly, when HisB10 is substituted by Asp which is also neutralized by protonation at pH 2, the 275-nm ellipticity equals that of 2Zn wild type insulin (Figure 3). It is very unlikely but not impossible that this could be due to a different spatial arrangement of the interacting chromophores, i.e., a different tertiary and not necessarily hexameric quaternary structure. Therefore, the hexameric state of [AspB10]insulin under these conditions has to be confirmed by molecular mass measurements. Lowering the pH to 2 also reduces the negative ellipticity around 250 nm of both wild-type insulin and, much more so, [AspB10]insulin. This effect may originate in a change of the dihedral angles of disulphide bonds, possibly A6 – A11 under the influence of protonated His B5.

It is thus plausible that with wild-type insulin the keystone effect fails. On the other hand, it seems that in the absence of electrostatic repulsion the scheme in Figure 1 basically applies. The two-contact concept has been followed up and applied successfully by Jeffrey and coworkers, though without emphasis on the keystone effect (Mark et al., 1987; Mark and Jeffrey, 1990; Nourse and Jeffrey, 1998). However, a question is left open that has intrigued us all the time: as any association beyond the classical dimer depends on the weak contact, how can tetramers form, whereas alternative dimers D′ have never really been shown to exist? – The reason occurred to us when reading an article by Shuker et al. (1996) on an apparently unrelated subject, entitled "Discovery of high-affinity ligands for proteins", which induced us to reconsider the scheme.

3. ASSOCIATION SCHEME BASED ON THREE SURFACES

3.1. General Description

The two-dimensional representation in Figure 1, the pie model in our jargon, is misleading as it withholds the fact that there are other association steps beyond keystone insertion that also involve more than just one contact at the same time. This becomes apparent when the third dimension is taken into account in the revised scheme which is based on the fact that the self-association of insulin actually employs three discrete surface areas of the monomer. In Figure 4 they are colour coded blue, red and green, respectively, which illustrates how the subunits fit together. As in the former scheme, only two types of contacts are possible between subunits, one strong, related to K_P, the other weak, related to K_Q. The weak contact, however, employs two different surfaces, a red and a green one. Neglect of this fact is the main shortcoming of the former two-dimensional scheme and is acknowledged not to be allowed. The complete list of oligomers with the number of their contacts and surfaces exposed, respectively, is given in Table 1. The contacts formed in any binary association step and the surfaces exposed before and after, respectively, are compiled in Tables 2 and 3. Now, the point of the three-dimensional scheme is that the classical dimer D formed by the strong contact is a species that exposes *two* surfaces *each* of the red and the green ones (see Table 1). Hence, when two classical dimers D associate to the tetramer Te, *two* of the weak red/green contacts are formed (*rather than one* according to the former scheme). This is exactly the situation that is artificially created and exploited in the strategy described by Shuker et al. (1996). In target-directed drug development, for instance, small molecules that bind to proximal subsites of a protein, each with only moderate affinity, are tethered together to become a high-affinity ligand. This explains why the formation of Te tramers is preferred to any other step that would establish a single weak contact only as, for instance, alternative dimerization. It has been considered that the overall association constant, e.g., for this reaction would be K_Q^2 and for any reaction in which more than one contact is formed would correspond to the product of the respective contact constants. A mathematical desription of the association scheme is the subject of Part II (Mülders and Wollmer, this volume). The many more steps that would also establish two weak or even two strong contacts (see Tables 2 and 3), however, are practically unimportant, as one or both of the oligomeric association partners involved scarcely form. The same applies to keystone insertions completing trimers.

The channelling of association through D dimers and Te tetramers is a consequence of the inequality of K_P and K_Q for wild-type insulin. Numerous derivatives and analogues have been designed with the aim to reduce association, i.e., increase diffusibility and, thus, to obtain fast acting insulins. Residues typically modified or replaced are those involved in the strong contact, i.e., preferentially residues of the C-terminal B-chain. Diffusion is, of course, expected to be fastest for

Table 1. Arrangement of subunits in oligomers and its degeneracy, existing contacts and exposed surfaces (compare Figure 4).

	Contacts			***Exposed surfaces***		
		blue/blue	red/green	blue	red	green
M		-	-	1	1	1
D	1/4=2/5=3/6	**1**	-	-	2	2
D′	1/2=2/3=3/1=4/5=5/6=6/4	-	**1**	2	1	1
Tr	1/4/2=2/5/3=3/6/1=1/4/5= 2/5/6=3/6/4	**1**	**1**	1	2	2
Tr′	1/2/3=4/5/6	-	**3**	3	-	-
Te	1/4/2/5=2/5/3/6=3/6/1/4	**2**	**2**	-	2	2
Te′	1/4/2/6=2/5/3/4=3/6/1/5	**1**	**2**	2	2	2
Te″	1/2/3/4=1/2/3/5=1/2/3/6= 1/4/5/6=2/4/5/6=3/4/5/6	**1**	**3**	2	1	1
P	1/2/3/4/5=1/2/3/5/6= 1/2/3/6/1=1/2/4/5/6= 2/3/4/5/6=3/1/4/5/6	**2**	**4**	1	1	1
H	1/2/3/4/5/6	**3**	**6**	-	-	-

Table 2. Binary association steps involving monomers: Contact formation and change in exposed surfaces.

	Exposed surfaces before association			Contacts formed upon association		Exposed surfaces after association		
	blue	red	green	blue/blue	red/green	blue	red	green
M + M = D	2	2	2	1	-	-	2	2
M + M = D′	2	2	2	-	1	2	1	1
M + D = Tr	1	3	3	-	1	1	2	2
M + D′ = Tr	3	2	2	1	-	1	2	2
M + D′ = Tr′	3	2	2	-	**2***	3	-	-
M + Tr = Te	2	3	3	1	1	-	2	2
M + Tr = Te′	2	3	3	-	1	2	2	2
M + Tr′ = Te″	4	1	1	1	-	2	1	1
M + Te = P	1	3	3	-	**2***	1	1	1
M + P = H	2	2	2	**1***	**2***	-	-	-

) Keystone-insertion steps completing a trimer or the hexamer, respectively

Table 3. Binary association steps involving oligomers: Contact formation and change in exposed surfaces.

	Exposed surfaces before association			Contacts formed upon association		Exposed surfaces after association		
	blue	red	green	blue/blue	red/green	blue	red	green
D + D = Te	-	4	4	-	2	-	2	2
D′ + D′ = Te	4	2	2	2	-	-	2	2
D′ + D′ = Te′	4	2	2	1	-	2	2	2
D + D′ = Te″	2	3	3	-	2*	2	1	1
D + Tr = P	1	4	4	-	3*	1	1	1
D′ + Tr′ = P	5	1	1	2	-	1	1	1
D + Te = H	-	4	4	-	4*	-	-	-
D′ + Te″ = H	4	2	2	2*	2*	-	-	-
Tr + Tr = H	2	4	4	1*	4*	-	-	-
Tr′ + Tr′ = H	6	-	-	3*	-	-	-	-

***)** *Keystone-insertion steps completing a trimer or the hexamer, respectively.*

monomers. The suitability of monomeric insulins as fast acting preparations for therapy, however, is compromised by insufficient physical and chemical stability with respect to the regulations for storage. Therefore, the design of fast acting insulins must take care that a residual ability to associate is maintained. This has been achieved, e.g., with [LysB28, ProB29]-, [AspB28]- and [LysB3, GluB29]insulin. These analogues can be stored as hexamers in the presence of zinc and phenolic compounds. Upon injection these low molecular weight ligands will quickly diffuse away and only then the increased propensity to dissociate will become effective.

Weakening of the strong contacts means that the two constants K_P and K_Q are closer to equivalence. Consequently, not only D dimers and Te tetramers, but also the other intermediates have a chance to form. The composition of the solution population as a function of the total concentration calculated for $K_P = K_Q$ on the basis of the revised scheme is shown in Part II (Mülders and Wollmer, this volume). As expected, the curves for the classical and the alternative dimers D and D′ coincide.

Hexamer formation is known to be enhanced by metal ions, e.g., Zn ions, and hexamer stability is known to be higher for R_6 than for T_6. This has raised the question about the smallest oligomeric species to which metal ions, e.g., Zn ions, and ligands interfering with T-R equilibria, e.g., phenol, can bind.

3.2. Interference of Zinc Ions

It is commonly accepted that Te tetramers are prepared to bind Zn ions (Coffman and Dunn, 1988) whereas the capability of D dimers is being questioned. The Te tetramer provides two thirds of each of the two prospective specific binding sites in the form of two times two properly preoriented HisB10 side chains. A residue recognized to also play a role in the acquisition of Zn ions is GluB13 (Coffman and Dunn, 1988, Wollmer et al., 1989). The smallest species that would display one rudimentary Zn binding site of the kind just described is the alternative dimer D′. Truncation of the C-terminal B-chain severely impairs or even abolishes the formation of the strong contact. Correspondingly, des-(B26-B30)-pentapeptide insulin (DPI) and particularly des-(B23-B30)-octapeptide insulin (DOI) in metal-free neutral solution are believed to be essentially monomeric even at elevated concentration. Nevertheless, they were reported to be partially hexameric in the presence of Zn ions (Jeffrey, 1986; Pittman and Tager, 1995, respectively).

In this case with the strong contact weakened it may well be that association passes through the alternative dimer D′. With more than one of the "weakened blue surfaces" (in the terminology of Figure 2) exposed and Zn ions in place, GluB13 repulsion might be overcome and association occur across the trimer/trimer interface. In principle these considerations could also apply to the fast acting analogues. In their case, however, CD spectroscopy shows that the strong contacts are still more effective than, e.g., in the truncated analogues.

3.3. Interference of Phenol

In the circular head-to-tail situation within trimers (see Figure 4) T↔R transition only involves the red surfaces whereas the green ones remain essentially unchanged. The trimers of 2-metal hexamers are known to exist in a ligand independent dynamic equilibrium between T_3 and R_3. The R-state is arrested by phenol binding. R_6 hexamers with 6 phenol molecules bound are known to be more stable than the T_6 hexamers owing to extra H-Bonding and van der Waals contacts provided by the phenol molecules, notwithstanding trapping of the metal ions. Hence, weak contacts can be reinforced by phenol, provided that the red surface converts towards the R-conformation. Theoretically all species containing weak contacts (see Table 1) should be potential candidates for undergoing the T→R transition and binding phenol. In accordance with this occurring, any equilibria involving weak contacts (see Tables 2 and 3) should be shifted to the associated state when phenol is present. The contribution of both the converted red surface and the unchanged green surface to the phenol binding pocket suggests that the existence of their contact is prerequisite for binding to occur. If however red surfaces could adopt the R-conformation also when freely exposed, would they be stabilized by phenol binding? Would this increase their binding affinity for green surfaces? In the end, could R-state monomers exist? The latter questions have been addressed by Pittman and Tager (1995).

Coming down from the hexamer, the next smaller oligomer containing weak contacts that has been identified is the Te tetramer. A NOE of the tetramers between HisB5 and a tyrosyl residue has been interpreted as an indication that tetramers can adopt the R-conformation (Kadima et al., 1992). Also the CD of metal-free wild-type insulin at 12 g/l concentration and pH 7.8, i.e., under conditions where it is assumed to be essentially tetrameric, indicates that a substantial fraction of it is in the R-state when a high molar excess of phenol is present (>20 M / M monomer; Krüger et al., 1990). At 2.8 g/l concentration, where metal-free wild-type insulin still is essentially tetrameric, 10 M phenol / M monomer is an excess insufficient to change its CD spectrum (not shown). If however the helix propensity in the N-terminal B-chain is increased by replacing PheB1 by a residue of high N-capping capacity, such as Asn or Asp, the CD clearly indicates an enhancement of the R-state under the same conditions (Figure 5). In any case, with or without substituted N-cap, addition of both phenol and Zn ions results in much more pronounced CD changes indicative of R_6 hexamer formation. Interestingly, this observation is also made with metal-free [AsnB1]DPI and even with [AsnB1]DOI (Figures 6 and 7, respectively). In any case with or without substituted N-cap, addition of both phenol and Zn ions results in much more pronounced CD changes indicative of R_6 hexamer formation. As the ability to form the strong contact is impaired or even abolished by the C-terminal truncation of the B-chain and cannot at all be influenced by the B1 substitution, it is concluded that in these cases association does occur via the weak contacts, starting with the alternative dimer.

4. CONCLUDING REMARKS

The decision that this book would be put together was an opportunity for me (A.W.) to include thoughts and views that still lack experimental confirmation and isolated results requiring flanking experiments with complementary techniques. These are some of the loose ends that remain as my retirement deprives me of my laboratory. I would be pleased if colleagues were interested in fixing them by falsifying or, preferentially of course, verifying them.

5. REFERENCES

Blundell T., Dodson G., Hodgkin D. and Mercola D. "Insulin: The Structure in the Crystal and Its Reflection in Chemistry and Biology." *Adv. Prot. Chem.* 26 (1972): 279-431.

Coffman F.D. and Dunn M.F. "Insulin-Metal Ion Interactions: The Binding of Divalent Cations to Insulin Hexamers and Tetramers and the Assembly of Insulin Hexamers." *Biochemistry* 27 (1988): 6179-6187.

Dodson G.G., Cutfield S., Hoenjet E., Wollmer A. and Brandenburg D. "Crystal Structure, Aggregation and Biological Potency of Beef Insulin Cross-linked at A1 and B29 by Diaminosuberic Acid." In: *Insulin: Chemistry, Structure and Function of Insulin and Related Hormones* (Brandenburg D. and Wollmer A., eds.), (1980), Walter de Gruyter, Berlin.

Hansen J.F. "The self-association of zinc-free human insulin and insulin analogue B13-glutamine." *Biophys. Chem.* 39 (1991): 107-110.

Hoenjet E. "Über den Einfluss von Effektoren, Liganden und chemischen Modifizierungen auf die Quartärstruktur von Hämoglobin und Insulin." *Doctoral thesis, RWTH Aachen* (1981).

Jeffrey P.D. "Self-association of Des-(B26-B30)-insulin." *Biol. Chem. Hoppe-Seyler* 367 (1986): 363-369.

Jeffrey P.D. and Coates J.H. "The Sedimentation Coefficient of Insulin in Acid Solution." *Biochim. Biophys. Acta* 109 (1965): 551-556.

Kadima W., Roy M., Lee R.W.-K., Kaarsholm N.C. and Dunn M.F. "Studies on the Association and

Conformational Properties of Metal-free Insulin in Alkaline Sodium Chloride Solutions by One- and Two-dimensional ^{1}H NMR." *J. Biol. Chem.* 267 (1992): 8963-8970.

Kadima W., Ogendal L., Bauer R., Kaarsholm N., Brodersen K., Hansen J.F. and Porting P. "The Influence of Ionic Strength and pH on the Aggregation Properties of Zinc-Free Insulin Studied by Static and Dynamic Laser Light Scattering." *Biopolymers* 33 (1993): 1643-1657.

Lin M. and Larive C.K. "Detection of Insulin Aggregates with Pulsed-Field Gradient Nuclear Magnetic Resonance Spectroscopy." *Anal. Biochem.* 229 (1995): 214-220.

Mark A.E. and Jeffrey P.D. "The Self-Association of Zinc-Free Bovine Insulin. Four Model Patterns and Their Significance." *Biol. Chem. Hoppe-Seyler* 371 (1990): 1165-1174.

Mark A.E., Nichol L.W. and Jeffrey P.D. "The self-association of zinc-free bovine insulin. A single model based on interactions in the crystal that describes the association pattern in solution at pH 2, 7 and 10." *Biophys. Chem.* 27 (1987): 103-117.

Mercola D.A. and Wollmer A. "The Crystal Structure of Insulin and Solution Phenomena: Use of the High-resolution Structure in the Calculation of the Optical Activity of the Tyrosyl Residues." In: *Structural Studies on Molecules of Biological Interest* (Dodson, G., Glusker, J.P. and Sayre, D., eds.), (1981): pp.557-582, Clarendon Press, Oxford.

Nourse A. and Jeffrey P.D. "A sedimentation equilibrium study of platypus insulin: the HB 10D mutant does not associate beyond dimer." *Biophys. Chem.* 71 (1998): 21-23.

Pekar A.H. and Frank B.H. "Conformation of Proinsulin. A Comparison of Insulin and Proinsulin Self-Association at Neutral pH." *Biochemistry* 11 (1972): 4013-4016.

Pittman I. IV and Tager H.S. "A Spectroscopic Investigation of the Conformational Dynamics of Insulin in Solution." Biochemistry 34 (1995): 10578-10590.

Shuker S.B., Hajduk P.J., Meadows R.P. and Fesik S.W. "Discovering High-affinity Ligands for Proteins: SAR by NMR." *Science* 274 (1996): 1531-1534.

Strickland E.H. and Mercola D.A. "Near ultraviolet tyrosyl circular dichroism of pig insulin monomers, dimers and hexamers. Dipole-dipole coupling calculations in the monopole approximation." *Biochemistry* 15 (1976): 3875-3883.

Wollmer A., Fleischhauer J., Strassburger W., Thiele H., Brandenburg D., Dodson G. and Mercola D. "Side-chain Mobility and the Calculation of Tyrosyl Circular Dichroism of Proteins." *Biophys. J.* 20 (1977): 233-243.

Wollmer A., Strassburger W. Hoenjet E., Glatter U., Fleischhauer J., Mercola D.A., de Graaf R.A.G., Dodson E.J., Dodson G.G., Smith G.D., Brandenburg D. and Danho W. "Correlation of Structural Details of Insulin in the Crystal and in Solution." In: *Insulin: Chemistry, Structure and Function of Insulin and Related Hormones.* (Brandenburg D. and Wollmer A., eds.), (1980), Walter de Gruyter, Berlin.

Wollmer A., Rannefeld B., Johansen B.R., Hejnaes K.R., Balschmidt P. and Hansen F.B. "Phenol Promoted Structural Transformation of Insulin in Solution." *Biol. Chem. - Hoppe-Seyler* 368 (1987): 903-911.

Wollmer A., Rannefeld B., Stahl J. and Melberg S.G. "Structural Transition in the Metal-free Hexamer of Protein-engineered [B13Gln]insulin." *Biol. Chem. Hoppe-Seyler* 370 (1989): 1045-1053.

6. FIGURES

(Figure 1 is integrated in the text.)

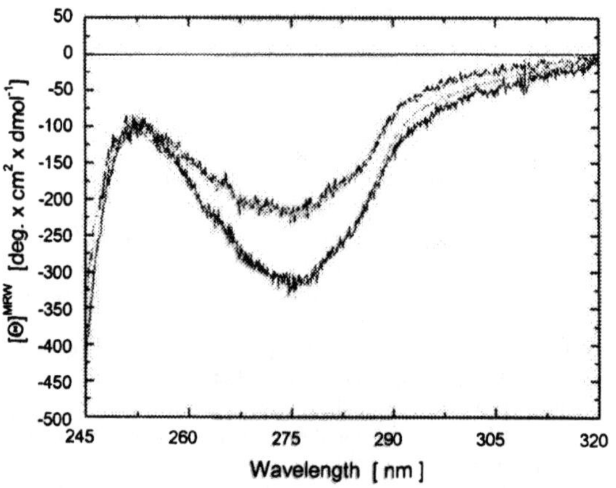

Figure 2. CD spectra of wild-type and [GlnB13]insulin in the near UV. Metal-free: Wild-type insulin, pH 7.5. Yellow: c = 10.4 g/l; 33 mM NaCl (M_{av} ≈24,000 Dalton; Kadima et al., 1993). • Cyan: c = 11.9 g/l; 100 mM NaCl (M_{av} ≈36,000 Dalton; Kadima et al., 1993). • Black: c = 30.5 g/l; 100 mM NaCl. • Red: [GlnB13]insulin; c = 12 g/l, 25 mM Tris·HCl, pH 7.8.
0.33 M Zn^{2+} / M monomer: Green: Wild-type insulin, c = 0.92 g/l, 25 mM Tris·HCl, pH 7.8.

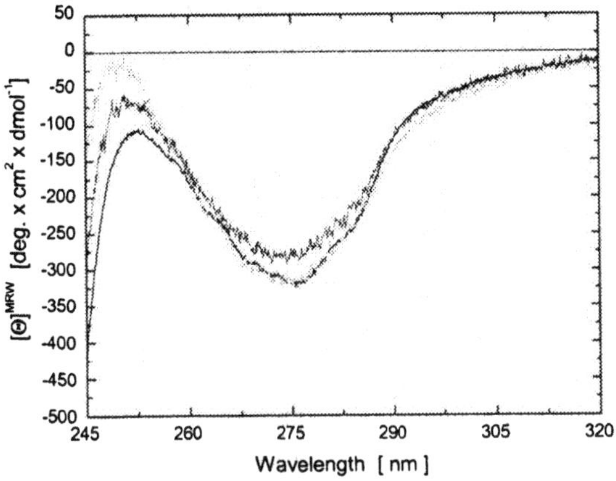

Figure 3. CD spectra in the near UV. Black: Wild-type insulin; c = 1 g/l; 0.33 M Zn^{2+} / M monomer; pH 7.8. • Red: Wild-type insulin; c = 9 g/l; metal-free , pH 2.0. • Green: [AspB10]insulin; c = 9 g/l; metal-free; pH 2.0.

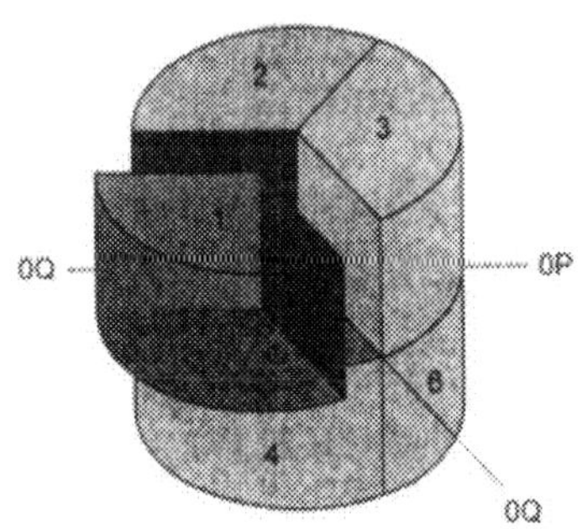

Figure 4. Schematic representation of an insulin pentamer. One monomer of the hexamer was removed to demonstrate the distribution of inter-subunit contacts. Strong contacts formed by two blue surfaces each exist between the subunits of the upper and lower trimer. Weak contacts are formed by a red and a green surface each between the subunits within the trimers.

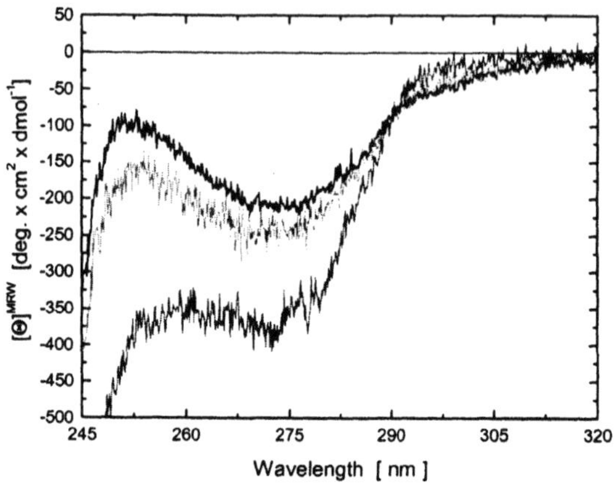

Figure 5. The effect of phenol on the near-UV CD spectrum of [AsnB1]insulin in the absence and presence of zinc ions at 2.8 g/l concentration and pH 7.8.
Black: Free of phenol and metal. • Green: 10 M phenol / M monomer. •
Red: 10 M phenol and 0.33 M Zn^{2+} / M monomer.

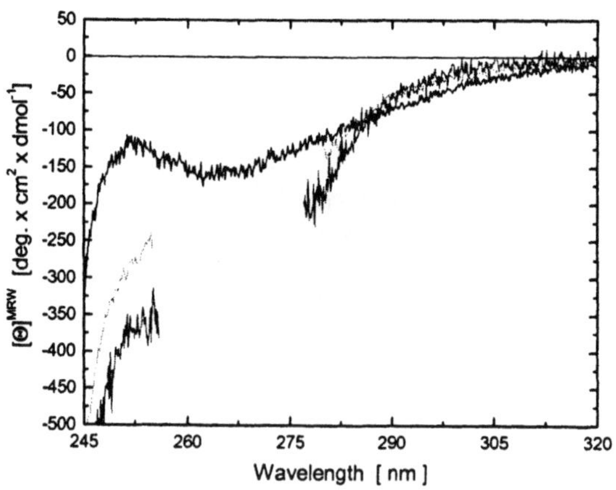

Figure 6. The effect of phenol on the near-UV CD spectrum of [AsnB1]des-(B26-B30)-pentapeptide insulin in the absence and presence of zinc ions at pH 7.8.
Black: Free of phenol and metal; c = 8.8 g/l. • Green: 65 M phenol / M monomer; c = 7.3 g/l. • Red: 20 M phenol and 0.33 M Zn^{2+} / M monomer; c = 2.8 g/l.
(The interruption of spectra is due to excessive phenol absorption.)

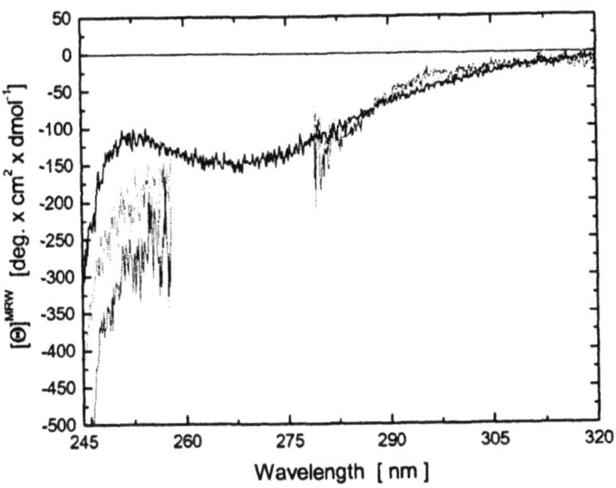

Figure 7. The effect of phenol on the near-UV CD spectrum of [AsnB1]des-(B23-B30)-octapeptide insulin in the absence and presence of zinc ions at pH 7.8. Black: Free of phenol and metal; c = 9.9 g/l. • Green: 65 M phenol / M monomer; c = 7.8 g/l. • Red: 20 M phenol and 0.33 M Zn^{2+} / M monomer; c = 7.4 g/l. (The interruption of spectra is due to excessive phenol absorption.)

T. MÜLDERS AND A. WOLLMER

SELF-ASSOCIATION OF INSULIN RECONSIDERED

II. Mathematical Description

Abstract. The association scheme for insulin introduced in Part I (Wollmer and Mülders, this volume) will be described mathematically in terms of surface-occupation probabilities. As a result numerical population solutions for the concentrations of the different oligomeric species are presented.

1. BASIC ASSUMPTIONS

In the following the association scheme for insulin introduced in Part I will be described mathematically in terms of surface-occupation probabilities. As to the arrangement of subunits in the oligomers and the colour coding of their contact surfaces the reader is referred to Table 1 and Figure 4 of Part I. This second part provides a mathematical description of the association scheme for insulin introduced in Part I (Wollmer and Mülders, this volume). The assumptions implied in the scheme on which the description is based are listed below:

- The monomer exposes three discrete contact surfaces, a blue, a red and a green one.
- Only two types of contacts can form, a strong contact by occupation of a blue surface by another blue one, and a weak contact by occupation of a red surface by a green one (or vice versa).
- At a given total concentration the probability of a specific surface to be occupied by a complementary one is independent of the molecular species exposing it and of the other surfaces in the same oligomer being occupied or not.

Certain association steps of the scheme stand out for establishing more than one inter-subunit contact at the same time. This implies an intrinsic strive of the self-association for completing the hexamer by insertion of the keystone with the simultaneous formation of up to four contacts (see Part I, Tables 2 and 3). All this is implicitly taken into account by assumptions 1 and 2. The main difference to the former scheme is that the focus is on the contact surfaces and their probability to be occupied by a partner surface rather than on the single oligomeric species. Therefore, explicit assumptions about the species involved in the formation of a specific oligomer are superfluous. As will be shown, the above assumptions are sufficient to calculate the populations of the different oligomers as functions of the total insulin concentration.

M. Federwisch, M. Leyck Dieken and P. De Meyts (eds.), Insulin & Related Proteins, 67–76.

2. DERIVATION OF THE EQUATIONS

In the following, the probability for a blue surface to be occupied (by another blue surface) will be denoted by P_b. Correspondingly, the probabilities for a green surface to be occupied by a red one or a red surface by a green one, are written P_g and P_r, respectively. The probabilities for a blue, green or red surface not to be occupied are then given by $(1-P_b)$, $(1-P_g)$ and $(1-P_r)$. Within this framework of occupation probabilities it suffices to simply count how many of the occupied and non-occupied surfaces characterizing a specific oligomer occur, in order to find the probability of the appearance of this specific oligomer among all oligomers. This results from the assumed independence of the occupation probabilities for a contact surface of the oligomer type, and therefore leads to a multiplicative behaviour for the probabilities. For instance, the probability to find an oligomer with n occupied blue surfaces is simply proportional to P_b^n. In an analogous way, counting the occupied and non-occupied surfaces in each oligomeric species leads to the probabilities of occurrence for the different oligomers, P_1, ..., P_6, i.e., the probabilities to find specific oligomeric species among all, which are listed in Table 1. Because there are always as many green as red surfaces, and since a green surface can only be occupied by a red one, and vice versa, it follows immediately that $P_g = P_r$, which has been used to obtain the equalities in Table 1.

Table 1. Probabilities of occurrence of the different oligomers.

Oligomer type	probability of occurrence
M	$P_1 = (1-P_b)(1-P_g)^2$
D	$P_{2a} = P_b^2\,(1-P_g)^4$
D'	$P_{2b} = (1-P_b)^2\,P_g^2\,(1-P_g)^2$
Tr	$P_{3a} = P_b^2\,(1-P_b)\,P_g^2\,(1-P_g)^4$
Tr'	$P_{3b} = (1-P_b)^3\,P_g^6$
Te	$P_{4a} = P_b^4\,P_g^4\,(1-P_g)$
Te'	$P_{4b} = P_b^2\,(1-P_b)^2\,P_g^6\,(1-P_g)^2$
Te"	$P_{4c} = P_b^2\,(1-P_b)^2\,P_g^4\,(1-P_g)^4$
P	$P_5 = P_b^4\,(1-P_b)\,P_g^8\,(1-P_g)^2$
H	$P_6 = P_b^6\,P_g^{12}$

In order to determine the concentration of a given oligomeric species, its probability of occurrence has to be multiplied by the total concentration of all oligomers, **C**, which is defined as the number of all oligomers divided by the volume. Expressed in formulas: $[M] = P_1 \times \mathbf{C}$, ... , $[H] = P_6 \times \mathbf{C}$. It remains to show how this total concentration of all oligomers, **C**, is related to the total insulin concentration, c_{tot}, and to the occupation probabiltites.

An explicit expression for the concentration **C** of all oligomers in terms of c_{tot} and the probabilties of occurrence $P_1, \ldots, P_6$ (explicitly given in Table 1) can be obtained by the following equalities:

$$c_{tot} = [M] + 2\,(\,[D] + [D']\,) + 3\,(\,[Tr] + [Tr']\,)$$

$$+\,4\,(\,[Te] + [Te'] + [Te'']\,) + 5\,[P] + 6\,[H]$$

$$= \mathbf{C} \times \{\; P_1 + 2\,(\,P_{2a} + P_{2b}\,) + 3\,(\,P_{3a} + P_{3b}\,)$$

$$+\,4\,(\,P_{4a} + P_{4b} + P_{4c}\,) + 5\,P_5 + 6\,P_6\;\} \qquad (1)$$

Here c_{tot} is the total insulin concentration in monomer units, and **C** denotes the concentration of all oligomers. Equation (1) is the desired relation connecting the concentration of all oligomers with the total insulin concentration c_{tot} and with the two (unknown) probabilities P_b and P_g. The equations in Table 1 yield explicitly

$$\mathbf{C}(c_{tot}, P_b, P_g) = c_{tot} / Y(P_b, P_g)\,,$$

where

$$Y(P_b, P_g) \quad - \;(1-P_b)\,(1-P_g)^2$$

$$+\,2\,(1-P_g)^2\,\big(\,P_b^{\,2}\,(1-P_g)^2 + (1-P_b)^2\,P_g^{\,2}\,\big)$$

$$+\,3\,(1-P_b)\,P_g^{\,2}\,\big(\,P_b^{\,2}\,(1-P_g)^4 + (1-P_b)^2\,P_g^{\,4}\,\big)$$

$$+\,4\,P_b^{\,2}\,P_g^{\,4}\,(1-P_g)^2\,\big(\,P_b^{\,2}\,(1-P_g)^2 + (1-P_b)^2\,\{\,P_g^{\,2} + (1-P_g)^2\,\}\,\big)$$

$$+\,5\,P_b^{\,4}\,(1-P_b)\,P_g^{\,8}\,(1-P_g)^2 + 6\,P_b^{\,6}\,P_g^{\,12}\,. \qquad (2)$$

It has to be noted that the occupation probabilities P_b and P_g are themselves functions of the total insulin concentration c_{tot}. As will be shown next, this can be expressed by their dependence on **C** (which in turn depends on c_{tot}) and on the equilibrium constants K_P and K_Q: $P_b = P_b(\mathbf{C}, K_P, K_Q)$, $P_g = P_g(\mathbf{C}, K_P, K_Q)$. In order to determine these two functions, the law of mass action involving the equilibrium constants K_P and K_Q for the formation of a strong and a weak contact, i.e., the association of monomers to classical and alternative dimers, respectively, can be used:

$$K_P = [D]\,[M]^{-2} = \mathbf{C} \times P_b^2\,(1-P_g)^4 \times (\mathbf{C} \times (1-P_b)\,(1-P_g)^2)^{-2}$$

$$= P_b^2 \times (\mathbf{C}(c_{tot}, P_b, P_g) \times (1-P_b)^2)^{-1}$$

$$(3)$$

$$K_Q = [D']\,[M]^{-2} = \mathbf{C} \times (1-P_b)^2\,P_g^2\,(1-P_g)^2 \times (\mathbf{C} \times (1-P_b)\,(1-P_g)^2)^{-2}$$

$$= P_g^2 \times (\mathbf{C}(c_{tot}, P_b, P_g) \times (1- P_g)^2)^{-1}$$

Inserting formula (2) in the equations (3) finally yields:

$$K_P \times c_{tot} \times (1-P_b)^2 - Y(P_b, P_g) \times P_b^2 = 0,$$

and (4)

$$K_Q \times c_{tot} \times (1-P_g)^2 - Y(P_b, P_g) \times P_g^2 = 0\,.$$

The explicit expression for the abbreviation $Y(P_b,P_g)$ is given in equation (2). Given fixed values for the total insulin concentration c_{tot} and the contact constants K_P and K_Q, the problem of determining the two unknwons P_b and P_g reduces to the standard task of numerically finding the zeros of the coupled equations (4). Once the solutions P_b and P_g to these equations have been found, all the concentrations of the different oligomeric species can be computed as
$[M] = P_1(P_b, P_g) \times \mathbf{C}(c_{tot}, P_b, P_g), \ldots, [H] = P_6(P_b, P_g) \times \mathbf{C}(c_{tot}, P_b, P_g)$,
compare also Table 1 and equation (2).

3. SOLVING EQUATION (4)

In order to solve the equations (4) for P_b and P_g the standard Newton method (Stoer,

1994) was used. This iterative method has to be started with an initial guess for P_b and P_g, which has for both occupation probabilities been chosen to be 1/2. Convergence was assumed to be reached if the solutions for the probabilities varied less than the predefined tolerance value of 10^{-6}. This iterative process has to be repeated for all total insulin concentrations of interest. When a solution belonging to one total insulin concentration has been found, it can be used as an initial guess for the next iterative solution process corresponding to another (either slightly increased or decreased) total insulin concentration. It was checked that the so obtained probabilities always stayed inside the interval [0,1], and that the result did not depend on whether the changes in the total insulin concentration were done in an increasing or decreasing manner starting either from the lowest or highest total insulin concentration of interest, respectively.

4. RESULTS AND DISCUSSION

The values for the association constants to be gathered from the literature are scattered over a wide range that extends from 2.5×10^4 l/mol to 7.5×10^5 l/mol for $K_{12} = K_P$ and from 220 l/mol to 2×10^4 l/mol for $K_{24} = K_Q$ (Pekar and Frank, 1972; Goldman and Carpenter, 1974; Jeffrey et al., 1976; Pocker and Biswas, 1981; Mark et al., 1987; Mark and Jeffrey, 1990; Kadima et al., 1993; Nourse and Jeffrey, 1998; Hassiepen et al., 1998). As it is not intended here to fit measured data, $K_P = 2 \times 10^5$ l/mol and $K_Q = 10^3$ l/mol were used as representative values for demonstrating the scheme. Figure 1 shows the numerical population solutions obtained with the model for $K_P = 2 \times 10^5$ l/mol and $K_Q = 10^3$ l/mol. In Figure 1 mainly monomers (M), classical dimers (D), tetramers (Te) and hexamers (H) are visible. All other oligomeric species lie constantly below a relative fraction of less than 0.03. During the very first increase in total concentration the monomer concentration rapidly falls off and classical D dimers are formed. Around a concentration of $c_{tot} \approx 1 \times 10^{-4}$ mol/l Te tetramers appear. Their population curve shows a maximum near $c_{tot} \approx 1.8 \times 10^{-4}$ mol/l. Around that concentration range there is a small window, where D dimers and Te tetramers are the predominant species. It is interesting to note that hexamers do not appear until sufficiently many Te tetramers are present to be completed by D dimer keystones. It has been pointed out in Part I that in the case of metal-free wild-type insulin the keystone effect is counteracted mainly by charge repulsion due to accumulation of GluB13 carboxylate groups in close proximity, so that self-association essentially ends with Te tetramers. Certainly, for oligomers of small size and/or of topological complexity, the 3rd assumption, i.e., that the occupation probability of a contact surface were independent of the oligomer type exposing it, is questionable. In the case of wild-type insulin, for instance, it neglects the influence of electrostatic repulsion which may counteract the association to higher oligomers, i.e., the probability for a surface to be occupied does in reality depend on how many charges have already been accumulated in the centre of the complex, and thus on the oligomer type to which a monomer (or another oligomer) is attached. Nevertheless, as pointed out in Part I, for some cases, e.g. [GlnB13]insulin, it may be considered as a working hypothesis.

Because of the close relationship between self-association, on the one hand, and pharmacokinetics as well as chemical and physical stability on the other, the contact surfaces of insulin have been subject to extensive modification. For instance, enhancement of dissociation to obtain fast acting insulins is achieved by impairment of the strong contact, whereas hexamers can be stabilised by reinforcement of the weak contacts by ligands such as phenol. The effect of such measures on the solution population as a function of total concentration is examined by variation of K_P and K_Q (see Figs. 2 and 3). Figure 2, $K_P = 2 \times 10^5$ l/mol, $K_Q = 10^4$ l/mol, and Figure 3, $K_P = 10^5$ l/mol and $K_Q = 10^4$ l/mol, show a picture qualitatively similar to Figure 1. The main difference is that the association processes already start at lower total insulin concentration and that Te tetramers are absorbed into hexamers much earlier, so that their population curves never exceed the one of the monomers.

A completely different view is offered by Figures 4 and 5, where $K_P = K_Q$, either equal 10^3 l/mol or 10^4 l/mol, respectively. Here, the population curves for classical (D) and alternative (D') dimers coincide, their sum, of course, corresponding to the total population of dimeric species. Furthermore, the role of association intermediate is with Tr' trimers. The reason for this change of the intermediate from Te tetramers to Tr' trimers is twofold: now that the two types of contacts are of equal strength, the number of surfaces for the red/green type-contact formation is twice that for blue/blue, and Tr' trimers benefit from the keystone effect whereas Te tetramers do not. There are examples where association seems to follow this route (see Part I).

Admittedly, the association behaviour of metal-free wild-type insulin is not correctly described by the present scheme as pointed out in Part I. This will also apply to other insulins with contacts 0P clearly stronger than 0Q and Glu in position B13. It is the accumulation of GluB13 carboxylate groups in close proximity that mainly counteracts the keystone effect by charge repulsion and can be overcome primarily in the presence of metal ions. Therefore, metal-free self-association essentially ends with Te tetramers. In Part I it is also shown that in the case of metal-free [GlnB13]insulin, i.e., in the absence of electrostatic repulsion the scheme apparently applies. It is challenging, but not straightforward to modify the present scheme appropriately that the influence of charge repulsion on the surface-occupation probabilities and, hence, on the populations of the different oligomeric species is taken into account.

5. REFERENCES

Goldman J. and Carpenter F.H. "Zinc Binding, Circular Dichroism and Equilibrium Sedimentation Studies on Insulin (Bovine) and Several of Its Derivatives." *Biochemistry* 13 (1974): 4566-4574.

Hassiepen U., Federwisch M., Mülders T., Lenz V.J., Gattner H.G., Krüger P. and Wollmer A. "Analysis of Protein Self-association at Constant Concentration by Fluorescence-Energy Transfer." *Eur. J. Biochem.* 255 (1998): 580-587.

Jeffrey P.D., Milthorp B.K. and Nichol L.W. "Polymerization Pattern of Insulin at pH 7.0." *Biochemistry* 15 (1976): 4660-4665.

Kadima W., Ogendal L., Bauer R., Kaarsholm N., Brodersen K., Hansen J.F. and Porting P. "The Influence of Ionic Strength and pH on the Aggregation Properties of Zinc-Free Insulin Studied by Static and Dynamic Laser Light Scattering." *Biopolymers* 33 (1993): 1643-1657.

Mark A.E. and Jeffrey P.D. "The Self-Association of Zinc-Free Bovine Insulin. Four Model Patterns and Their Significance." *Biol. Chem. Hoppe-Seyler* 371 (1990): 1165-1174.

Mark A.E., Nichol L.W. and Jeffrey P.D. "The self-association of zinc-free bovine insulin. A single model based on interactions in the crystal that describes the association pattern in solution at pH 2, 7 and 10." *Biophys. Chem.* 27 (1987): 103-117.
Nourse A. and Jeffrey P.D. "A sedimentation equilibrium study of platypus insulin: the HB10D mutant does not associate beyond dimer." *Biophys. Chem.* 71 (1998): 21-23.
Pekar A.H. and Frank B.H. "Conformation of Proinsulin. A Comparison of Insulin and Proinsulin Self-Association at Neutral pH." *Biochemistry* 11 (1972): 4013-4016.
Pocker Y. and Biswas S.B. "Self-association of Insulin and the Role of Hydrophobic Bonding: A Thermodynamic Model of Insulin Dimerization." *Biochemistry* 20 (1981): 4354-4361.
Stoer J., *Numerische Mathematik 1*, 7. Auflage, Springer-Verlag Berlin, (1994): p. 288.

6. FIGURES

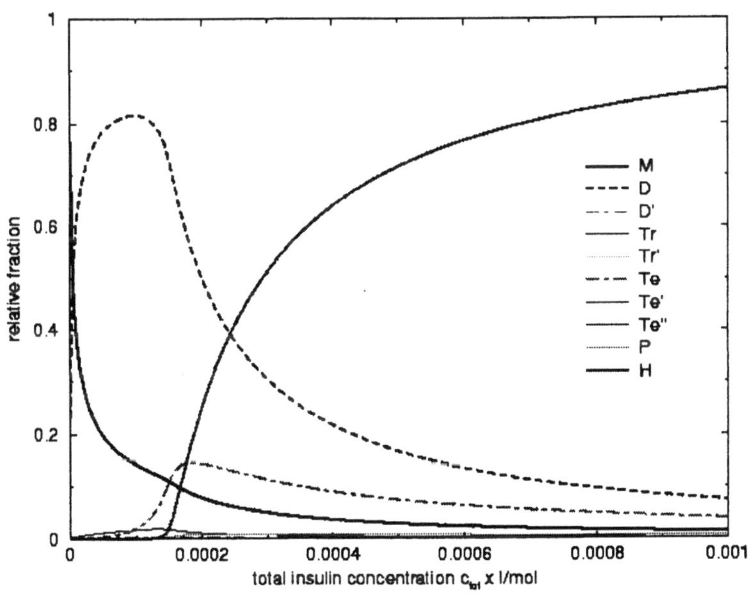

Figure 1. Dependence of the distribution of oligomers on the total concentration of insulin for $K_P = 2 \times 10^5$ l/mol and $K_Q = 10^3$ l/mol.
(For the structural identity of oligomers see Table 1 and Table 1 of Part I.)

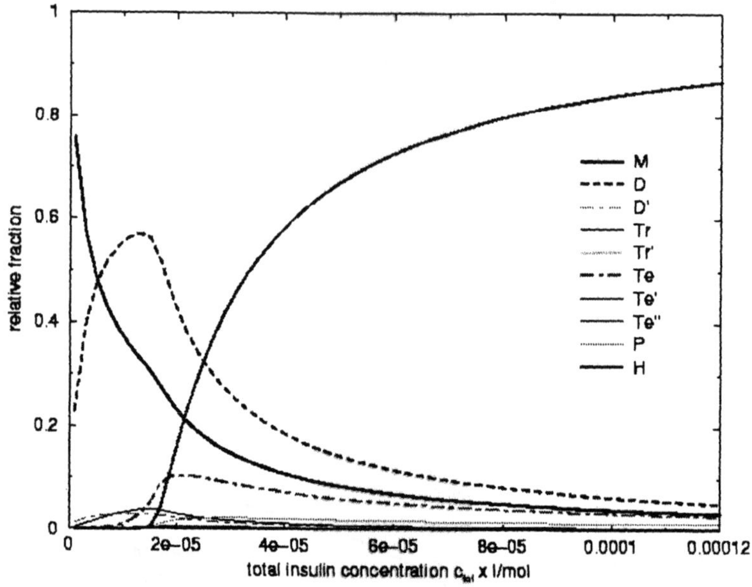

Figure 2. Dependence of the distribution of oligomers on the total concentration of insulin for $K_P = 2 \times 10^5$ l/mol and $K_Q = 10^4$ l/mol.
(For the structural identity of oligomers see Table 1 and Table 1 of Part I.)

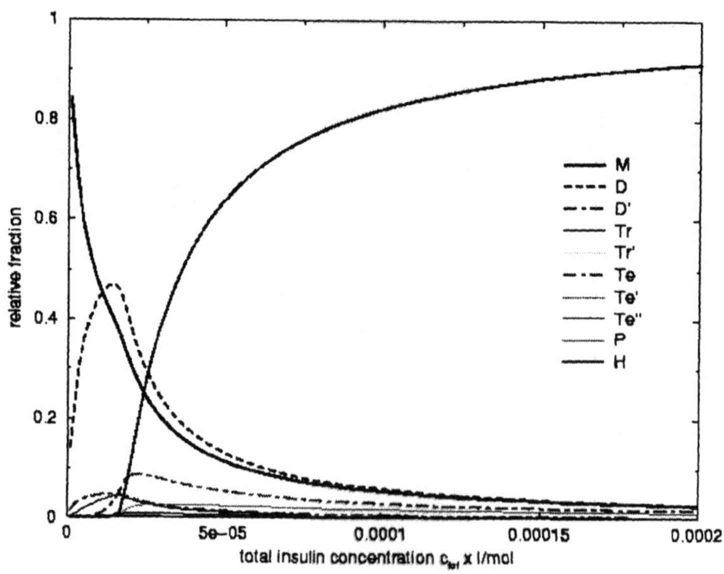

Figure 3 (see opposite page). Dependence of the distribution of oligomers on the total concentration of insulin for $K_P = 10^5$ l/mol and $K_Q = 10^4$ l/mol (For the structural identity of oligomers see Table 1 and Table 1 of Part I.)

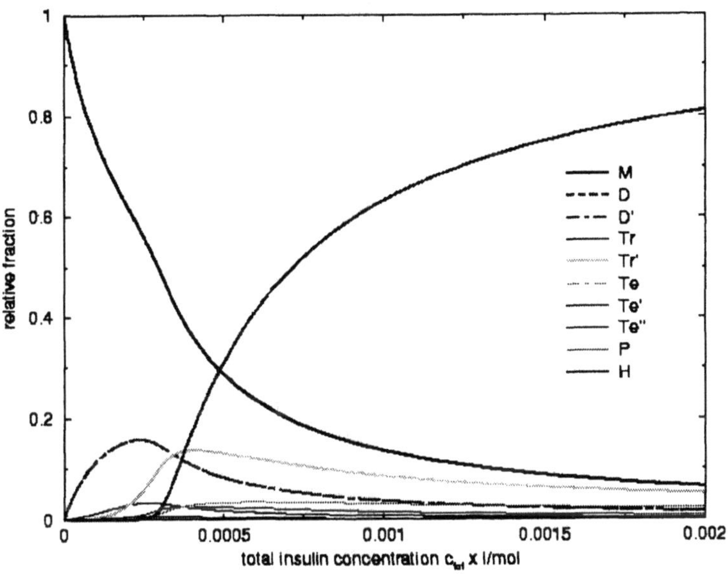

Figure 4. Dependence of the distribution of oligomers on the total concentration of insulin for $K_P = K_Q = 10^3$ l/mol. The curves for the classical and alternative dimers coincide. (For the structural identity of oligomers see Table 1 and Table 1 of Part I.)

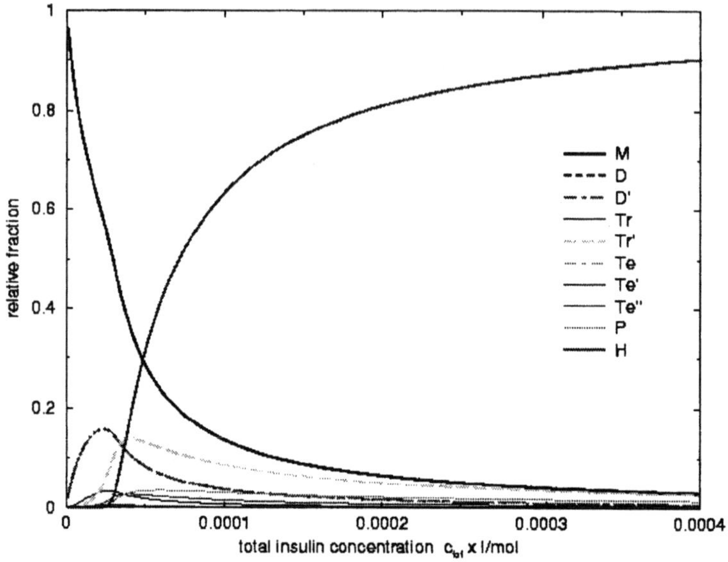

Figure 5. *Dependence of the distribution of oligomers on the total concentration of insulin for* $K_P = K_Q = 10^4$ *l/mol. The curves for the classical and alternative dimers coincide. (For the structural identity of oligomers see Table 1 and Table 1 of Part I.)*

A. WOLLMER

T-R TRANSITION

Abstract. Insulin hexamers crystallize in essentially three different structural states that have all been defined by high-resolution X-ray analysis. In solution these structural states are related by dynamic equilibria, the transitions between them being ligand controlled. The history of the detection and investigation of the solution structural transitions is outlined and a few recent results are presented. Meanwhile insulin has been established as an outstanding model for the study of allosterism and cooperativity, respectively.

1. INTRODUCTION

The origin of the theme treated here can be seen in the observation by Jørgen Schlichtkrull (1958) that when insulin is crystallized in the presence of 7% NaCl, its zinc content rises from 2 to 4 atoms per hexamer. After the 2Zn insulin crystal structure had been solved (Adams et al., 1969) it was observed that presence of NaCl leads to a crystal modification which surprised by asymmetric hexamers. Whereas in one trimer the conformation was the same as in the 2Zn structure, it was strikingly different in the other: residues B1-B8 of all three subunits, normally extended, had adopted a helical conformation so that the pre-existing central B-helix B9-B19 continued to the very N-terminus of the B-chain. The zinc-binding site of this trimer on the threefold axis had dissolved and three off-axial sites had formed instead, each involving one reoriented side chain each of HisB10 and HisB5, which accounts for the total content of 4 zinc ions per hexamer (Bentley et al., 1976). Crystallization from a NaCl containing mother liquor is not the only way to obtain the 4Zn structure. It also forms when a 2Zn insulin crystal is subsequently imbibed in a NaCl containing solution; notwithstanding its extent, the structural rearrangement can occur in the existing crystal without destroying it. Because of the increase in helicity connected with the transformation, we set out to investigate by CD spectroscopy whether NaCl would have a helix promoting effect also in solution. The first structural evidence for the 2Zn→4Zn transition in solution was presented at the Sheffield Meeting of the British Diabetes Association in 1975. We took the subject up again in a systematic study only many years later (Renscheidt, 1982). At the time, high resolution insight into the solution structure of proteins, much later provided by two-dimensional NMR spectroscopy, was not possible. Therefore, the aim of our studies was to validate the structural information obtained (and obtainable only) by crystal X-ray analysis also for the protein in solution. For this to be achieved the demonstration that, for instance, the secondary structural composition deduced from a static CD spectrum roughly matches the amounts of secondary structural elements identified in the X-ray structure is insufficient. In the case of insulin the structural

M. Federwisch, M. Leyck Dieken and P. De Meyts (eds.), Insulin & Related Proteins, 77–89.

alternatives in the crystal offered a unique possibility to also include dynamic behaviour such as the 2Zn→4Zn transition. Not only the evidence that the very same conditions, for instance use of Zn ions, permitting the anion induced transformation to occur in the crystal permit it to occur equally in solution, but also the counter-evidence produced with the essentially non-permissive Ni ions enabled us to postulate a solution equivalence of the 2Zn→4Zn transformation of insulin in the crystal. Adoption of the 4Zn state neither requires the three off-axial Zn-binding sites to be occupied nor to exist: the 2Zn→4Zn also occurs with only 2 Zn ions per hexamer present and with an analogue in which the off-axially coordinating HisB5 is replaced by an alanine residue, respectively (Renscheidt et al., 1984).

In 1985 we started to examine whether phenolic excipients extensively used in insulin preparations for therapeutic application, might have an influence on the solution structure of the hormone. Interestingly, the phenolic compounds, though chemically incomparable to inorganic anions, elicited CD spectral changes that were qualitatively the same (Wollmer et al., 1986), but double in amplitude (Wollmer et al., 1987, 1989). While monoclinic crystals grown in the presence of phenol were still under analysis we were informed (Dodson, G.G. and Smith, G.D., personal communication) that they consisted of symmetrical hexamers with the N-terminal B-chains of all subunits in helical conformation such as first encountered in one trimer of 4Zn hexamers. Six phenol molecules were accommodated in pockets between any two adjacent subunits of each trimer, close to the HisB5 side chains and H-bonded to the CysA11 carbonyls so that off-axial Zn binding is no longer possible. Therefore, our CD spectroscopic observations have provided evidence also of a 2Zn→monoclinic transition in solution. As the phenol-binding sites do not exist in 2Zn insulin, it follows that the structure must convert first before a phenol molecule can bind. Furthermore, as mixed trimers of converted and unconverted sub-units are sterically forbidden, the subunits of a trimer must undergo the structural transition in a cooperative process. As a consequence, when one site in a trimer is occupied the other ones are kept open for two further phenols to bind. This positive cooperativity was revealed by the sigmoidal onset of the CD spectroscopic titration curve (Krüger et al., 1990). The curve also exhibited two distinct steps of roughly the same amplitude indicating that the hexamer is transformed trimer-wise. Hence, assembling as a trimer of dimers, the insulin hexamer, once it exists, behaves as a dimer of trimers, more precisely a dimer of positively cooperative trimers which are related by negative cooperativity.

As the 2Z→4Zn and 2Zn→monoclinic terminology originally geared to emphasizing the correspondence between crystal and solution, turned out to be unpractical and inappropriate (e.g., with respect to metal stoichiometry) it was replaced by the allosteric T/R nomenclature of hemoglobin following a suggestion of Kaarsholm et al. in 1989.

While the key features had been outlined before, the T→R transition was studied in great detail by X-ray crystallographic and spectroscopic techniques as well as theoretical methods during the 90ies. This research has essentially gone on in the labs of Guy Dodson in York, David Smith in Buffalo, Michael Dunn in Riverside,

Michael Weiss in Chicago, in the industrial research labs of Novo Nordisk and Lilly as well as in my own lab. About 40 articles on the subject were published in the last decade which are compiled under References to give credit also to those not explicitly discussed and/or cited in the text. Meanwhile insulin has been established as an outstanding allosteric model system.

2. RESULTS AND DISCUSSION

Since the most striking aspect of the T→R transition is helix formation in the N-terminal B-chain, a question that suggested itself was whether increased helix propensity in the sequence there would enhance the R-state. Strangely, there were no sequence modifications we knew of that had been carried out under this point of view when we started in 1996. Because, lacking recombinant technology, we had to apply Edman degradation and fragment condensation, our modifications were restricted to positions B1, B2 or N-terminal extensions with special emphasis on N-capping. We refrained from replacements deeper in the chain, because upon removal also of AsnB3, GlnB4, cyclisation would have caused extremely low yields. According to CD spectroscopic titrations with phenol, 12 out of 15 analogues synthesized in total underwent transition more readily than wild-type insulin (for an example see Figure 1).

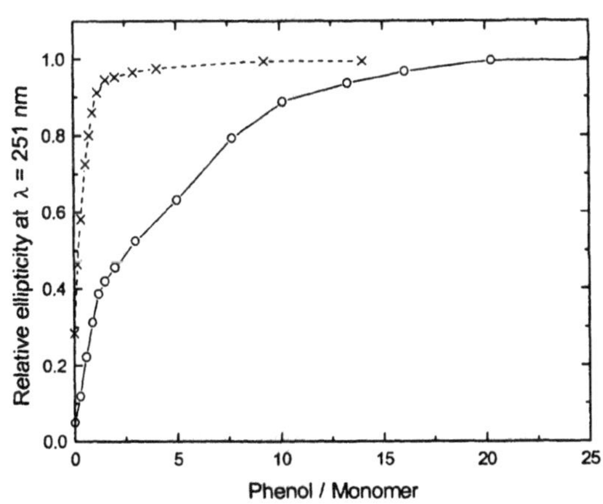

Figure 1. CD spectroscopic titration of [AsnB1]insulin, c = 10.0 g/l (crosses/dashed line) and wild-type insulin. c = 11.0 g/l (open circles / solid line) in the presence of 0.33 M Zn^{2+} / M monomer with phenol; solvent: 25mM Tris/HCl, pH 7.8 . (Reproduced from Shneine et al., 2000, with kind permission).

The 2Zn hexamers of 4 analogues ([AsnB1,GluB0], [AsnB0,GluB1], [AsnB0,AlaB1] and [AspB1]) turned out to be 70-80% in the T_3R_3-state already in the absence of any ligand. This however was not the case with Co; its preference for octahedral coordination suffices to withstand the increased helix propensitiy of the modified sequence (Figure 2).

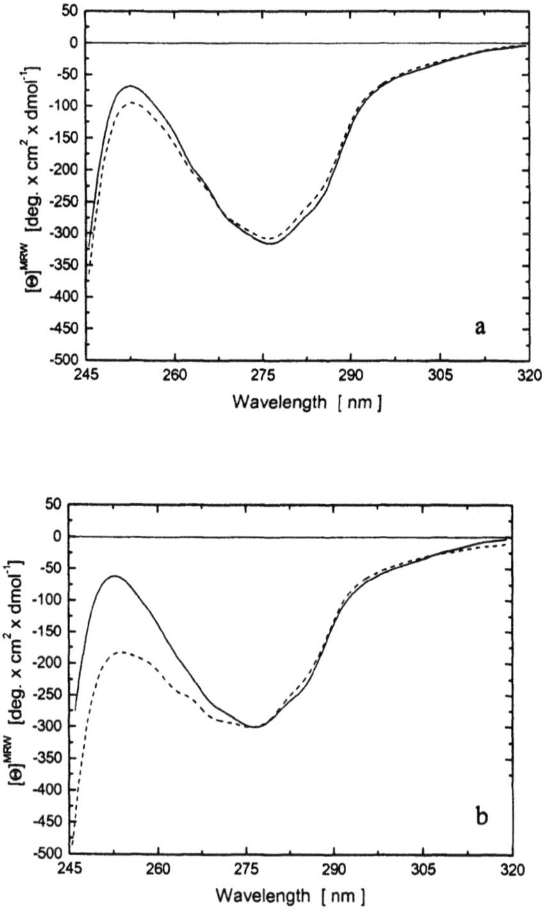

Figure 2. The effect of 0.33 M metal ions / M monomer on the CD spectrum of wild-type insulin (a) and [AsnB1]insulin (b) in the near UV. Solvent: 25mM Tris/HCl, pH 7.8. (a) Wild-type insulin: c = 7 g/l; Zn^{2+} (dashed line). c = 7 g/l; Co^{2+} (solid line). (b) [AsnB1]insulin: c = 1 g/l; Zn^{2+} (dashed line). c = 7 g/l; Co^{2+} (solid line). More negative ellipticity around 251 nm is indicative of T→R transition.

On the other hand, the phenol titration curves for both 2Zn and 2Co hexamers are so steep that the two-step character indicative of negative cooperativity between trimers is no longer discernible, and saturation is achieved with near-stoichiometric amounts of phenol (see Figure 1). In their case transformation by SCN ions also exceeds the T_3R_3-state, its normal limit (Figure 3).

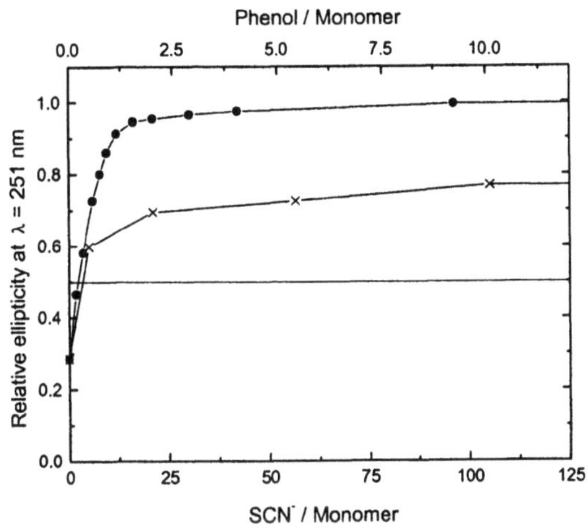

Figure 3. CD spectroscopic titration of [AsnB1]insulin in the presence of 0.33 M Zn^{2+} / M monomer with KSCN (crosses; lower abscissa) and phenol (filled circles; upper abscissa). c = 11.0 g/l, in both cases; solvent: 25mM Tris/HCl, pH7.8 . (Reproduced from Shneine et al., 2000, with kind permission).

Another approach to stabilising the R-structure was recognized simply by inspection: the side chains closest to the threefold axis are those of AsnB3. For an analogue with AsnB3 substituted by His the T→R transition could be expected to result in a spatial arrangement of the HisB3 side chains similar to the HisB10 zinc binding sites and, hence, in the possibility to stabilise the R-state by metal binding to the two extra HisB3 sites in the hexamer (Figure 4).

In this case degradation was continued to include GlnB4 and severe losses in yield had to be tolerated. As the desired extra binding sites only exist in the R-state, the introduction of His was combined with that of helix promoting residues in the other positions of the tetrapeptide that had to be synthesized anyway for condensation. The resulting analogue [AsnB1,AlaB2,HisB3,GluB4]insulin (NAHE insulin) was obtained only in small amounts. [HisB3]insulin was prepared independently also by J. Ertl, R. Obermeier and G. Seipke at Aventis Pharma Deutschland GmbH and the results described below were agreed to be shared.

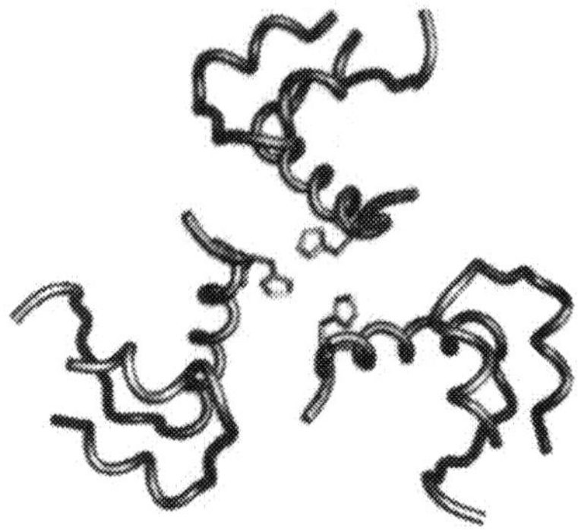

Figure 4. Projection of the main chain model structure of a [HisB3]insulin R_3 trimer in the direction of the three-fold axis. Only HisB3 side chains are drawn. (Coordinates of R_6 insulin kindly provided by G.G. Dodson and J.L. Whittingham.)

The CD spectrum of NAHE insulin as a function of concentration is very much the same as that of wild-type insulin (Figure 5a,b).

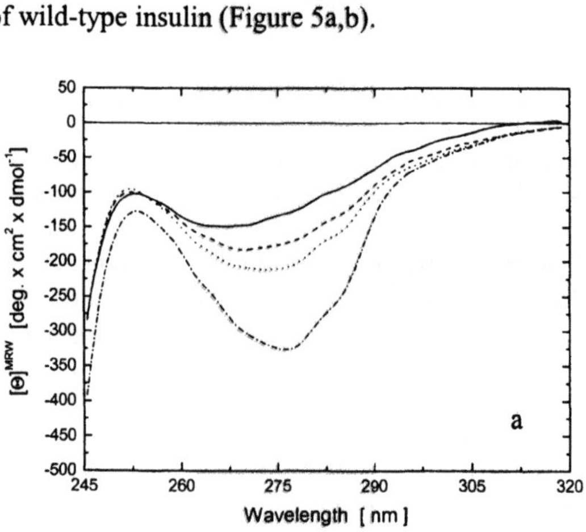

Figure 5a. The CD spectrum of NAHE insulin in the near UV as a function of concentration and in the presence of Zn ions. Solvent: 25mM Tris/HCl, pH 7.8. Metal-free: c = 0.02 g/l (solid line); c = 0.20 (dashed line); c = 0.94 g/l (dotted line). 0.33 M Zn^{2+} / M monomer; c = 0.90 g/l (dashed/dotted line).

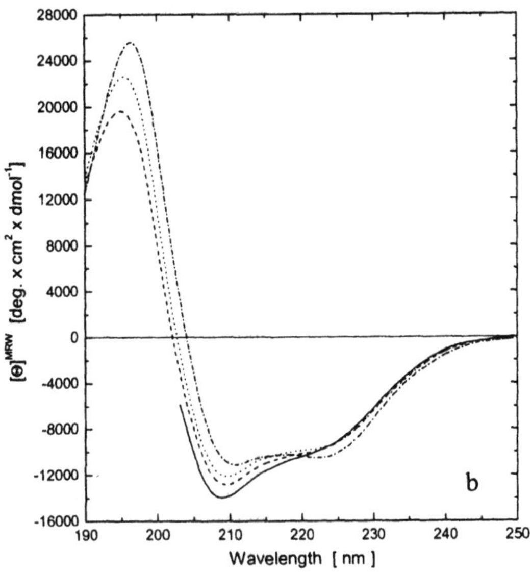

Figure 5b. The CD spectrum of NAHE insulin in the far UV as a function of concentration and in the presence of Zn ions. Solvent: 25mM Tris/HCl, pH 7.8. Metal-free: c = 0.02 g/l (solid line); c = 0.20 (dashed line); c = 4.0 g/l (dotted line). 0.33 M Zn^{2+} / M monomer; c = 3.8 g/l (dashed/dotted line).

Contrary to the example shown in Figure 2 there is hardly any increase in the ellipticity around 251 nm upon addition of 2 Zn per hexamer, i.e., in this case the helix promoting substitutions do not enhance the T→R transition. Addition of 2 Co^{2+} to the 2Zn hexamer, however, results in CD spectral changes in both the near and far UV characteristic of the T→R transition. These changes are increased further by addition of phenol (Figure 6), but not to the extent known for wild-type R_6.

It seems that indeed the HisB3 side chains are pulled together for cobalt coordination whereas, on the other hand, the coordination geometry and/or the sequence modification in B1, B2 and B4 may not allow the R_6 structure to be fully adopted. Interestingly, the time required for the final spectrum with 2 Zn and 2 Co ions per hexamer to establish strongly depends on the order in which the metal ions are added. Whereas upon addition of Co^{2+} to 2ZnNAHE insulin the changes are immediate, they are taking hours when Zn is added to 2CoNAHE insulin. This means that ion sorting occurs across a few Angstroms' distance ending up with the Zn ions in the B10 and Co ions in the B3 sites even though they are equal in charge and radius. Cobalt oxidation to Co^{3+} by H_2O_2 (Storm and Dunn, 1985) could have

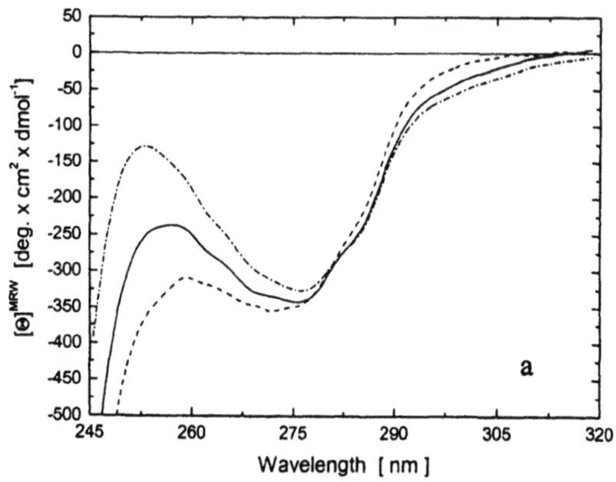

Figure 6a. The combined effect of Zn and Co ions on the CD spectrum of NAHE insulin in the near UV. Solvent: 25mM Tris/HCl, pH 7.8 . 0.33 M Zn^{2+} / M monomer, c = 0.90 g/l (dashed/dotted line); 0.33 M Zn^{2+} + 0.33 M Co^{2+} / M monomer, c = 3.6 g/l (solid line); 0.33 M Zn^{2+} + 0.33 M Co^{2+} + 5 M phenol / M monomer, c = 3.3 g/l (dashed line).

been an informative experiment which was not done because of lack of material. That treatment, by the way, may lead to partial oxidation of the disulfide groups to –S(O)–S– (as indicated by reduced recovery of Cys residues in the amino acid analysis). This could be the reason of the positive ellipticity of $2Co^{3+}$ insulin around 251 nm (Thomas and Wollmer, 1989).

No spectral effects are observed when 2 Zn ions are added to 2ZnNAHE insulin. It remains to be shown whether under these conditions Zn binding to HisB3 at all occurs or not and whether it would occur when the R-state is pre-formed by phenol. Co ions are hard to remove even by gel chromatography with EDTA.

R-state stabilisation is not the only motivation for creating extra HisB3 metal binding sites, the other being stabilisation of the hexamer, e.g., for prolonged action. At a concentration of 0.02 mg/ml = 3.4 μM there is no effect of metal ions on the CD spectrum of wild-type insulin whatsoever, no matter how the solution is prepared. The CD of NAHE insulin does not change either when 0.33 M Zn^{2+} / M monomer are added to a 0.02 mg/ml = 3.4 μM solution of NAHE insulin. However, the spectrum increases upon further addition of 0.33 M Co^{2+} / M monomer to this solution. Moreover, when a solution of NAHE insulin with 2 Zn^{2+} and 2 Co^{2+} per hexamer is prepared at 0.9 mg/ml = 155 μM and subsequently diluted to 0.02 mg/ml = 3.4 μM the CD spectrum of the hexamer is largely maintained (Figure 7). This clearly demonstrates that hexamer stability is substantially increased by the extra HisB3 sites with Co ions bound.

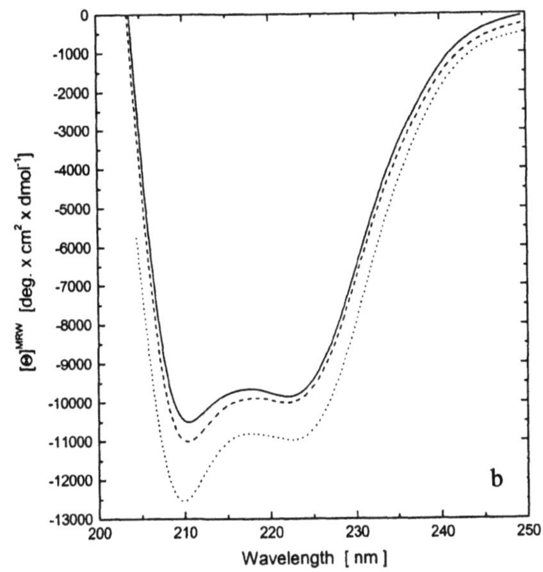

Figure 6b. The combined effect of Zn and Co ions on the CD spectrum of NAHE insulin in the far UV. Solvent: 25mM Tris/HCl, pH 7.8 . 0.33 M Zn^{2+} / M monomer, c = 0.94 g/l (solid line); 0.33 M Zn^{2+} + 0.33 M Co^{2+} / M monomer, c = 0.94 g/l (dashed line); 0.33 M Zn^{2+} + 20 M phenol / M monomer, c = 2.7 g/l (dotted line).

It seems noteworthy that addition of 2 Co ions to 2ZnNAHE insulin hexamers also entails changes in the isotropic absorption spectrum similar to those caused by light scattering (not shown). In the R_6 hexamers the HisB3 side chains might be close enough to the surface that inter-hexamer cobalt coordination could be a possible explanation (though a highly speculative one). Regrettably, the amount of material was much too small to allow for molecular weight measurements or characterization of the metal complexes.

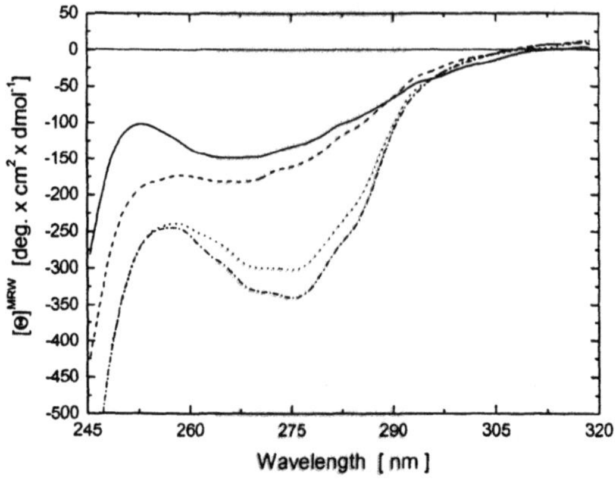

Figure 7. The combined effect of Zn and Co ions on the CD spectrum of NAHE insulin in the near UV at different concentrations and pH 7.8.
NAHE insulin at c = 0.02 g/l, metal-free = wild-type insulin at c = 0.02 g/l, metal-free = wild-type insulin at c = 0.02 g/l + 0.33 M Zn^{2+} / M monomer = wild-type insulin + 0.33 M Zn^{2+} / M monomer, prepared at c = 1 g/l and diluted to 0.02 g/l (all solid line).
NAHE insulin at c = 0.02 g/l + 0.33 M Zn^{2+} + 0.33 M Co^{2+} / M monomer (dashed line).
NAHE insulin at c = 0.9 g/l + 0.33 M Zn^{2+} + 0.33 M Co^{2+} / M monomer (dashed/dotted line).
Same after dilution to c = 0.02 g/l (dotted line).

3. REFERENCES

Adams M.J., Blundell T.L., Dodson E.J., Dodson G.G., Vijayan M., Baker E.N., Harding M.M., Hodgkin D.C., Rimmer B. and Sheat S. *Nature* 224 (1969): 491-???

Bentley G.A., Brange J., Derewenda Z., Dodson E.J., Dodson G.G., Wilkinson A.J., Wollmer A. and Xiao B. "Role of B13 Glu in Insulin Assembly. The hexamer Structure of Recombinant Mutant [B13 Glu → Gln]insulin." *J. Mol. Biol.* 228 (1992): 1163-1176.

Bentley G.A., Dodson G.G., Dodson E.J., Hodgkin D.C. and Mercola D.A. "Structure of insulin in 4-zinc insulin." *Nature* 261 (1976): 166-168 (1976)

Bentley G.A., Dodson G.G., Dodson E.J., Hodgkin D.C., Mercola D.A. and Wollmer A. "Structural Rearrangement in Insulin; a Comparison of 2 and 4 Zinc Insulin Structures." Paper presented at the Spring Meeting of the British Diabetic Association, (1975), Sheffield, UK.

Bentley G., Dodson G. and Lewitova A. "Rhombohedral Insulin Crystal Transformation." *J. Mol. Biol.* 126 (1978): 871-875.

Berchtold H. and Hilgenfeld R. "Binding of Phenol to R_6 Insulin Hexamers." *Inc. Biopoly* 51 (1999): 165-172.

Birnbaum D.T., Kilcomons M.A., DeFelippis M.R. and Beals J.M. "Assembly and Dissociation of Human Insulin and $Lys^{B28}Pro^{B29}$-Insulin Hexamers: A Comparison Study." *Pharmaceutical Research* 14 (1997): 25-36.

Bloom C.R., Choi W.E., Brzovic P.S., Ha J.J., Huang S.-T., Kaarsholm N.C. and Dunn M.F. "Ligand Binding to Wild-type and E-B13Q Mutant Insulins: A Three-state Allosteric Model System Showing Half-site Reactivity." *J. Mol. Biol.* 245 (1995): 324-330.

Bloom C.R., Heymann R., Kaarsholm N.C. and Dunn M.F. "Binding of 2,6- and 2,7-

Dihydroxynaphthalene to Wild-Type and E-B13Q Insulins: Dynamic, Equilibrium, and Molecular Modeling Investigations. *Biochemistry* 36 (1997): 12746-12758.

Bloom C.R., Kaarsholm N.C., Ha J. and Dunn M.F. "Half-Site Reactivity, Negative Cooperativity, and Positive Cooperativity: Quantitative Considerations of a Plausible Model." *Biochemistry* 36 (1997): 12759-12765.

Bloom C.R., Wu N., Dunn, A., Kaarsholm, N.C. and Dunn, M.F. "Comparison of the Allosteric Properties of the Co(II)- and Zn(II)-Substituted Insulin Hexamers." *Biochemistry* 37 (1998): 10937-10944.

Brader M.L. and Dunn M.F. "Insulin hexamers: new conformations and applications." *TIBS* 16 (1991): 341-345.

Brader M.L., Kaarsholm N.C., Lee R. W.-K. and Dunn M.F. "Characterization of the R-State Insulin Hexamer and Its Derivatives. The Hexamer Is Stabilized by Heterotropic Ligand Binding Interactions." *Biochemistry* 30 (1991): 6636-6645.

Brzovic P.S., Choi W.E., Borchardt D., Kaarsholm N.C. and Dunn M.F. "Structural Asymmetry and Half-Site Reactivity in the T to R Allosteric Transition of the Insulin Hexamer." *Biochemistry* 33 (1994): 13057-13069.

Choi W.E., Brader M.L., Aguilar V., Kaarsholm N.C. and Dunn M.F. "The Allosteric Transition of the Insulin Hexamer Is Modulated by Homotropic and Heterotropic Interactions." *Biochemistry* 32 (1993): 11638-11645.

Choi W.E., Borchardt D., Kaarsholm N.C., Brzovic P.S. and Dunn M.F. "Spectroscopic Evidence for Preexisting T- and R-State Insulin Hexamer Conformations." *Proteins: Structure, Function, and Genetics* 26 (1996): 377-390.

Chothia C., Lesk A.M., Dodson G.G. and Hodgkin D.C. "Transmission of conformational change in insulin." *Nature* 302 (1983): 500-505.

Ciszak E. and Smith G.D. "Crystallographic Evidence for Dual Coordination areound Zinc in the T_3R_3 Human Insulin Hexamer." *Biochemistry* 33 (1994): 1512-1517.

Derewenda U., Derewenda Z., Dodson E.J., Dodson G.G., Reynolds C.D., Smith G.D., Sparks C. and Swenson D. "Phenol stabilizes more helix in a new symmetrical zinc insulin hexamer." *Nature* 338 (1989): 594-596.

Derewenda U., Derewenda Z., Dodson G.G., Hubbard R.E. and Korber F. "Molecular structure of insulin: The insulin monomer and its assembly." *British Medical Bulletin* 45 (1989): 4-18.

Derewenda U., Derewenda Z.S., Dodson G.G. and Hubbard R.E. "Insulin Structure." In: *Handbook of Experimental Pharmacology*, Vol 92 (P. Cuatrecasas and S. Jacobs, eds.), (1990) Springer Verlag, Berlin, pp. 23-39.

Dodson E.J., Dodson G.G., Hubbard R.E. and Reynolds C.D. "Insulin's Structural Behavior and Its Relation to Activity." *Biopolymers* 22 (1983): 281-291.

Engels M., Jacoby E., Krüger P., Schlitter J. and Wollmer A. "The T↔R structural transition of Insulin; pathways suggested by targeted energy minimization." *Protein Engineering* 5 (1992): 669-677.

Fabris D. and Fenselau C. "Characterization of Allosteric Insulin Hexamers by Electrospray Ionization Mass Spectrometry." *Anal. Chem.* 71 (1999): 384-387.

Gross L. and Dunn M.F. "Spectroscopic Evidence for an Intermediate in the T6 to R6 Allosteric Transition of the Co(II)-Substituted Insulin Hexamer." *Biochemistry* 31 (1992): 1295-1301.

Hardaway L.A., Brems D.N., Beals J.M. and MacKenzie N.E. "Amide hydrogen exchange of the central B-chain helix within the T- and R-states of insulin hexamers." *Biochim. Biophys. Acta* 1208 (1994): 101-103.

Huang S.T., Choi W.E., Bloom C., Leuenberger M. and Dunn M.F. "Carboxylate Ions Are Strong Allosteric Ligands for the HisB10 Sites of the R-State Insulin Hexamer." *Biochemistry* 36 (1997): 9878-9888.

Jacoby E., Krüger P., Karatas Y. and Wollmer A. "Distinction of Structural Reorganisation and Ligand Binding in the T↔R Transition of Insulin on the Basis of Allosteric Models." *Biol. Chem. Hoppe-Seyler* 374 (1993): 877-885.

Kaarsholm N.C., Ko H.-C. and Dunn M.F. *Biochemistry* 28 (1989): 4427-4435.

Kadima W. *Biochemistry* 38 (1999): 13443-13452.

Kadima W., Roy M., Lee R. W.-K., Kaarsholm N.C. and Dunn M.F. "Studies of the Association and Conformational Properties of Metal-free Insulin in Alkaline Sodium Chloride Solutions by One- and

Two-dimensional ^{1}H NMR." *J. Biol. Chem.* 267 (1992): 8963-8970.
Karatas Y., Krüger P. and Wollmer A. "Kinetic Measurements of T→R Structural Transitions in Insulin." *Biol. Chem. Hoppe-Seyler* 372 (1991): 1035-1038.
Kim Y. and Shields J.E. "pH Dependent Conformational Changes in the T- and R-States of Insulin in Solution: Circular Dichroic Studies in the pH Range of 6 to 10." *Biochem. Biophys. Res. Commun.* 186 (1992): 1115-1120.
Krüger P., Gilge G., Cabuk Y. and Wollmer A. "Cooperativity and Intermediate States in the T→R-Structural Transformation of Insulin." *Biol. Chem. Hoppe-Seyler* 371 (1990): 669-673.
Markussen J., Andersen A.S., Kaarsholm N.C., Kjeldsen T., Olsen O.H. and Schäffer L. "Insulin and Its Receptor." In: *Frontiers in Insulin Pharmacology* (M. Berger and F.A. Gries, eds.), (1993) Georg Thieme Verlag, Stuttgart, pp. 1-16.
Nakagawa S.H. and Tager H.S. "Implications of Invariant Residue Leu^{B6} in Insulin-Receptor Interactions." *J. Biol. Chem.* 266 (1991): 11502-11509.
O'Donoghue S.I., Chang X., Abseher R., Nilges M. and Led J.J. "Unraveling the symmetry ambiguity in a hexamer: Calculation of the R_6 human insulin structure." *J. Biomol. NMR* 16 (2000): 93-108.
Pittman I. and Tager H.S. "A Spectroscopic Investigation of the Conformational Dynamics of Insulin in Solution." *Biochemistry* 34 (1995): 10578-10590.
Rahuel-Clermont S., French C.A., Kaarsholm N.C. and Dunn M.F. "Mechanisms of Stabilization of the Insulin Hexamer through Allosteric Ligand Interactions." *Biochemistry* 36 (1997): 5837-5845.
Ramesh V. and Bradbury J.H. "^{1}H n.m.r. studies of insulin." *Int. J. Peptide Protein Res.* 28 (1986): 146-153.
Renscheidt H. (1982) Diploma Dissertation, RWTH Aachen.
Renscheidt H., Straßburger W., Glatter U., Wollmer A., Dodson G.G. and Mercola D.A. "A solution equivalent of the 2Zn→4Zn transformation of insulin in the crystal." *Eur. J. Biochem.* 142 (1984): 7-14.
Roy M., Brader M.L., Lee R.W.-K., Kaarsholm N.C., Hansen J.F. and Dunn M.F. "Spectroscopic Signatures of the T to R Conformational Transition in the Insulin Hexamer." *J. Biol. Chem.* 264 (1989): 19081-19085.
Schlichtkrull J. (1958) Dissertation, University of Copenhagen, Munskgaard
Schlitter J., Engels M., Krüger P. Jacoby E. and Wollmer A. "Targeted Molecular Dynamics Simulation of Conformational Change – Application to the T↔R Transition in Insulin." *Molecular Simulation* 10 (1993): 291-308.
Shneine J., Voswinkel M., Federwisch M. and Wollmer A. "Enhancing the T→R Transition of Insulin by Helix-Promoting Sequence Modifications at the N-Terminal B-Chain." *Biol. Chem.* 381 (2000): 127-133.
Smith G.D. "The phenolic binding site in $T_3R_3^f$ insulin[1]." *J. Molecular Structure* 469 (1998): 71-80.
Smith G.D. and Ciszak E. "The structure of a complex of hexameric insulin and 4'-hydroxyacetanilide." *Proc. Natl. Acad. Sci. USA* 91 (1994): 8851-8855.
Smith G.D. and Dodson G.G. "The Structure of a Rhombohedral R_6 Insulin Hexamer That Binds Phenol." *Biopolymers* 32 (1992): 441-445.
Smith G.D., Ciszak E. and Pangborn W. "A novel complex of a phenolic derivative with insulin: Structural features related to the T→R transition." *Protein Science* 5 (1996): 1502-1511.
Smith G.D., Swenson D.C., Dodson E.J., Dodson G.G. and Reynolds C.D. "Structural stability in the 4-zinc human insulin hexamer." *Proc. Natl. Acad. Sci. USA* 81 (1984): 7093-7097.
Smith G.D., Ciszak E. Magrum L.A., Pangborn W.A. and Blessing R.H. "R_6 hexameric insulin complexed with *m*-cresol or resorcinol." *Acta Cryst.* D56 (2000): 1541-1548.
Storm M.C. and Dunn M.F. "The Glu(B13) Carboxylates of the Insulin Hexamer Form a Cage for Cd^{2+} and Ca^{2+} Ions." *Biochemistry* 24 (1985): 1749-1756.
Tang L., Whittingham J.L., Verma C.S., Caves L.S.D. and Dodson G.G. "Structural Consequences of the B5 Histidine → Tyrosine Mutation in Human Insulin Characterized by X-Ray Crystallography and Conformational Analysis." *Biochemistry* 38 (1999): 12041-12051.
Thomas B. and Wollmer A. "Cobalt Probing of Structural Alternatives for Insulin in Solution." *Biol. Chem. Hoppe-Seyler* 370 (1989): 1235-1244.
Wang D., Zeng Z., Hu Y. and Markussen J. "Oligomerization in vitro of the extracellular domain of the insulin receptor by non-covalent interactions is accompnied by increased binding affinity, decreased

binding capacity and curvilinear Scatchard plots." In: *Peptides* (Y. Du, J.P. Tam and Y. Zhang, eds.), (1993) ESCOM, Leiden, pp. 241-244.

Whittingham J.L., Chaudhuri S., Dodson E.J., Moody P.C.E. and Dodson G.G. "X-ray Crystallographic Studies on Hexameric Insulins in the Presence of Helix-Stabilizing Agents, thiocyanate, Methylparaben, and Phenol." *Biochemistry* 43 (1995): 15553-15563.

Whittingham J.L., Edwards D.J., Antson A.A., Clarkson J.M. and Dodson G.G. "Interactions of Phenol and m-Cresol in the Insulin Hexamer, and Their Effect on the Association Properties of B28 Pro→Asp Insulin Analogues." *Biochemistry* 37 (1998): 11516-11523.

Williamson K.L. and Williams R.J.P. "Conformational Analysis by Nuclear Magnetic Resonance: Insulin." *Biochemistry* 18 (1979): 5966-5972.

Wollmer A., Led J.J., Geismar H., Balschmidt P. and Hansen F.B. "The effects of phenolic preservatives on the structure of insulin in solution." Paper presented at the Second Assisi International Symposium on Advanced Models for the Therapy of Insulin-dependent Diabetes, Assisi, Italy, April 20-23 (1986).

Wollmer A., Rannefeld B., Johansen B.R., Hejnaes K.R., Balschmidt P. and Hansen F.B. "Phenol-Promoted Structural Transformation of Insulin in Solution." *Biol. Chem. Hoppe-Seyler* 368 (1987): 903-911.

Wollmer A., Rannefeld B., Stahl J. and Melberg S.G. "Structural Transition in the Metal-Free Hexamer of Protein-Engineered [B13 Gln]Insulin." *Biol. Chem. Hoppe-Seyler* 370 (1989): 1045-1053.

4. ACKNOWLEDGEMENTS

It is not surprising that I am contributing a chapter entitled T→R transition, with insulin my main interest for many years. However, still another motivation for the choice of this title is even more obvious: Which title could be more to the point for the contribution of a retiring person than this? - So it is my turn to go from tense - provided I have ever been that way - to relaxed - provided I shall ever learn to be that way. I wouldn't undergo the transition without extending my warmest thanks to all friends and colleagues with whom I have enjoyed collaboration and exchanges during my working life.

P. KRÜGER

COMPARISON OF CONFORMATIONAL TRANSITIONS IN PROTEINS

Aspects of pathway calculations

Abstract. The T→R transition in insulin is a well known example for a conformational change in a protein. Many other proteins undergo a conformational transition, often associated with functional aspects. The investigation of large transitions using experimental methods usually exhibits macroscopic properties. In insulin the T→R transition has been investigated by simulation techniques. A method named 'Targeted Molecular Dynamics' has been developed in this context in order to describe the changes at the atomic level. Meanwhile this method has been applied to a number of other proteins. A comparison of simulated conformational changes in various systems underlines the necessity of a detailed description of the pathways at the atomic level.

1. INTRODUCTION

The T→R transition in insulin can mainly be described as a change of the N-terminal part of the B-chain, which is extended in T and helical in R. This transition was investigated extensively by a number of experiments (e.g. see this volume) and has also stimulated theoretical work in order to be able to study the details of the conformational transition. As a result, a simulation method named 'Targeted Molecular Dynamics' (TMD) was developed (Schlitter et al., 1993). This method combines the flexibility of a molecular dynamics simulation with the information on the limiting states of the conformational transition. Meanwhile, this method has been applied to a number of other proteins having at least two conformational states. The proteins are for instance chymotrypsin (Wroblowsky et al., 1997; Verheyden, 2000), Ha-*ras* p21 (Diaz et al., 1997; Ma and Karplus, 1997; Krüger et al., 2000a), aspartate transcarbamylase (Roche et al., 1999), plasminogen activator inhibitor (Krüger et al., 2000b) and recently the GroEL system (Ma et al., 2000). Most of these examples show quite different types of movements: the formation of secondary structure, the shift of loops, insertion and shear motions.

At the beginning, when the TMD development started, it was unknown which kind of conformational transition could be tackled. Questions were: Is the T→R transition already too large? Is it possible to include solvent molecules? Can TMD be applied to other systems? How close is the calculated pathway to reality? Many questions could already be answered: even larger conformational changes than the one seen in insulin could be simulated. The inclusion of water is not only possible but also helpful. Recently a publication has shown that even a folding event could be modelled using TMD (Ferrara et al., 2000). Despite these possibilities the main

M. Federwisch, M. Leyck Dieken and P. De Meyts (eds.), Insulin & Related Proteins, 91–101.

question remains, which concerns the quality of the calculated pathway. A global search using the ordinary simulation methods to find pathways with the lowest free energy is not possible in most cases. Therefore some 'helping hands' had to be introduced. In TMD this 'helping hand' is a constraint which is defined between an initial and a target structure. This guidance introduces limits to the exploration of the conformational space. One outcome is the dependence of the pathway on the direction. This means, it is not the same starting from the target structure and moving to the final one or performing the simulation in the opposite direction. Even if the simulation is successful, i.e. reaches the target, only a few possible alternatives for the pathway(s) can be found.

The ultimate test for the quality of the pathway is the comparison with the experiment, e.g. the introduction of mutations which can serve as a probe. This is a very extensive task. Estimates on the possible pathways can be obtained by looking at the conformational transition in a simplified way. A simplified view is for instance to consider the conformational transition as a movement of rigid bodies. Recently a number of groups have worked out methods along this line (Hayward et al., 1997; Wriggers and Schulten, 1997; Verbitsky et al., 1999; Whisstock et al., 2000). The 'rigid body' description gives already an idea of what can be expected from a more detailed simulation using for instance TMD. If an overall description is possible, why is it still useful to go for a more elaborate approach like TMD? In the following, a few examples from our own work are discussed, showing different aspects of conformational transitions. These examples emphasise the necessity to investigate conformational transitions with a method that takes the full flexibility of the proteins into account.

2. COMPUTATION

The TMD method establishes a distance constraint between two structures. For further details of the method see Schlitter et al. (1993). In order to perform a simulation, two structures are necessary, e.g. an initial and a final structure. The pathway from the initial to the final structure describes the direction the simulation takes. It can be the same direction that is followed macroscopically by experiments but this is not necessarily the case. In each step of the simulation, first an unconstrained step of a molecular dynamics (MD) simulation is performed. After the free simulation step the distance between the initial and target structure has to be reduced. This is realised by distance constraints defined as the sum of distances of all atom positions between the initial and final structure. It is a time dependent constraint that reduces the distance in each step of the simulation. Usually a target structure can be reached within a time of 500 ps to 1 ns. The fact that the overall distance between the initial and the final structure is only reduced by a small step after each free MD step, gives some flexibility to the system. Because the constraint is defined through the target structure, the calculated pathway is dependent on the direction taken, i.e. it is different performing a simulation from an initial to a target structure or vice versa (defining the target as new initial structure). The force field of a simulation program guaranties that reasonable energies are found, otherwise the simulation stops. The energy criterion can be used to discriminate between successful and unsuccessful pathways. It can happen that obstacles block the pathway and the potential energy raises. Therefore the TMD method, if successful, delivers pathways with no high energy barriers between the steps. The pathways found can be influenced by the simulation conditions like temperature, starting velocities, etc.. The simulations that are described here were done with the TMD method implemented in the Gromos program package (version 87 and 96) (van Gunsteren et al., 1996; Schlitter et al., 1993; Swegat, 1996).

3. SYSTEMS THAT UNDERGO A CONFORMATIONAL TRANSITION

3.1. Insulin

The simulation of the conformational transition of insulin has been performed using the hexamer structure. In the following some aspects of the transition are discussed based on the pathway calculation between the hexamer in T-state (T6) and in R-state (T3R3) (Jacoby et al., 1995) The organisation of the subunits are displayed in Figure 1. The simulation started from T6. In one trimer of this hexamer the protomers are transformed from T to R. The constraint in the simulation was placed on all subunits simultaneously that means all protomers start their transition at the same time.

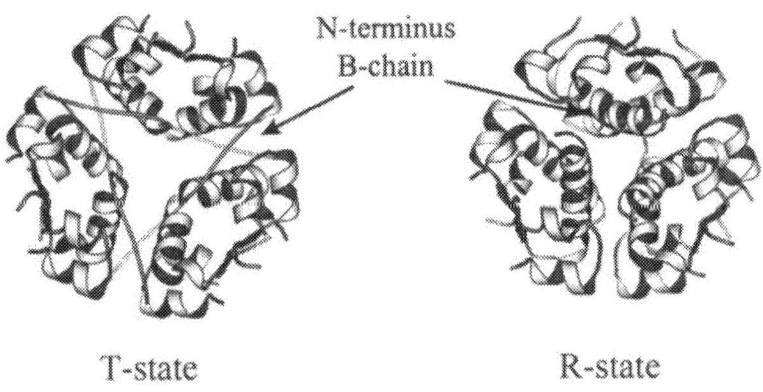

Figure 1. T and R state insulin hexamer. The N-terminus of the B-chain is extended in T and helical in R. The transformation to the R state can be induced by treatment with phenolic reagents. The figures of the proteins structures were made with the program Molscript (Kraulis, 1991).

The simulation trajectory can be displayed in a simplified sketch shown in Figure 2. Here only the change in the monomer is outlined. The two pathways (forward and backward) display possible alternatives of the motion. This motion can be seen as a movement of rigid bodies. First, a swivel movement of the extended N-terminus of the B-chain to the position of the helix, and afterwards the winding up of the extended part to a helical conformation. In the backward direction, there is first a kink of the helix, which is stretched afterwards to the extended situation. An explicit analysis in terms of a rigid body motion has not been made yet, but the cartoon shows, both cases are quite obvious alternatives. This sketch characterises only the change in the monomer. In case of hexamer insulin having quaternary interaction, the movement of the other subunits has to be taken into account as well. One example is the calculation from the T_3R_3 state to the T_6 state that shows a monomer which could not be transformed, because the pathway to the target structure was blocked by an other monomer of the trimer.

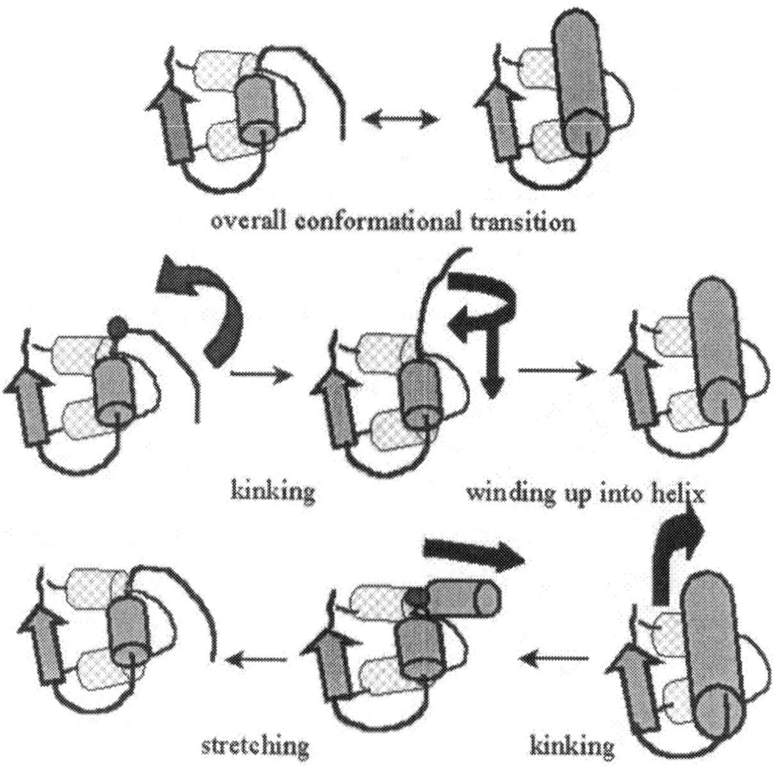

Figure 2. Simplified scheme of the T-R transition in the forward and backward direction. It shows the two possible alternatives for the movement of the N-terminal B-chain corresponding to its secondary structural change.

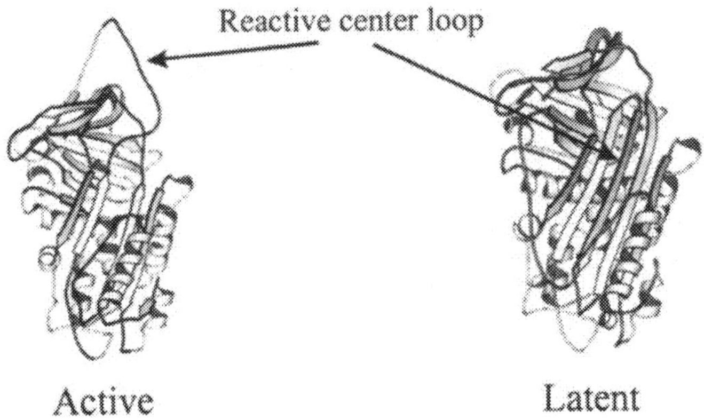

Figure 3. Plasminogen Activator Inhibitor Type I: The transition of the active into the latent structure mainly consists in the insertion of the 'reactive center loop' into the β–sheet A.

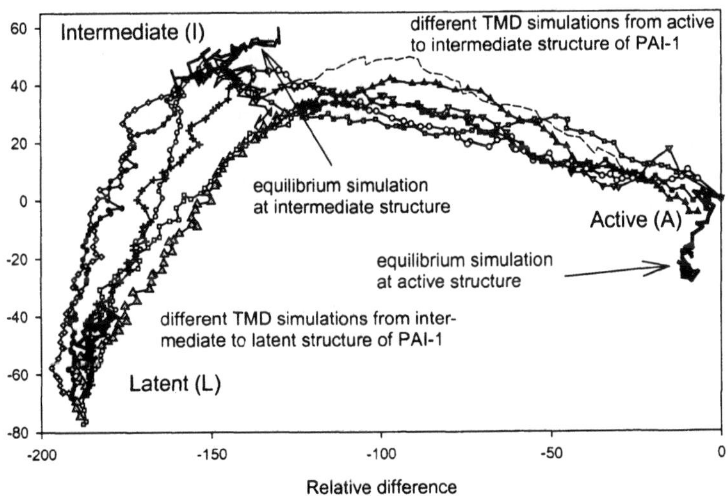

Figure 4. Two-dimensional representation of the conformational space in PAI-1. This projection of the conformational space is calculated according to Diamond, 1974 and Levitt, 1983.

3.2. Plasminogen Activator Inhibitor

Plasminogen Activator Inhibitor 1 (PAI-1) is a molecule that can inhibit proteases. A conformational transition takes place in which a large extended loop, the so-called 'Reactive Centre Loop' (RCL) is inserted into a β–sheet. The structure with RCL extended is the 'active' structure, the one with the RCL inserted the 'latent' one. The latter one cannot inhibit a protease.

In the case of PAI-1 the conformational transition was too large in order to perform a TMD simulation in one step. Not only the size of the conformational transition is essential here, but also the fact that the RCL has to pass through a gate formed by loop 1 and 2. In an ordinary TMD simulation this obstacle hinders the movement of the RCL to the final structure.

In order to enable the simulation from the active structure to latent, a rigid body movement of the RCL was performed. An intermediate was modelled by rotating the RCL around an axis through its hinge points. This new structure was called the intermediate still having clashes with the rest of the molecule. Nevertheless this intermediate had a geometry that made it possible to surround the obstacle. The simulations were performed between the active and the intermediate structure as well as between the intermediate and the latent structure in both directions.

The trajectories of the transition are shown in Figure 4. The plot, which is a two-dimensional projection of the multi-dimensional conformational space, shows a continuous pathway from the active to the latent structure via an intermediate. The computation of the pathway from the active to intermediate structure exhibits an overlap of the different trajectories. In order to circumvent the clashes which are present in the modelled intermediate, the TMD simulation stopped before reaching the intermediate target. This end structure of the TMD simulation had no clashes and showed reasonable potential energies. This structure was used as a new initial structure to start the simulation in order to reach the latent form. Equilibrium simulation, for instance starting from the trajectory close to the intermediate were used to analyse the contacts between the RCL and the loops forming the gate.

3.3. Chymotrypsin

In the case of the system chymotrypsinogen/chymotrypsin the conformational change is associated with the formation of a mature active site and substrate binding pocket. For the activation, residues are cleaved by proteolysis and afterwards the molecule consists of 3 chains A,B,C connected by disulfide bridges. During the conformational transition from chymotrypsinogen to chymotrypsin the N-terminus of the B-chain has to move. The residue Ile 16 is one of the residues that participates in the active site formation. Its structural change is indicated in Figure 5.

In Figure 6 a projection of the conformational space in two dimensions, similar to the one in Figure 4, is displayed. The conformational transition was calculated between the active and the inactive molecule, based on the X-ray structures 4CHA and 2CGA of bovine chymotrypsin, respectively chymotrypsinogen. The active structures can be seen in the left bottom part and the inactive structures in the upper right of the plot. An initial equilibrium MD simulation was performed in order to refine the starting structures. This was done for the molecule with and without the A-chain for the active and the inactive structure. Starting from these equilibrated structures, a TMD simulation was performed between the active and the inactive structure with the A-chain in both directions, and the same was done for the molecules without the A-chain. In case of the inactive form, the cleavage of the A-chain leads to a drift in the overall conformation which is comparable in size to the differences between the active and inactive structure. This is indicated by the end points of the conformational transition and the distances between the inactive structure with and without the A-chain, as well as by the end points of the free equilibrium simulation.

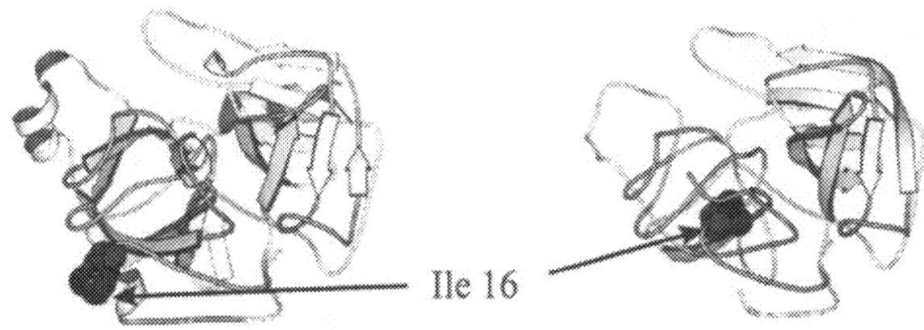

Figure 5. Chymotrypsinogen/chymotrypsin: after cleavage there is a movement of the residue Ile 16 towards the active site.

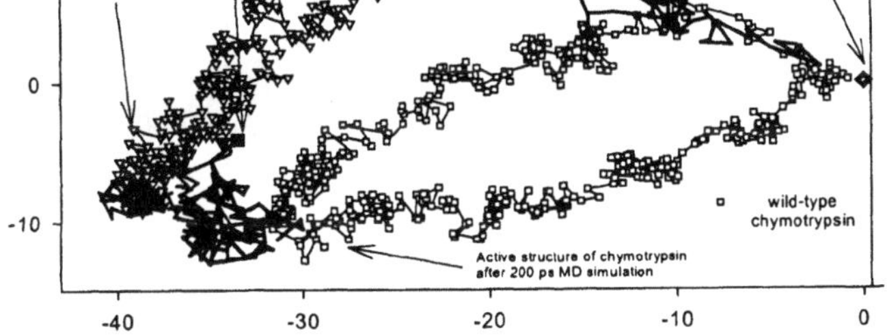

Figure 6. Two-dimensional representation of the conformational space of chymotrypsin in the active and inactive conformation.

3.4. Ha-ras p21

The molecule Ha-*ras* p21 plays an important role for signal transduction in the cell. The triggering mechanism is associated with the cleavage of GTP to GDP and a subsequent conformational change. Mutants can have an influence on this conformational transition.

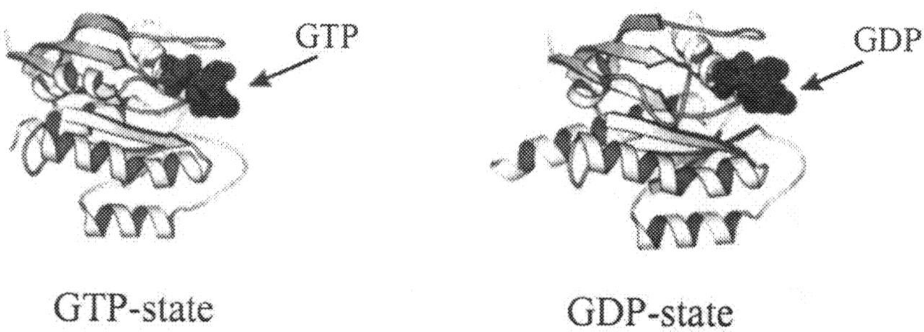

Figure 7. Ha-ras-p21 in the GTP and GDP conformation. After cleavage of the phosphate group the conformation changes especially in loops surrounding the ligand.

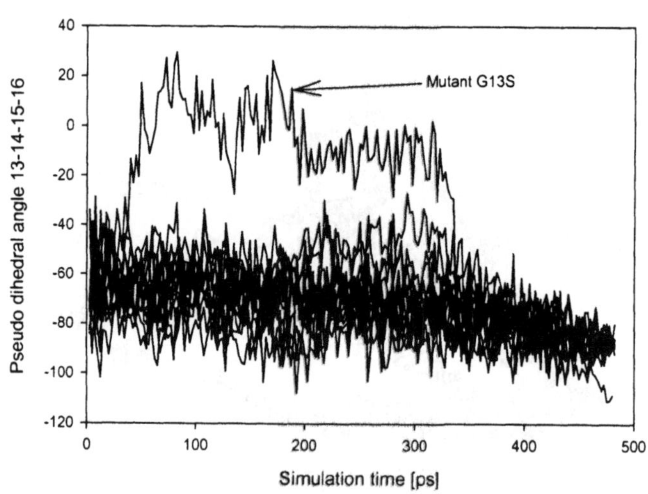

Figure 8. Pseudo dihedral angle (C_α-atoms 13-14-15-16) trajectory of the simulation of one series with 15 mutants of Ha-ras-p21.

The pathway of the conformational transition in Ha-*ras*-p21 was calculated using the structures of the GDP (1Q21) and the GTP (P121) form (Krüger et al., 2000). 15

mutants of Ha-*ras*-p21 were modelled based on the X-ray structure. Three simulation series with the mutants were performed with different ligands (GTP and GDP) and different directions of the pathways (GTP $\rightarrow$ GDP and vice versa). The main aspect of the simulation series was to investigate whether the mutants can deliver different pathways.

An analysis of the trajectories of the rms-positional differences of C_α-atoms demonstrated differences between the various mutants. This was especially the case for functional important segments that move during the conformational change (not depicted). The mutants were for instance designed to affect the main-chain flexibility. An example of a transition in a region in which a mutation is introduced is shown in Figure 8. Here the pseudo dihedral angle between the C_α-atoms i, i+1, i+2, i+3 (13-14-15-16) was used to monitor the transition. The mutant G13S shows a significant different pathway compared to the other simulations. This can be seen also in the other simulation series, but less pronounced.

4. CONCLUSION

The four simulations discussed here show different aspects of the conformational transition. In insulin the simulation of a monomer molecule can be regarded in a simple way as to be composed of two different kinds of rigid body motions. Both alternatives are found in the simulation in both directions. Although this simple description corresponds to the simulation, it becomes less obvious if the transition is seen in context with quaternary interaction. A concerted movement of the three subunits is necessary for a successful conformational transition otherwise the pathways for one subunit can lead to a cul-de-sac. In case of PAI-1 the simple picture of a rigid body motion has been used initially in order to get a first guess for an intermediate. This intermediate is needed to perform a large and complex conformational transition. It has still clashes with the remaining protein body. Free equilibrium simulations show the flexibility of the RCL. These free simulations can be used to analyse the contacts between the RCL and the obstacle formed by loop1 and 2. In chymotrypsin the simulations show an influence of remote parts of the molecule (here the A-chain) on the stability and therefore the pathway of the molecule. The simulations of Ha-*ras* p21 show clearly an influence of mutants on the calculated pathway. This corresponds to the experimental situation, in which pronounced differences between the mutants are found, although a straightforward comparison with the simulation cannot be made yet.

All these examples show the complex influences on a conformational transition: quaternary interaction, sequence specific interaction, impacts due to the molecule stability. On top of this the plasticity of the whole molecule is important. In the case of PAI-1 a simple rigid body motion can neither describe a successful pathway nor does it give a manner to circumvent the obstacle loops on the pathway. Even if we do not know which calculated alternatives come close to the real pathway, the complexity of the possible alternatives can be unravelled using TMD. Simplified models to describe the conformational transitions are valuable because they give first insight how the conformational transition might look like, but many important features are not visible from there.

5. REFERENCES

Diamond R. "Real space refinement of the structure of hen-white lysozyme." *J. Mol. Biol.* 82 (1974): 371-391.

Díaz J.F., Schlitter J. and Engelborghs Y. "Calculation of pathways for the conformational transition between GTP and GDP bound states of the Ha-ras-p21 protein." *Proteins, Structure, Function and Genetics* 28 (1997): 434-451.

Ferrara P., Apostolakis J. and Caflisch A. "Computer simulations of protein folding by targeted molecular dynamics." *Proteins* 39 (2000):252-260.

Hayward S., Kitao A. and Berendsen H.J.C. "Model-Free Methods of Analyzing Domain Motions in Proteins from Simulation: A Comparison of Normal Mode Analysis and Molecular Dynamics Simulation of Lysozyme." *Proteins, Structure, Function and Genetics* 27 (1997): 425-437.

Jacoby E., Krüger P., Schlitter J. and Wollmer A. "Simulation of a complex protein structural change: The T↔R transition in the insulin hexamer." *Protein Engineering* 9 (1995): 113-125.

Krüger P., Diaz J.F., Kuppens S. and Engelborghs Y. "Impact of Mutations on a Conformational Transition. Ha-ras p21 as an Example." In preparation.

Krüger P., Verheyden S, Declerck J.P. and Engelborghs Y. "Extending the Capabilities of Targeted Molecular Dynamics: Simulation of a Large Conformational Transition in Plasminogen Activator Inhibitor 1." *Prot. Sci. 10* (2001): 798-808.

Kraulis J.P. "MOLSCRIPT: A program to produce both detailed and schematic plots of protein structures." *J. Appl. Crystallogr.* 24 (1991): 946-950.

Levitt M. "Molecular dynamics of native protein. II. Analysis and nature of motion." *J. Mol. Biol.* 168 (1983): 621-637.

Ma J., and Karplus K. "Molecular switch in signal transduction: reaction paths of the conformational changes in ras p21." *Proc. Natl. Acad. Sci.* 94 (1997):11905-11910.

Ma J., Sigler P.B., Xu Z. and Karplus M.A. "A Dynamic Model for the Allosteric Mechanism of GroEL." *J. Mol. Biol.* 302 (2000):303-313.

Roche O., and Field M.J. "Simulations of the T↔R conformational transition in aspartate transcarbamylase." *Protein Eng.* 12 (1999): 285-295.

Schlitter J., Engels M., Krüger P., Jacoby E. and Wollmer A. "Targeted molecular dynamics simulation of conformational change - application to the T - R transition in insulin." *Molecular Simulation* 10 (1993): 291-308.

Swegat W., RWTH-Aachen and ETH Zürich, unpublished, 1996

van Gunsteren W.F., Billeter S.R., Eising A.A., Hünenberger P.H., Krüger P., Mark A.E., Scott W.R. and Tironi I.G. "Biomolecular Simulation: The GROMOS96. Manual and User Guide." Zürich, Groningen: VDF Hochschulverlag AG and BIOMOS b.v. an der ETH-Zürich. 1996.

Verbitsky G., Nussinov R. and Wolfson H. "Flexible structural comparison allowing hinge-bending, swiveling motions." *Proteins* 34 (1999): 232-254.

Verheyden G. "Theoretische en experimentele karakterisering van de conformatieveranderingen die optreedt bij activering van chymotrypsinogeen tot chymotrypsine." Thesis University Leuven. 2000.

Whisstock, J.C., R. Skinner, R.W. Carrell, and A.M. Lesk. "Conformational Changes in Serpins: I. The Native and Cleaved Conformations of α_1-Antitrypsin." *J. Mol. Biol.* 295 (2000): 651-665.

Wriggers W. and Schulten K. "Protein domain movements: detection of rigid domains and visualization of hinges in comparisons of atomic coordinates." *Proteins* 29 (1997): 1-14.

Wroblowski B., Díaz J.F., Schlitter J. and Engelborghs Y. "Modeling pathways of α-chymotrypsin activation and deactivation." *Protein Eng.* 10 (1997): 1163-1174.

6. ACKNOWLEDGEMENTS

I would like to thank Dr. Gert Verheyden for the permit to cite results from his thesis prior to publication. I also would like to thank Prof. Dr. Yves Engelborghs, Steven Kuppens, Stefan Verheyden and Jo Verkammen for helpful discussions. Thanks go to Priv. Doz. Dr. J. Schlitter for long standing cooperation on the development of the TMD method. PK is a recipient of the fellowship F/97/115 and the research is supported by grant OT/97/19 from the research council of the University of Leuven.

M.A. WEISS, Q.-X. HUA, W. JIA, S.H. NAKAGAWA, Y.-C. CHU, P.G. KATSOYANNIS

ACTIVITIES OF INSULIN ANALOGUES AT POSITION A8 ARE UNCORRELATED WITH THERMODYNAMIC STABILITY

Abstract. The biological activity and thermodynamic stability of human insulin are enhanced by substitution of Thr^{A8} by arginine or histidine. These substitutions on the surface of insulin stabilize the N-terminal α-helix of the A chain, a key element of hormone-receptor recognition. Does enhanced stability necessarily imply enhanced activity? Here, we test a proposed correlation between the hormone's thermodynamic stability and biological activity by comparison of Gln^{A8} and Glu^{A8} analogues. To circumvent possible confounding effects of insulin self-association, these A-chain substitutions were combined with a monomeric B-chain variant containing three substitutions, [Asp^{B10}, Lys^{B28}, Pro^{B29}] ("DKP"). Two analogues, Gln^{A8}–DKP-insulin and Glu^{A8}–DKP-insulin, were obtained by chain combination with native yield. ^{1}H-NMR spectra of the analogues are essentially identical to those of DKP-insulin, indicating a correspondence of structures. Biological activities were determined by competitive displacement of ^{125}I-human insulin from insulin receptors in human placental membranes. Under conditions wherein DKP- insulin has 150 ± 23% activity relative to native human insulin, Gln^{A8}–DKP-insulin exhibits 122 ± 1% activity and Glu^{A8}–DKP- insulin, 61 ± 2% activity. Thermodynamic stabilities were measured by circular dichroism: protein unfolding was monitored as a function of guanidine concentration. Surprisingly, activity correlates inversely with stability: inferred values of ΔG_u are 4.8 ± 0.2 kcal/mole (DKP-insulin), 5.0 ± 0.1 kcal/mole (Gln^{A8}–DKP-insulin) and 5.9 ± 0.1 kcal/mole (Glu^{A8}–DKP-insulin). We suggest that the relative biological activities of these A8 analogues reflect electrostatic features of the hormone-receptor interface rather than the free hormone's thermodynamic stability.

1. INTRODUCTION

The functional surface of insulin has long been the object of speculation (Adams et al., 1969; Blundell et al., 1972; Pullen et al., 1976; De Meyts et al., 1978; Chothia et al., 1983; Baker et al., 1988) Despite many years of investigation by mutagenesis (Katsoyannis et al., 1973; Cosmatos and Katsoyannis, 1975; Danho et al., 1975; Inouye et al., 1978; Tager et al., 1979; Inouye et al., 1981; Kobayashi et al., 1982; Okada et al., 1981; Shoelson et al., 1983; Kitagawa et al., 1984; Kobayashi et al., 1986; Nakagawa and Tager, 1986; Nanjo et al., 1986; Nakagawa and Tager, 1987; Casaretto et al., 1987; Schwartz et al., 1989; Mirmira and Tager, 1989; Wang et al., 1991; Mirmira and Tager, 1991; Nakagawa and Tager, 1992; Hu et al., 1993; Kaarsholm et al., 1993; Wollmer et al., 1994; Weiss et al., 2000), nuclear magnetic resonance (NMR)[1] spectroscopy (Hua and Weiss, 1990; Hua et al., 1991; Weiss et al., 1991; Knegtel et al., 1991; Hua et al., 1992; Hua et al., 1993; Hua et al., 1993; Sørensen and Led, 1994; Hua et al., 1995; Jørgensen et al., 1996; Hua et al., 1996; Olsen et al., 1996; Ludvigsen et al., 1998; Olsen et al., 1998; Kurapkat et al., 1999)

M. Federwisch, M. Leyck Dieken and P. De Meyts (eds.), Insulin & Related Proteins, 103–119.

and X-ray crystallography (Adams et al., 1969; Peking Insulin Structure Group, 1971; Bentley et al., 1978; Bi et al., 1984; Liang et al., 1985; Dai et al., 1987; Derewenda et al., 1989; Badger et al., 1991; Derewenda et al., 1991), structures of insulin and insulin analogues do not consistently predict relative potencies. These observations suggest that a change in structure may occur on receptor binding (Nakagawa and Tager, 1987; Mirmira and Tager, 1989, 1991; Hua et al., 1991; Hua et al., 1993; Derewenda et al., 1991). This intriguing but controversial hypothesis (Ludvigsen et al., 1998) has motivated examination of the relationship between the thermodynamic stability of insulin analogues and their biological activity. Do mutations that stabilize the native insulin T state[2] (Figure 1) also enhance biological activity, and if so, how can such effects be compatible with an induced-fit mechanism of receptor binding? A subset of analogues with enhanced potency has been found by Kaarsholm and coworkers (Kaarsholm et al., 1993) to exhibit enhanced stability (Table 1). This influential study focused on the substitution of Thr^{A8} in human insulin by alanine, histidine or arginine. The A8 site defines one edge of insulin's classical receptor-binding surface (Pullen et al., 1976; De Meyts et al., 1978; Baker et al., 1988) and demarkates the A1-A8 A-chain α-helix (asterisk in Fig. 1). The enhanced activity of His^{A8} and Arg^{A8} analogues was ascribed to thermodynamic stabilization of this critical helix by a combination of local and long-range electrostatic interactions (Kaarsholm et al., 1993; Olsen et al., 1998). To test the generality of this conclusion, we have prepared an A8 analogue containing a negative charge at position A8 (Glu^{A8}). Surprisingly, the substitution stabilizes the free hormone but destabilizes the hormone-receptor complex. These observations suggest that, contrary to the hypothesis of Kaarsholm et. al. (Kaarsholm et al., 1993), effects of mutations on insulin's thermodynamic stability are uncorrelated with effects on biological activity.

Our experimental design employs an engineered insulin monomer (Brange et al., 1988) as a framework for A8 substitutions. The monomer, designated DKP-insulin (Weiss et al., 1991; Brems et al., 1992; Bakaysa et al., 1996), contains three substitutions in the B chain ([Asp^{B10}, Lys^{B28}, Pro^{B29}]). Its solution structure under physiological conditions (Hua et al., 1996) resembles the crystallographic T state (Fig. 1). Two substitutions (interchange of Pro^{B28} and Lys^{B29}) were originally designed by R. DiMarchi and colleagues at Eli Lilly and Company (Brems et al., 1992; Bakaysa et al., 1996) by analogy to the sequence of insulin-like growth factor I (IGF-I). These substitutions weaken the hormone's classical dimer contact without altering activity (Hua et al., 1992). The third substitution, His^{B10}→Asp, disrupts the trimer interface and Zn^{2+}-binding site utilized in crystallographic hexamers. The B10 substitution, originally described in a patient with hyperproinsulinemia and diabetes mellitus (Gruppuso et al., 1984; Chan et al., 1987; Carroll et al., 1988), enhances the

---

[2] The usual crystal forms of insulin are hexameric (T_6, T_3R_3 and R_6). The structure of an isolated dimer has been determined in a Zn-free cubic lattice (Badger et al., 1991). The crystal structure of a native insulin monomer has not been obtained. However, the structure of proteolytic fragment des-pentatpeptide-insulin (DPI), lacking a portion of the dimerization surface, has been determined in two crystal forms (Bi et al., 1984; Dai et al., 1987).

hormone's stability (Kaarsholm et al., 1993) and affinity for the insulin receptor (Schwartz et al., 1989). The resulting "DKP" B chain may be combined with variant A chains to enable investigation of A-chain substitutions in a monomeric framework (Weiss et al., 1991). We describe here the synthesis and characterization of two analogues of DKP-insulin. Glu^{A8}-DKP-insulin was prepared to test the relationship between the hormone's activity and stability. The substitution places a negative charge in a seemingly unfavorable environment: near another negative charge (Glu^{A4}) and at the negative end of an α-helical dipole. A corresponding Gln^{A8} analogue, containing a polar but uncharged side chain isosteric to glutamic acid, was prepared as control. The structures of the analogues, as probed by circular dichroism (CD) and ^{1}H-NMR spectroscopy, are similar to those of the parent monomer. Surprisingly, the Glu^{A8} analogue is more stable than either the Gln^{A8} analogue or the parent monomer. Further and as a seeming paradox, relative affinities correlate inversely with thermodynamic stabilities. We propose that variant A8 side chains contact the periphery of the insulin receptor and in so doing can introduce favorable or unfavorable electrostatic interactions irrespective of their effects on the thermodynamic stability of the free hormone.

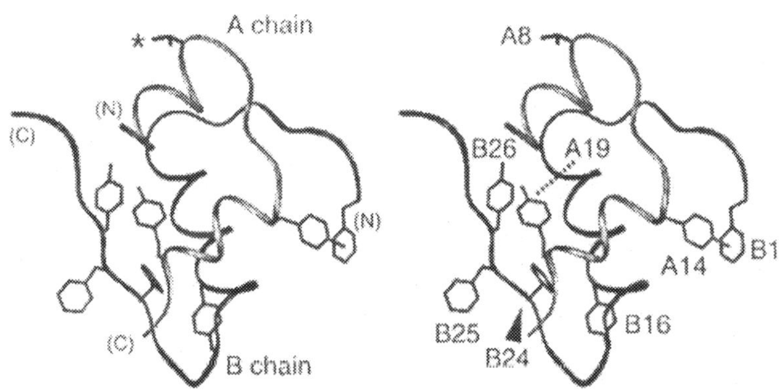

Figure 1. Stereo ribbon representation of the porcine insulin monomer, as inferred from the crystal structure of 2-Zn (T_6) porcine insulin (molecule 1; Chinese nomenclature). The B chain is shown in black and A chain in grey. Asterisk in left-hand panel indicates A8 side chain (threonine). For reference the four tyrosine and two phenylalanine side chains are also shown; these provide informative NMR probes as shown in Figure 4. NMR structures of engineered insulin monomers in solution (Hua et al., 1996; Olsen et al., 1996) resemble the crystallographic T-state protomer (Blundell et al., 1972; Adams et al., 1969; Peking Insulin Structure Group, 1971)

Table 1. Past Studies of A8 Analogues of Human Insulin[a]

analogue	potency	ΔG_u (kcal/mole)	C_{mid} (M)[b]
human insulin	100	3.76	4.37
His^{A8}-human	309	5.52	5.41
Ala^{A8}-human	112	3.92	4.51
Arg^{A8}-human	302	5.01	4.77

[a] *Values are taken from Kaarsholm and coworkers (Kaarsholm et al., 1993).*
[b] *C_{mid} is defined as that molar concentration of guanidine-HCl associated with fifty percent protein unfolding.*

2. METHODS

2.1. Materials

4-Methylbenzhydrylamine resin (0.6 mmol of amine/g; Star Biochemicals, Inc.) was used as solid support for synthesis of the A-chain analog; (N-butoxy-carbonyl, O-benzyl)-threonine-PAM resin, (0.56 mmol/g; Bachem, Inc.) was used as solid support for synthesis of the B-chain analog. *tert*-Butoxycarbonyl-amino acids and derivatives were obtained from Bachem and Peninsula Laboratories; N,N'-dicyclohexylcarbodiimide and N-hydroxybenzotriazole (recrystallized from 95% ethanol) from Fluka. Amino-acid analyses of synthetic chains and insulin analogues were performed after acid hydrolysis; protein determinations were carried out by the Lowry method using native insulin as standard. Chromatography resins were pre-swollen microgranular carboxymethylcellulose (CM-cellulose; Whatman CM52), DE53 cellulose (Whatman) and Cellex E (Ecteola cellulose; Sigma); solvents were HPLC grade.

2.2. Peptide Synthesis

The general protocol for solid-phase synthesis is as described (Stewart and Young, 1984). The C-terminal Asn in the synthesis of the A chain was incorporated into solid support by coupling *tert*-butoxycarbonyl aspartic acid-α-benzyl ester with 4-methylbenz-hydrylamine resin. After the final deprotection the Asp residue was converted to an Asn residue. (i) *Synthetic A-Chain S-Sulfonate.* From 606 mg of Glu^{A8} peptidyl resin, after deblocking, sulfitolysis and chromatographic purification, approximately 116 mg of purified S-sulfonated A-chain were obtained; from 602 mg of Gln^{A8} peptidyl resin, after analogous processing, 125 mg of purified S-sulfonated A-chain were obtained. (ii) *Synthetic B-Chain S-Sulfonate.* After deblocking, sulfitolysis and chromatographic purification 610 mg of peptidyl resin yielded approximately 125 mg of purified S-sulfonated B-chain. Amino-acid analyses were in agreement with expected values.

2.3. Peptide Purification

Crude S-sulfonated A chain was purified by chromatography on a Cellex E column (1.5 x 47 cm) as described (Hu et al., 1993; Hua et al., 1996), dialyzed against distilled water, and lyophilized to yield the purified Glu^{A8} or Gln^{A8} A-chain S-sulfonate. Crude S-sulfonated B-chain was likewise purified on a cellulose DE53 column (1.5 x 47 cm), dialyzed and lyophilized to yield the [Asp^{B10}, Lys^{B28}, Pro^{B29}] B-chain S-sulfonate.

2.4. Chain Recombination

Chain recombination employed S-sulfonated A- and B chains (approximately 2:1 by weight) in 0.1 M glycine (pH 10.6) in presence of dithiothreitol (Hu et al. 1993; Hua et al. 1996). The insulin analogue was isolated from the combination mixture as described (Hu et al. 1993) and purified on a 0.9 x 23 cm CM-cellulose chromatography and rp-HPLC on a Vydac 218 TP column (0.46 x 25 cm); the latter used a flow rate of 0.5 ml/min with 20-80% linear gradient of 80% aqueous acetonitrile containing 0.1% trifluoroacetic acid over 80 min. Re-chromatography of this material on rp-HPLC under the same conditions gave a single sharp peak. Amino-acid analyses and mass spectrometry gave expected values.

2.5. Receptor Binding Studies

Radio-labelled [^{125}I-Tyr^{A14}]-human insulin was purchased from Amersham (Piscataway, N.J.). Receptor binding assays of insulin analogues were performed as described (Cara et al., 1990) with minor modifications. Human placental cell membranes were prepared as described (Marshall et al., 1974), stored at -80 °C in small aliquots and thawed prior to use. Membrane fragments (0.025 mg protein/tube) were incubated with radio-labelled insulin (approximately 30,000 cpm) in the presence of selected concentrations of unlabelled peptide for 18 hours at 4 °C in a final volume of 0.25 ml of 0.05 M Tris-HCl and 0.25% (w/v) bovine serum albumin at pH 8. Subsequent to incubation, each mixture was diluted with 1 ml of ice-cold buffer and centrifuged (10,000 x g) for 5 min at 4 °C. The supernatant was then removed by aspiration, and the membrane pellet was counted for radioactivity. Data were corrected for nonspecific binding (amount of radioactivity remaining membrane-associated in the presence of 1 μM human insulin). Each determination was performed with three or four replicates (see Table 2); data are reported as the mean and standard deviation.

2.6. Spectroscopy

^{1}H-NMR spectra were obtained at 600 MHz at 25 °C in 50 mM potassium phosphate (pH 7); the protein concentration was 1.5 mM. Aromatic resonance assignment for DKP-insulin was accomplished as described (Hua et al., 1996) and extended by inspection to analogues. Spectra in H_2O were obtained using pulse-field gradients and laminar shaped pulses. CD spectra were obtained using an Aviv spectropolarimeter equipped with thermister temperature control and automated titration unit for guanidine denaturation studies. CD samples contained 25-50 μM insulin or analogue in 50 mM potassium phosphate (pH 7); samples were diluted to 5 μM for equilibrium denaturation studies (Fig. 2C). Data were fitted by non-linear least squares to a two-state model as described (Hua et al., 2000).

3. RESULTS

Glu^{A8}-DKP-insulin and Gln^{A8}-DKP-insulin were prepared by total synthesis (see *Methods*). The analogues thus contain one substitution in the A chain and three substitutions in the B chain. Use of the monomeric DKP template simplifies biophysical analysis by avoiding confounding effects of insulin self-association (Weiss et al., 1991; Hua et al., 1996). Chain combination yielded native disulfide pairing with an efficiency similar to that observed in the total synthesis of native insulin. Non-native disulfide isomers (Hua et al., 1995) were not observed.

3.1. Analogues Exhibit Native Structure

CD spectra of the analogues at 4 °C exhibit similar helix context relative to the parent DKP-insulin monomer (Fig. 2A). Similar results are obtained at 25 °C (not shown). One-dimensional ^{1}H-NMR spectra of the analogues at pH 7 and 25 °C are similar to that of DKP-insulin (Fig. 3); each exhibits non-random dispersion of chemical shifts characteristic of the T-state structure (Fig. 1). Two-dimensional total correlation (TOCSY) ^{1}H-NMR spectra reveal essentially identical spin systems with corresponding chemical shifts. This correspondence is illustrated in the aromatic region (Fig. 4), which provides intrinsic probes of the hormone's major structural elements. An example is provided by the chemical shifts of Phe^{B24} and Tyr^{A19}, whose large secondary shifts reflect their distinct environments in the native hydrophobic core. Such correspondence of chemical shifts provides compelling evidence of structural similarity; accordingly, detailed analysis of nuclear Overhauser effects (Hua et al., 1996) was not performed.

3.2. Glu^{A8} Substitution Impairs Receptor Binding

Specific binding to the insulin receptor was measured relative to binding of ^{125}I-labelled human insulin to a human placental membrane fraction (see *Methods*). Results are given in Table 2. Whereas the activity of GlnA8-DP-insulin is similar to that of DKP-insulin (within experimental error), the Glu^{A8} analogue is significantly less active (approximately 61% relative to human insulin). The activities of Gln^{A8} and Glu^{A8} analogues differ by a factor of two.

3.3. Glu^{A8} Substitution Enhances Thermodynamic Stability

The native DKP monomer and A8 analogues exhibit similar temperature-dependent changes in mean residue ellipticity at 222 nm in the range 4-50 °C (Fig. 2B). Analysis of protein unfolding in the denaturant guanidine-HCl at 4 °C demonstrates significant differences in thermodynamic stability (Fig. 2C) Whereas DKP-insulin exhibits a cooperative and apparent two-state transition with ΔG_u 4.8 ± 0.2 kcal/mole, Glu^{A8}-DKP-insulin exhibits enhanced stability (ΔG_u 5.9 ± 0.1 kcal/mole; see Table 2). The extent of stabilization ($\Delta\Delta G_u$ 1.1 ± 0.3 kcal/mole) is thus similar to that reported for His^{A8} ($\Delta\Delta G_u$ 1.76 kcal/mole) and Arg^{A8} ($\Delta\Delta G_u$ 1.25 kcal/mole; see Table 1). Although qualitative inspection of the unfolding curves strongly suggests that Gln^{A8}-DKP-insulin is slightly more stable than DKP-insulin in accord with its greater C_{mid} guanidine concentration (Table 2), the inferred increment in stability is imprecisely determined ($\Delta\Delta G_u$ 0.2 ± 0.3 kcal/mole).

Table 2. Potencies and Guanidine Denaturation of DKP Analogues[a]

analogue	potency[b]	ΔG_u	$\Delta\Delta G_u$	C_{mid}
insulin	100	4.4 ± 0.1	--	5.3 ± 0.1 M
DKP-insulin	150 ± 23 (4)	4.8 ± 0.2	--	5.8 ± 0.1 M
Gln^{A8}-DKP-insulin	122 ± 1 (3)	5.0 ± 0.1	0.2 ± 0.3	6.4 ± 0.1 M
Glu^{A8}-DKP-insulin	61 ± 2 (3)	5.9 ± 0.1	1.1 ± 0.3	6.6 ± 0.2 M

[a]Values of ΔG_u and $\Delta\Delta G_u$ are given in kcal/mole. [b]Potency was measured by relative affinity for the insulin receptor; number of assays is given in parenthesis with standard deviations provided. Under these conditions the K_d for native insulin is 0.48 ± 0.06 nM. C_{mid} is defined as that molar concentration of guanidine-HCl associated with fifty percent protein unfolding.

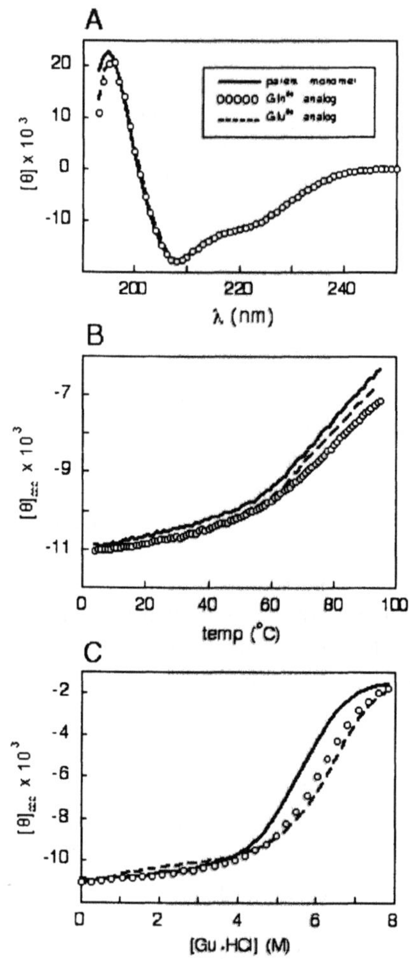

Figure 2. Circular dichroism (CD) studies of DKP-insulin (solid lines) and analogues Gln^{A8}–DKP-insulin (open circles) and Glu^{A8}–DKP-insulin (red dashed lines). (A) Far-ultraviolet CD spectra of A8 analogues are similar to that of DKP-insulin. (B) Thermal unfolding studies demonstrate similar thermal stabilities in the range 4-50 °C. For clarity curves of Gln^{A8}–DKP-insulin and Glu^{A8}–DKP-insulin are offset downward. At higher temperatures relative thermal stabilities follow the order Gln^{A8}–DKP-insulin > Glu^{A8}–DKP-insulin > DKP-insulin. Chemical degradation of analogues at elevated temperatures is possible and cannot be excluded. (C) Denaturant unfolding studies of analogues at 4 °C demonstrate marked differences in thermodynamic stabilities. Mean residue ellipticity is shown as a function of [guanidine-HCl]. The most stable analogue is Glu^{A8}–DKP-insulin, which is least active. Inferred values of ΔG_u are given in Table 2.

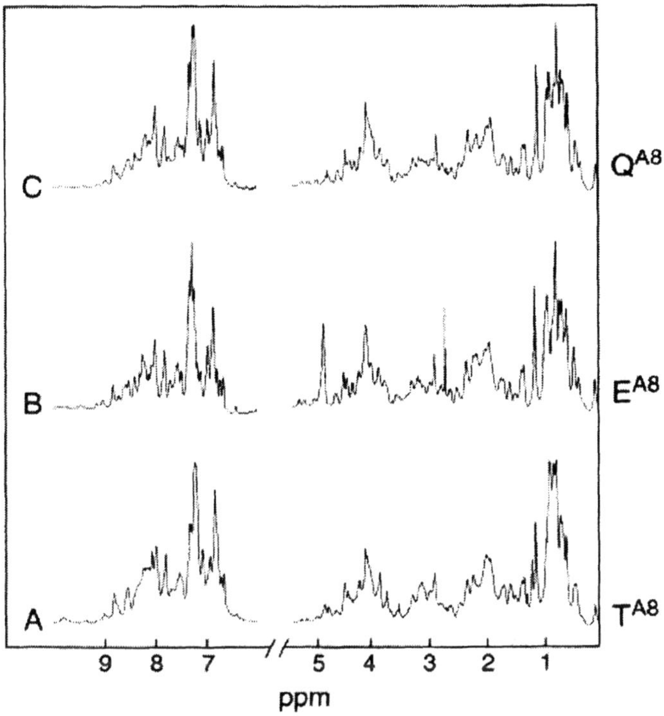

Figure 3. Overall structural similarities are demonstrated by one-dimensional ^{1}H-NMR spectroscopy. Spectra of DKP-insulin (A), Glu^{A8}–DKP-insulin (B), and Gln^{A8}–DKP-insulin (C) at 25 °C and pH 7.0. Spectra at right were obtained in D2O (pD 6.6, direct meter reading). Chemical shifts are referenced to 3-Trimethylsilyl propionate sodium salt at 0 ppm.

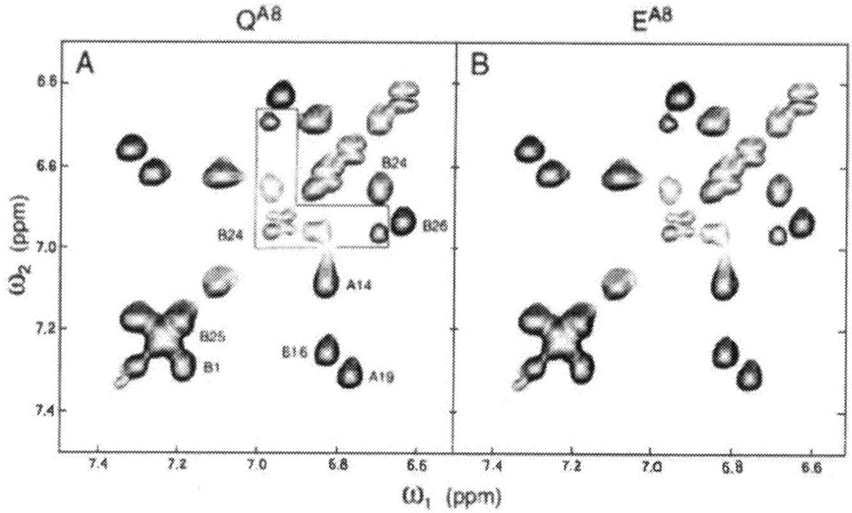

Figure 4. Detailed correspondence of chemical shifts is observed in aromatic region (and elsewhere) of two-dimensional TOCSY spectra of (A) Gln^{A8}–DKP-insulin and (B) Glu^{A8}–DKP-insulin. Assignments are as indicated in panel A. The spectrum of DKP-insulin is essentially identical. Locations of aromatic rings in structure of insulin are as shown in Figure 1. Conditions were as described in the caption to Figure 3; the TOCSY mixing time was 55 ms. The unusual Phe^{B24} spin system is boxed at left.

4. DISCUSSION

The relationship between insulin's thermodynamic stability and its bioactivity is intimately connected to the nature of the hormone's active structure. Should a change in conformation be required for receptor binding, then destabilizing mutations at key positions may enhance binding by facilitating this conformational change (Mirmira and Tager, 1989, 1991; Hua et al., 1991). Conversely, should the native structure dock within the receptor as a rigid body, then stabilizing mutations may enhance binding by reducing the entropic cost of association. In fact, we imagine that the active structure of insulin will be found to consist of a mixture of native and induced elements, leading to no consistent relationship between the thermodynamic effects of mutations and their consequences for biological activity. Whereas the C-terminal A-chain α-helix appears to function as a preformed recognition element (Weiss et al., 2000; Hua et al., 1996), considerable speculation has focused on conformational changes involving the C-terminal β-strand of the B chain (Baker et al., 1988; Kobayashi et al., 1982; Mirmira and Tager, 1989, 1991; Hua et al., 1991; Ludvigsen et al., 1998; Derewenda et al., 1991). That the native B-chain β-sheet, as seen in crystal structures of insulin, represents an inactive structural element was suggested by the paradoxical finding of native structure in an inactive single-chain analogue ("mini-proinsulin") (Derewenda et al., 1991). The single-chain analogue, which contains a peptide bond between the C-terminal β-strand of the B-chain (Lys^{B29}) and N-terminal α-helix of the A-chain (Gly^{A1}), is highly stable yet presumably constrained from reorganizing on receptor binding (Markussen et al., 1985). This hypothesis raises the intriguing possibility that detachment or destabilization of the C-terminal β-strand could enhance activity. This process would be expected to facilitate contacts between the receptor and the conserved side chains of Ile^{A2} and Val^{A3}. Aliphatic substitutions at these sites, although seemingly conservative, markedly impair receptor binding (Baker et al., 1988; Okada et al., 1981; Shoelson et al., 1983; Nakagawa and Tager, 1992). The proposed detachment of the C-terminal B-chain β-strand would rationalize the enhanced activity of D-amino-acid substitutions at position B24 (Kobayashi et al., 1982).

Evidence that the N-terminal α-helix of the A-chain functions as a preformed recognition element has been obtained through the study of a two-disulfide analogue of DKP-insulin in which cystine A6-A11 is substituted by serines (Hua et al., 1996). The analogue exhibits segmental unfolding of the A1-A8 α-helix in an otherwise native-like subdomain. Despite overall preservation of the insulin fold and retention of putative receptor contacts, the analogue is essentially inactive. Why cannot the two-disulfide analogue readily refold on receptor binding as a consequence of induced fit? Because cystine A6-A11 is buried in the hormone's hydrophobic core, the analogue's low activity has plausibly been ascribed to the unfavorable polarity of serine when forced into a non-polar environment. This model predicted that the corresponding [Ala^{A6}, Ala^{A11}] two-disulfide analogue would be markedly more active than the [Ser^{A6}, Ser^{A11}] analogue. This prediction has recently been verified (Weiss et al., 2000). The structure of the alanine analogue exhibits similar segmental destabilization, indicating that the function of thee native A6-A11 disulfide bridge is to stabilize the α-helical folding of the A1-A8 segment. Together, these findings demonstrate that this helix is not a cooperative element of structure; *i.e.*, local folding or unfolding can occur independently of the remainder of the native state.

The critical importance of the N-terminal α-helix of the A-chain has focused attention on the structural details of its stabilization. The C-terminal residue is threonine (Thr^{A8}), not the most favorable residue as a peptide C-cap (Doig and Baldwin, 1995; Chakrabartty et al., 1993). The presence of glutamic acid at A4 raised the possibility of designing a more stable peptide helix by introduction of a postively charged side chain at A8, creating an (i, i+4) salt bridge as a "capping box" (Motoshima et al., 1997). Additional stabilization might occur by long-range electrostatic interactions with charges in the B chain. Accordingly, Kaarsholm and coworkers (Kaarsholm et al., 1993) studied an Arg^{A8} analogue together with His^{A8}, previously known to enhance activity in the context of porcine insulin (Baker et al., 1988). The results were in accord with the above electrostatic model (Table 1). This correlation suggested that the introduced interactions stabilized the native structure of the hormone and thereby enhanced its biological activity. The present study addresses the possibility that the consistency of the then-available data and proposed model does not reflect a causal relationship. It seemed possible to us that the correlation observed between enhanced thermodynamic stability and enhanced binding (Table 1) was spurious. This possibility was raised by the absence of characterization of a negatively charged substitution at A8. The C-cap model predicted that such an analogue would exhibit both decreased stability and decreased receptor binding. To test this prediction, we have synthesized and characterized Glu^{A8}-DKP-insulin in relation to Gln^{A8}-DKP-insulin and the parent monomer.

Our results (Table 2) demonstrate that introduction of a negative charge at A8 is associated with an increase in stability almost as large as introduced by Arg^{A8} or His^{A8}. The structural basis for this is enigmatic. CD and ^{1}H-NMR spectra suggest that the overall structure of the analogue is essentially the same as that of DKP-insulin. The net thermodynamic effect of the Glu^{A8} substitution presumably reflects the sum of favorable and unfavorable terms; the latter include the helix-dipole effect (Doig and Baldwin, 1995; Chakrabartty et al., 1993; Motoshima et al., 1997) and possible repulsion between the side chains of Glu^{A4} and Glu^{A8}. Favorable interactions may include a strengthened salt bridge between the ε-NH_3^+ group of Lys^{B28} and Glu^{A4}. Such an interaction could stabilize the native configuration of the C-terminal β-strand of the B-chain and thus hinder its proposed detachment on receptor binding. Regardless of structural mechanism, these results illustrate the importance of testing a variety of side-chain functionalities before inferring cause-and-effect relationships. Indeed, the complexity of protein structures and interactions implies that changes in physical or functional properties often reflect small differences between large entropic and enthalpic driving forces, including those associated with solvation. Even in the presence of high-resolution crystal structures it is usually difficult to dissect the physical origins of changes in free energies of stability or interaction.

Functional effects of A8 substitutions may in principle reflect a variety of mechanisms. One possible mechanism, non-local changes in the structure of insulin, is excluded by the available data (Olsen et al., 1998). Kaarsholm and coworkers have proposed, by contrast, that effects on receptor binding are mediated by the local structure or stability of the A1-A8 helical segment in the free hormone (Kaarsholm et al., 1993). The following reasoning suggests that this effect is minor. We hypothesize that variations in segmental helical stability or configuration can readily be "repaired" by the receptor by local induced fit. Because the A8 side chain itself lies at the periphery at the receptor-binding surface (Pullen et al., 1976; Baker et al., 1988), this hypothesis predicts that diverse side chains can be accomodated at A8 without significant change in net binding affinity. An upper limit to the cost of induced fit is estimated by the residual activity of an [Ala^{A6}, Ala^{A11}] two-disulfide analogue in which the A1-A8 segment is disordered (Weiss et al., 2000). Despite disorganization of this critical recognition element, the two-disulfide analogue exhibits relative binding of 5%, yielding a thermodynamic "repair bill" (ΔΔG of binding) of approximately 1.6 kcal/mole. Because its stimulation of lipogenesis is identical to that of native porcine insulin at saturating concentrations of the hormone, the variant hormone-receptor complex achieves native potency and so presumably native structure. This example illustrates an extreme case of induced fit – a disordered coil-to-helix transition – and yet it is associated with only a small thermodynamic penalty. We thus expect that subtle changes in the helical structure of the free hormone (as may be associated with A8 substitutions) would in themselves not significantly aid or impair receptor binding; *i.e.*, induced fit should readily be able to override such local structural variations. This reasoning suggests that whether or not the native carboxylate of Glu^{A4} makes a salt bridge with Arg^{A8} or hydrogen bond with His^{A8} in the unbound hormone (Olsen et al., 1998) is unrelated to the biological effects of the latter substitutions. Analogous reasoning pertains to the proposed correlation between biological activity and the global thermodynamic stability of the free hormone. Global stabilities (ΔG_u) reflect an overall equilibrium between folded and unfolded states of a protein. A mutational increase in ΔG_u from 4 kcal/mole to 5.4 kcal/mole, for example, would correspond to an increase in the fraction of folded insulin molecules in an ensemble from 0.9990% to 0.9999%. In itself such a marginal change in the availability of the folded state would be expected to have a negligible effect on receptor binding. Activity of A8 analogues could nonetheless be enhanced or impaired by three possible residue-specific mechanisms: (i) the mutation could cause a local or non-local structural changes in the free hormone (as considered above); (ii) the mutation could facilitate or hinder a change in the hormone's conformation on receptor binding; or (iii) the mutation could stabilize or destabilize the hormone-receptor interface irrespective of its effects in the free hormone. The first mechanism was considered above. The second mechanism is unlikely as the A1-A8 appears to function as a preformed recognition helix whose segmental folding is not tightly coupled to the remainder of the structure of the insulin monomer (Weiss et al., 2000; Hua et al., 1991). This leaves the third possibility.

We propose that the variable activities of A8 insulin analogues reflect primarily the chemistry of the hormone-receptor interface. In accord with classical models of the hormone's receptor-binding surface (Adams et al., 1969; Blundell et al., 1972; Pullen et al., 1976; De Meyts et al., 1978; Chothia et al., 1983; Baker et al., 1988), we imagine that Thr^{A8} is positioned at the edge of the receptor and that this location has a negative electrostatic potential. Differences in activity between A8 analogues would then arise from electrostatic interactions within the complex and not within the free hormone. This model predicts that substitution of A8 by non-standard amino acids that carry a positive charge but destabilize α-helices, such as diamino-butyric acid, would destabilize insulin but not perturb the hormone-receptor complex. By enlarging the repertoire of structure-activity relationships, such non-standard mutagenesis (Mendel et al., 1995) is a strength of the total chemical synthesis of proteins (Dawson and Kent, 2000). As exemplified by the elegant studies of Professor A. Wollmer and his colleagues (Casaretto et al., 1987; Wollmer, et al., 1994; Kurapkat et al., 1999), non-standard mutagenesis of insulin promises to illuminate the forces and interactions that contribute to the biological activity of this fascinating hormone.

5. REFERENCES

Adams M.J., Blundell T.L., Dodson E.J., Dodson G.G., Vijayan M., Baker E.N., Hardine M.M., HodgkinD.C., Rimer B. and Sheet S. "Structure of rhombohedral 2 zinc insulin crystals." *Nature (London)* 224 (1969): 491-495.

Badger J., Harris M.R., Reynolds C.D., Evans A.C., Dodson E.J., Dodson G.G. and North A.C. "Structure of the pig insulin dimer in the cubic crystal." *Acta Crystallogr. B* 47 (Pt 1) (1991): 127-136.

Bakaysa D.L., Radziuk J., Havel H.A., Brader M.L., Li S., Dodd S.W., Beals J.M., Pekar A.H. and Brems D.N. "Physicochemical basis for the rapid time-action of LysB28ProB29- insulin: dissociation of a protein-ligand complex." *Protein Sci.* 5 (1996): 2521-2531.

Baker E.N., Blundell T.L., Cutfield J.F., Cutfield S.M., Dodson E.J., Dodson G.G., Hodgkin D.M., Hubbard R.E., Isaacs N.W., Reynolds C.D. et al. "The structure of 2Zn pig insulin crystals at 1.5 A resolution." *Philos. Trans. R. Soc. Lond. B. Biol. Sci.* 319 (1988): 369-456.

Bentley G., Dodson G. and Lewitova A. "Rhombohedral insulin crystal transformation." *J. Mol. Biol.* 126 (1978): 871-875.

Bi R.C., Dauter Z., Dodson E.J., Dodson G., Giordano F. and Reynolds C.D. "Insulin structure as a modified and monomeric molecule." *Biopolymers* 23 (1984): 391-395.

Blundell T.L., Dodson G.G., Hodgkin D.C. and Mercola D.A. "Insulin: the structure in the crystal and its reflection in chemistry and biology." *Adv. Protein. Chem.* 26 (1972): 279-402.

Brange J., Ribel U., Hansen J.F., Dodson G., Hansen M.T., Havelund S., Melberg S.G., Norris F., Norris K., Snel L. et al. "Monomeric insulins obtained by protein engineering and their medical implications." *Nature* 333 (1988): 679-682.

Brems D.N., Alter L.A., Beckage M.J., Chance R.E., DiMarchi R.D., Green L.K., Long H.B., Pekar A.H., Shields J.E. and Frank B.H. "Altering the association properties of insulin by amino acid replacement." *Protein Eng.* 5 (1992): 527-533.

Cara J.F., Mirmira R.G., Nakagawa S.H. and Tager H.S. "An insulin-like growth factor I/insulin hybrid exhibiting high potency for interaction with the type I insulin-like growth factor and insulin receptors of placental plasma membranes." *J. Biol. Chem.* 265 (1990): 17820-17825.

Carroll R.J., Hammer R.E., Chan S.J., Swift H.H., Rubenstein A.H. and Steiner D.F. "A mutant human proinsulin is secreted from islets of Langerhans in increased amounts via an unregulated pathway." *Proc. Natl. Acad. Sci. U.S.A.* 85 (1988): 8943-8947.

Casaretto M., Spoden M., Diaconescu C., Gattner H.G., Zahn H., Brandenburg D. and Wollmer A. "Shortened insulin with enhanced in vitro potency." *Biol. Chem. Hoppe Seyler* 368 (1987): 709-716.

Chakrabartty A., Doig A.J. and Baldwin R.L. "Helix capping propensities in peptides parallel those in proteins." *Proc. Natl. Acad. Sci.. U.S.A.* 90 (1993): 11332-11336.

Chan S.J., Seino S., Gruppuso P.A., Schwartz R. and Steiner D.F. "A mutation in the B chain coding region is associated with impaired proinsulin conversion in a family with hyperproinsulinemia." *Proc. Natl. Acad. Sci. U.S.A.* 84 (1987): 2194-2197.

Chothia C., Lesk A.M., Dodson G.G. and Hodgkin D.C. "Transmission of conformational change in insulin." *Nature* 302 (1983): 500-505.

Cosmatos A. and Katsoyannis P.G. "Elongation of the interchain disulfide bridges of insulin. A synthetic analog." *J. Biol. Chem.* 250 (1975): 5315-5321.

Dai J.-B., Lou M.-Z., You J.-M. and Liang D.-C. "Refinement of the structure of despentapeptide (B26-B30) insulin at 1.5 A resolution." *Sci. Sin.* 30 (1987): 55-65.

Danho W.O., Gattner H.G., Nissen D. and Zahn H. "B-chain shortening of matrix-bound insulin with pepsin, II. Preparation and properties of camel des-pentapeptide (B26-30)- and des-PheB1-des-pentapeptide (B26-30)- insulin." *Hoppe Seylers Z. Physiol. Chem.* 356 (1975): 1405-1412.

Dawson P.E. and Kent S.B. "Synthesis of Native Proteins by Chemical Ligation." *Annu. Rev. Biochem.* 69 (2000): 923-960.

De Meyts P., Van Obberghen E. and Roth J. "Mapping of the residues responsible for the negative cooperativity of the receptor-binding region of insulin." *Nature* 273 (1978): 504-509.

Derewenda U., Derewenda Z., Dodson E.J., Dodson G.G., Reynolds C.D., Smith G.D., Sparks C. and Swenson D. "Phenol stabilizes more helix in a new symmetrical zinc insulin hexamer." *Nature* 338 (1989): 594-596.

Derewenda U., Derewenda Z., Dodson E.J., Dodson G.G., Bing X. and Markussen J. "X-ray analysis of the single chain B29-A1 peptide-linked insulin molecule. A completely inactive analogue." *J. Mol. Biol.* 220 (1991): 425-433.

Doig A.J. and Baldwin R.L. "N- and C-capping preferences for all 20 amino acids in alpha-helical peptides." *Protein Sci.* 4 (1995): 1325-1336.

Gruppuso P.A., Gorden P., Kahn C.R., Cornblath M., Zeller W.P. and Schwartz R. "Familial hyperproinsulinemia due to a proposed defect in conversion of proinsulin to insulin." *N. Engl. J. Med.* 311 (1984): 629-634.

Hu S.Q., Burke G.T., Schwartz G.P., Ferderigos N., Ross J.B. and Katsoyannis P.G. "Steric requirements at position B12 for high biological activity in insulin." *Biochemistry* 32 (1993): 2631-2635.

Hua Q.X. and Weiss M.A. "Toward the solution structure of human insulin: sequential 2D ^{1}H NMR assignment of a des-pentapeptide analogue and comparison with crystal structure." *Biochemistry* 29 (1990): 10545-10555.

Hua Q.X., Shoelson S.E., Kochoyan M. and Weiss M.A. "Receptor binding redefined by a structural switch in a mutant human insulin." *Nature* 354 (1991): 238-241.

Hua Q.X., Shoelson S.E. and Weiss M.A. "Nonlocal structural perturbations in a mutant human insulin: sequential resonance assignment and ^{13}C-isotope-aided 2D-NMR studies of [PheB24-- >Gly]insulin with implications for receptor recognition." *Biochemistry* 31 (1992): 11940-11951.

Hua Q.X., Ladbury J.E. and Weiss M.A. "Dynamics of a monomeric insulin analogue: testing the molten-globule hypothesis." *Biochemistry* 32 (1993): 1433-1442.

Hua Q.X., Shoelson S.E., Inouye K. and Weiss M.A. "Paradoxical structure and function in a mutant human insulin associated with diabetes mellitus." *Proc. Natl. Acad. Sci. U.S.A.* 90 (1993): 582-586.

Hua Q.X., Gozani S.N., Chance R.E., Hoffmann J.A., Frank B.H. and Weiss M.A.. "Structure of a protein in a kinetic trap." *Nat. Struct. Biol.* 2 (1995): 129-138.

Hua Q.X., Hu S.Q., Frank B.H., Jia W., Chu Y.C., Wang S.H., Burke G.T., Katsoyannis P.G. and Weiss M.A. "Mapping the functional surface of insulin by design: structure and function of a novel A-chain analogue." *J. Mol. Biol.* 264 (1996): 390-403.

Hua Q.X., Zhao M., Narayana N., Nakagawa S.H., Jia W. and Weiss M.A. "Diabetes-associated mutations in a beta-cell transcription factor destabilize an antiparallel "mini-zipper" in a dimerization interface." *Proc. Natl. Acad. Sci. U.S.A.* 97 (2000): 1999-2004.

Inouye K., Watanabe K., Morihara K., Tochino Y., Kanaya T., Emura J. and Sakakibara S. "Enzyme-assisted semisynthesis of human insulin." *J. Am. Chem. Soc.* 101 (1978): 751-752.

Inouye K., Watanabe K., Tochino Y., Kobayashi M. and Shigeta Y. "Semisynthesis and properties of some insulin analogs." *Biopolymers* 20 (1981): 1845-1858.

Jørgensen A.M., Olsen H.B., Balschmidt P. and Led J.J. "Solution structure of the superactive monomeric des-[Phe(B25)] human insulin mutant: elucidation of the structural basis for the monomerization of des-[Phe(B25)] insulin and the dimerization of native insulin." *J. Mol. Biol.* 257 (1996): 684-699.

Kaarsholm N.C., Norris K., Jorgensen R.J., Mikkelsen J., Ludvigsen S., Olsen O.H., Sorensen A.R. and Havelund S. "Engineering stability of the insulin monomer fold with application to structure-activity relationships." *Biochemistry* 32 (1993): 10773-10778.

Katsoyannis P.G., Okada Y. and Zalut C. "Synthesis of a biologically active insulin analog lacking the intrachain cyclic system." *Biochemistry* 12 (1973): 2516-2525.

Kitagawa K., Ogawa H., Burke G.T., Chanley J.D. and Katsoyannis P.G. "Critical role of the A2 amino acid residue in the biological activity of insulin: [2-glycine-A]- and [2-alanine-A]insulins." *Biochemistry* 23 (1984): 1405-1413.

Knegtel R.M., Boelens R., Ganadu M.L. and Kaptein R. "The solution structure of a monomeric insulin. A two-dimensional ^{1}H-NMR study of des-(B26-B30)-insulin in combination with distance geometry and restrained molecular dynamics." *Eur. J. Biochem.* 202 (1991): 447-458.

Kobayashi M., Ohgaku S., Iwasaki M., Maegawa H., Shigeta Y. and Inouye K. "Supernormal insulin: [D-PheB24]-insulin with increased affinity for insulin receptors." *Biochem. Biophys. Res. Commun.* 107 (1982): 329-336.

Kobayashi M., Takata Y., Ishibashi O., Sasaoka T., Iwasaki T.M., Shigeta Y. and Inouye K. "Receptor binding and negative cooperativity of a mutant insulin, [Leu^{A3}]-insulin." *Biochem. Biophys. Res. Commun.* 137 (1986): 250-257.

Kurapkat G., Siedentop M., Gattner H.G., Hagelstein M., Brandenburg D., Grotzinger J. and Wollmer A. "The solution structure of a superpotent B-chain-shortened single- replacement insulin analogue." *Protein Sci.* 8 (1999): 499-508.

Liang D.-C., Stuart D., Dai J.-B., Todd R., You J.-M. and Luo M.-Z. "X-ray studies of des-pentapeptide (B26-B30) insulin at 2.4 A resolution: The DPI molecule and its structural relationship with insulin." *Sci. Sin.* 28 (1985): 472-484.

Ludvigsen S., Olsen H.B. and Kaarsholm N.C. "A structural switch in a mutant insulin exposes key residues for receptor binding." *J. Mol. Biol.* 279 (1998): 1-7.

Markussen J., Jorgensen K.H., Sorensen A.R. and Thim L. "Single chain des-(B30) insulin. Intramolecular crosslinking of insulin by trypsin catalyzed transpeptidation." *Int. J. Pept. Protein Res.* 26 (1985): 70-77.

Marshall R.N., Underwood L.E., Voina S.J., Foushee D.B. and Van Wyk J.J. "Characterization of the insulin and somatomedin-C receptors in human placental cell membranes." *J. Clin. Endocrinol. Metab.* 39 (1974): 283-292.

Mendel D., Cornish V.W. and Schultz P.G. "Site-directed mutagenesis with an expanded genetic code." *Annu. Rev. Biophys. Biomol. Struct.* 24 (1995): 435-462.

Mirmira R.G. and Tager H.S. "Role of the phenylalanine B24 side chain in directing insulin interaction with its receptor. Importance of main chain conformation." J. Biol. Chem. 1989 Oct 15;29: 17613. *J. Biol. Chem.* 264 (1989): 6349-6354.

Mirmira R.G. and Tager H.S. "Disposition of the phenylalanine B25 side chain during insulin-receptor and insulin-insulin interactions." *Biochemistry* 30 (1991): 8222-8229.

Motoshima H., Mine S., Masumoto K., Abe Y., Iwashita H., Hashimoto Y., Chijiiwa Y., Ueda T. and Imoto T. "Analysis of the stabilization of hen lysozyme by helix macrodipole and charged side chain interaction." *J. Biochem. (Tokyo)* 121 (1997): 1076-1081.

Nakagawa S.H. and Tager H.S. "Role of the phenylalanine B25 side chain in directing insulin interaction with its receptor. Steric and conformational effects." *J. Biol. Chem.* 261 (1986): 7332-7341.

Nakagawa S.H. and Tager H.S. "Role of the COOH-terminal B-chain domain in insulin-receptor interactions. Identification of perturbations involving the insulin mainchain." *J. Biol. Chem.* 262 (1987): 12054-12058.

Nakagawa S.H. and Tager H.S. "Importance of aliphatic side-chain structure at positions 2 and 3 of the insulin A chain in insulin-receptor interactions." *Biochemistry* 31 (1992): 3204-3214.

Nanjo K., Sanke T., Miyano M., Okai K., Sowa R., Kondo M., Nishimura S., Iwo K., Miyamura K., Given B.D. et al. "Diabetes due to secretion of a structurally abnormal insulin (insulin Wakayama). Clinical and functional characteristics of [Leu^{A3}] insulin." *J. Clin. Invest.* 77 (1986): 514-519.

Okada Y., Chanley J.D., Burke G.T. and Katsoyannis P.G. "[A2-Norleucine] Insulin. An analogue with unanticipated biological properties." *Hoppe Seylers Z.. Physiol.. Chem.* 362 (1981): 629-638.

Olsen H.B., Ludvigsen S. and Kaarsholm N.C. "Solution structure of an engineered insulin monomer at neutral pH." *Biochemistry* 35 (1996): 8836-8845.

Olsen H.B., Ludvigsen S. and Kaarsholm N.C. "The relationship between insulin bioactivity and structure in the NH_2- terminal A-chain helix." *J. Mol. Biol.* 284 (1998): 477-488.

Peking Insulin Structure Group. "Insulin's crystal structure at 2.5 A resolution." *Peking Rev.* 40 (1971):11-16.

Pullen R.A., Lindsay D.G., Wood S.P., Tickle I.J., Blundell T.L., Wollmer A., Krail G., Brandenburg D., Zahn H., Gliemann J. and Gammeltoft S. "Receptor-binding region of insulin." *Nature* 259 (1976): 369-373.

Schwartz G.P., Burke G.T. and Katsoyannis P.G. "A highly potent insulin: des-(B26-B30)-[Asp^{B10},Tyr^{B25}-NH_2]insulin(human)." *Proc. Natl. Acad. Sci. U.S.A.* 86 (1989): 458-461.

Schwartz G.P., Burke G.T. and Katsoyannis P.G. "A superactive insulin: [B10-aspartic acid]=insulin (human)." *Proc. Natl. Acad. Sci. U.S.A.* 84 (1989): 6408-6411.

Shoelson S., Haneda M., Blix P., Nanjo A., Sanke T., Inouye K., Steiner D., Rubenstein A. and Tager H. "Three mutant insulins in man." *Nature* 302 (1983): 540-543.

Sørensen M.D. and Led J.J. "Structural details of Asp(B9) human insulin at low pH from two-dimensional NMR titration studies." *Biochemistry* 33 (1994): 13727-13733.

Stewart J.M. and Young J.D. "Solid Phase Peptide Synthesis." *Raven Press: New York* (1984).

Tager H., Given B., Baldwin D., Mako M., Markese J., Rubenstein A., Olefsky J., Kobayashi M., Kolterman O. and Poucher R. "A structurally abnormal insulin causing human diabetes." *Nature* 281 (1979): 122-125.

Wang S.H., Hu S.Q., Burke G.T. and Katsoyannis P.G. "Insulin analogues with modifications in the beta-turn of the B-chain." *J. Protein. Chem.* 10 (1991): 313-324.

Weiss M.A., Hua Q.X., Lynch C.S., Frank B.H. and Shoelson S.E. "Heteronuclear 2D NMR studies of an engineered insulin monomer: assignment and characterization of the receptor-binding surface by selective ^{2}H and ^{13}C labeling with application to protein design." *Biochemistry* 30 (1991): 7373-7389.

Weiss M.A., Hua Q.X., Jia W., Chu Y.C., Wang R.Y. and Katsoyannis P.G. "Insulin's intrachain disulfide bridge tethers a recognition alpha-helix." *Biochemistry* (2000, submitted).

Wollmer A., Gilge G., Brandenburg D. and Gattner H.G. "An insulin with the native sequence but virtually no activity." *Biol. Chem. Hoppe Seyler* 375 (1994): 219-222.

6. ACKNOWLEDGEMENTS

We thank B. Frank and R. Chance for biosynthetic human insulin and advice; G.G. Dodson, D.F. Steiner and the late H.S. Tager for helpful discussion and communication of results prior to publication. This work was supported in part by grants from the National Institutes of Health to MAW and PGK and by the Diabetes Research & Training Center at the University of Chicago.

7. ABBREVIATIONS

CD, circular dichroism; DG, distance geometry; IGF-I, insulin-like growth factor I; NMR, nuclear magnetic resonance; rp-HPLC, reverse-phase high-performance liquid chromatography; SA, simulated annealing; TOCSY, total correlated spectroscopy; 1D or 2D-NMR, one- or two-dimensional NMR. Amino acids are designated by standard three-letter code.

R.H. JONES AND F. SHOJAEE-MORADIE

RELATIONSHIPS BETWEEN THE STRUCTURE OF INSULIN AND ITS PHYSIOLOGICAL EFFECTS

Thyronine Insulin Analogues

Abstract. Elucidation of the structure of insulin has provided opportùnities to explain its physiological properties. Following secretion directly into the hepatic portal vein, which flows directly to the liver, it acts initially to modulate hepatic glucose output, an effect primarily responsible for glucose homeostasis. Only 50% of secreted insulin passes from the liver to the other tissues where it has a role in controlling lipolysis and glucose uptake particularly after meals. In evolutionary terms selection pressure may have acted to optimize the affinity of the insulin to insulin receptor interaction in order to define the most appropriate relative hepatic to peripheral ratio of insulin action. Therapeutically insulin is given subcutaneously. This unphysiological route results in relative under-exposure of the liver to insulin with peripheral hyperinsulinaemia. By exploiting the peripheral capillary endothelium as a molecular sieve it is proving possible to design insulin analogues which compensate for this imbalance.

1. INTRODUCTION

When Banting and Best first isolated insulin in 1921 its chemical composition was unknown. It was to remain in that sense an enigmatic albeit 'unspeakably wonderful' treatment for diabetes for 34 years until Sanger published the primary sequence in 1955 (Ryle et al., 1955). With the solution by Dorothy Hodgkin (Hodgkin, 1972) of the crystal structure in 1969 it became possible to address questions that relate the molecular properties of insulin to its physiological effects.

2. INSULIN METABOLISM

Insulin is the major regulator of glucose homeostasis with important effects on both glucose production or 'Rate of appearance' (Ra) and glucose utilization or 'disposal' (Rd). In mammals including man insulin secreted from the pancreas passes directly and in its entirety to the liver in a portal vein. There it influences a number of biochemical pathways via initial binding to a specific receptor on the surface of hepatocytes. With respect to glucose metabolism it reduces glucose release from glycogen and inhibits synthesis from gluconeogenic precursors. Thus a fall in plasma glucose for example during exercise results in a subtle and precise reduction in portal insulin levels to allow release of sufficient quantities of glucose to maintain concentrations within a narrow and safe range. Approximately 50% of the insulin reaching the liver is degraded without passing on to the general, systemic circulation to the other tissues of the body (Chap et al., 1987). In normal subjects insulin is secreted briskly in response to a meal. This serves rapidly to reduce endogenous

M. Federwisch, M. Leyck Dieken and P. De Meyts (eds.), Insulin & Related Proteins, 121–130.

glucose Ra and at the concentrations reached sufficient insulin passes through the liver to facilitate glucose uptake and storage in muscle and fat. The result of this homeostatic response is that plasma concentration of glucose remains between 3.5-7 mmol/l through periods of fasting and feasting (Bolli et al., 1999) (Figure 1).

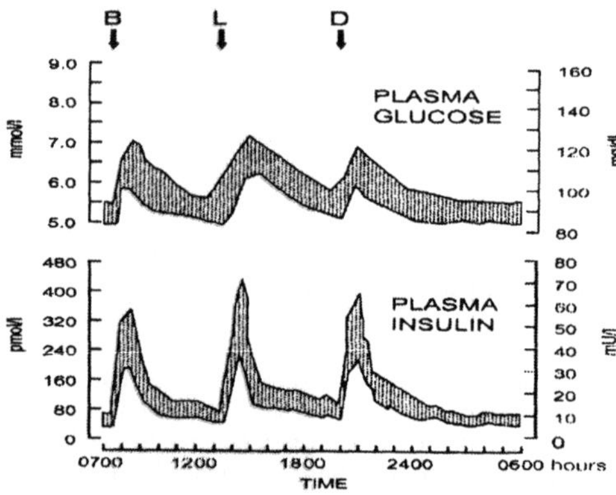

Figure 1. Plasma insulin and glucose profiles after consumption of meals in 8 normal, non-diabetic people. Means ± 2SD. B=Breakfast, L=Lunch and D= Dinner. Original data were adapted from Ciofetta et al., Diabetes Care 22:795-800, and reprinted from Diabetologia, Insulin analogues and their potential in the management of diabetes mellitus, Bolli et al., 42: 1151-1167, figure 1, 1999., © Springer-Verlag.

When glucose concentrations begin to fall it is essential that the insulin signal can be reduced rapidly to avoid dangerous levels of hypoglycaemia. Insulin metabolism is therefore a critically important process. Intravenous infusion of insulin with frequent assay of resultant plasma levels allows quantitative description of insulin metabolism in-vivo (Sönksen et al., 1973). In studies in which primed continuous infusions of insulin are applied in a stepwise manner, with a doubling of the infusion rate every 30 min the rate of metabolism (metabolic clearance rate or MCR) can be calculated over a range of plasma concentrations. The results show that as the concentration of insulin increases, even within the physiological range, the MCR shows a tendency to fall, i.e. the process shows saturation kinetics (Jones, 1984).

In a further study applying a similar infusion protocol, equimolar infusion rates of native insulin and insulin analogues (proinsulin and insulin variously modified at A1-glycine and B29-lysine) were used to investigate the relationships between MCR and effects on glucose metabolism. Previous studies with these analogues had suggested discrepancies between results obtained from in-vivo and in-vitro experiments. The biological activities in-vivo were unexpectedly greater than the biological responses and receptor binding affinities derived from in-vitro experiments (Jones et al., 1976). During the constant infusion studies the concentrations of the low affinity analogues achieved in the circulation were much higher than those observed with insulin itself. The converse is also true in that Turkey insulin, which exhibits an in-vitro potency of approximately 200%, has a higher MCR in dogs than dog insulin itself.

For all these analogues the MCR value was directly related to biological potency and receptor binding affinity in in-vitro systems (Jones et al., 1976). We concluded that in-vivo both the initiation of metabolic responses and access to pathways of insulin degradation occurred via the same receptor, a conclusion later supported by the discovery of receptor mediated endocytosis (Carpentier et al., 1979). One outcome of this phenomenon is that the biological effectiveness of an insulin analogue is largely independent of its receptor affinity over a wide range of values. It is of interest to consider therefore what evolutionary forces could apply to optimize this binding affinity.

3. INSULIN DISTRIBUTION

Pancreatic insulin is released into the hepatic portal circulation and delivered directly to the liver. Thus freshly secreted insulin both exerts direct hormonal effects on the liver and is metabolized by the hepatocytes. In this 'first-pass' the liver extracts up to 60% of the insulin delivered to it (Field, 1973) and the remainder enters the larger systemic plasma volume before redistribution through the arterial tree. In consequence after physiological insulin release hepatocytes are exposed to insulin concentrations approximately three times higher than exist in the capillaries of the other major target organs for insulin, muscle and adipose tissue. This can be clearly demonstrated if simultaneous samples are taken from hepatic portal and peripheral vessels before and after an oral glucose load (Chap et al., 1986; Jones, 1988) (Figure 2).

When insulin analogues are infused into dogs with appropriately placed catheters the sites of metabolism can be defined. Low affinity insulin analogues exhibit markedly reduced first pass hepatic extraction as expected from the fact that the liver is a major site of insulin clearance and such analogues exhibit reduced values for whole body MCR (Chap et al., 1987). Here then may be a clue to the 'ideal' value for insulin/insulin receptor binding affinity in evolutionary terms. Selection pressure could act to define the most appropriate distribution of insulin action between hepatic and peripheral targets. It may be relevant to this discussion that Turkey and other bird insulins which retain a histidine residue at A8 have very high receptor affinities. With mammalian circulatory anatomy such insulins would exhibit very high first-pass hepatic extraction such that a markedly reduced quantity could reach the peripheral tissues.

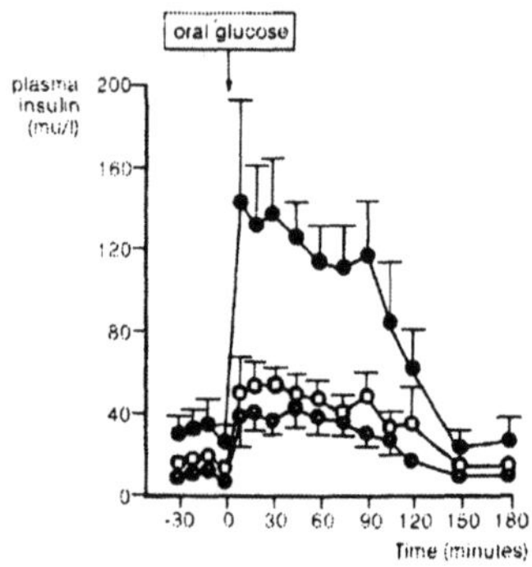

Figure 2. Plasma insulin concentrations before and after oral glucose (1g/kg body weight) in normal dogs. Portal vein (•), hepatic vein (o) and hepatic artery (◐). Original data were adapted from Chap et al., J. Clin. Invest. 78 (1986): 1355-1361, reprinted from Clinical Diabetes: An Illustrated Text, Jones RH, Insulin action and Metabolism, 5.1-5.16, 1988, © Publisher Mosby.

In this context it is of potential relevance that in birds there exists an anatomical shunt between the hepatic portal and systemic circulation which allows the liver to be bypassed. We suggest that when this shunt was lost in evolution of the mammalian line insulin/insulin receptor affinity was reduced by natural selection to allow continued adequate insulinization of muscle and adipose tissue.

4. INSULIN THERAPY

In the treatment of diabetes, insulin is injected into a subcutaneous depot from which it is absorbed into the peripheral circulation via which it is distributed at approximately equal concentrations throughout the body. As a result the normal hepatic/peripheral insulin concentration gradient is lost such that in insulin treated diabetes there is inevitable under-insulinisation of the liver and relative peripheral hyperinsulinaemia (Horwitz et al., 1975). The implications of this profound physiological imbalance are not yet clear. However patients on regimes of subcutaneous insulin therapy continue to exhibit a variety of metabolic abnormalities including excessive glycaemic fluctuations, dyslipidaemias and reduced plasma insulin-like growth factor I (IGF-I) concentrations with elevated plasma levels of growth hormone (GH) (The DCCT Group, 1993; Sönksen et al., 1993; Amiel et al., 1984).

Insulin present in plasma has to reach its receptors on the outer surface of target cells to initiate its effects on metabolism and the translocation of the complex to the inside of the cell where insulin is metabolised. In the liver there are no barriers between the vascular space and the surface of hepatocytes, which are fully exposed to all plasma constituents. In other tissues, notably the primary peripheral targets for insulin action, muscle and adipose tissue the endothelial lining of the capillaries acts as a molecular sieve limiting access to the extracellular space (Figure 3). Small molecules diffuse freely but large proteins remain in the plasma. In the absence of this semi-permeable membrane the tissues would become waterlogged under the influence of intra-vascular pressures generated by gravity and the beating of the heart. The osmotic gradient maintained by the endothelium between intra-vascular and extra-vascular spaces prevents this. It can be shown experimentally that at MW 6000 Da monomeric insulin concentration in lymph (extra-vascular fluid) is lower than the simultaneous values in plasma indicating that there is at least some limitation of diffusion through capillary membranes even for this relatively small protein. It follows that in addition to site of administration (portal versus peripheral) the influence of the peripheral capillary barrier could influence the balance of insulin action between the liver on the one hand and muscle and fat on the other. It turns out that this is indeed the case. Proinsulin has a MW of 9000 Da approx. Infusion studies with proinsulin in normal subjects (Glaubler et al., 1986) have shown that suppression of glucose Ra (liver effect) achieved in response to proinsulin (M_r 9000 Da) was similar to that of equimolar infusion of insulin (M_r 5807 Da). The effect on the stimulation of glucose Rd (effect in the periphery) was lower with proinsulin than with insulin.

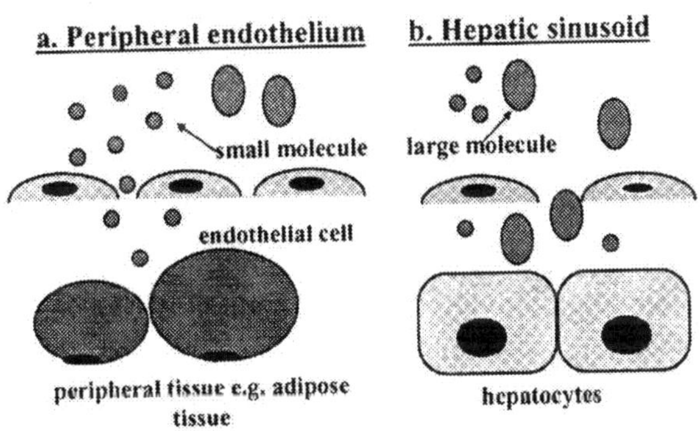

Figure 3. Endothelial sieve.

Infusion studies with insulin and covalently linked insulin dimers showed a similar result (Figure 4) (Shojaee-Moradie et al., 1995). Thus it is possible to redistribute insulin action within an intact mammal by manipulation of the molecular dimensions of the active substance. We have therefore explored whether this capacity to convey relative hepatoselectivity to insulin can be exploited to restore a more normal balance of action to insulin therapy without direct access to the hepatic portal vein.

Our approach has been to design and synthesize active analogues of insulin which include an additional moiety capable of binding to normal large molecular weight protein constituents of plasma.

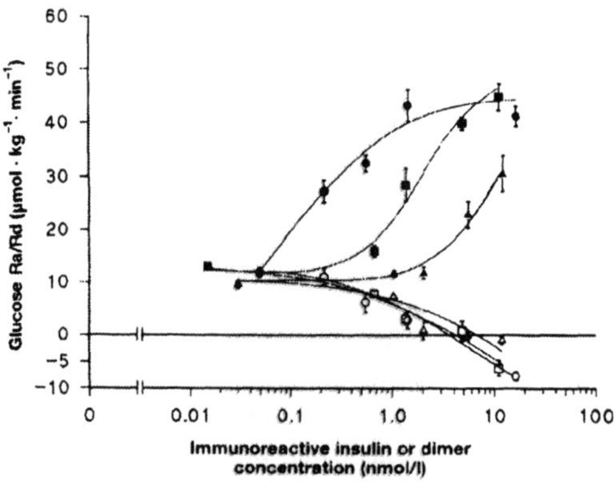

Figure 4. Dose response relationships of glucose Rd (closed symbols) and glucose Ra (open symbols) to serum concentrations of insulin (• o), B1-B'1 D (■ □) and B29-B'29 D (▲△). Reprinted from Diabetologia, Demonstration of a relatively hepatoselective effect of covalent insulin dimers on glucose metabolism in dogs. Shojaee-Moradie et al 38:1007-1013, figure3, 1995, © Springer-Verlag.

Several such analogues have now been prepared and studied.

5. THYRONINE INSULIN ANALOGUES

Here we report the in vitro biological activity of insulin covalently linked to thyroxine: $N^{\alpha B1}$ L-thyroxyl-insulin (T4-Ins) and reverse 3', 5', 5 triiodothyronine: $N^{\alpha B1}$ L-rT3-insulin (rT3-Ins) (Figure 5) and in vivo biological activity of B1-T4-Ins. These analogues have the capacity to bind both to the insulin receptor via the insulin moiety and via the thyroxyl moiety, to thyroid hormone binding proteins (THBPs) i.e. thyroxine binding protein (TBG), transthyretin and human serum albumin (HSA). The size of the bound complex (i.e. THBP bound B1-T4-Ins) limits access of the analogue to peripheral receptors but not to the more exposed hepatocytes.

Using a calibrated fast performance liquid chromatography (FPLC) system, we have shown that these analogues bind to the binding proteins (Figure 6).

In vitro binding experiments using rat liver plasma membrane (LPM) as a source of insulin receptor, $[^{125}I\text{-}Tyr^{A14}]$insulin as the tracer and in the absence of binding THBPs demonstrated that there were no significant differences between insulin and either of the analogues indicating that the addition of the thyroid moiety to the B1 position had not reduced the capacity for normal association with the insulin receptor (at least at equilibrium) (Telfer et al., 1998). Dynamic experiments will be required to confirm normal on and off rates and to explore possible effects of the substitution on negative co-operativity (De Meyts, 1994).

Further experiments were designed to test whether thyroxyl insulin (T4-Ins) with restricted access to receptor sites in peripheral tissues could display relative hepatoselectivity.

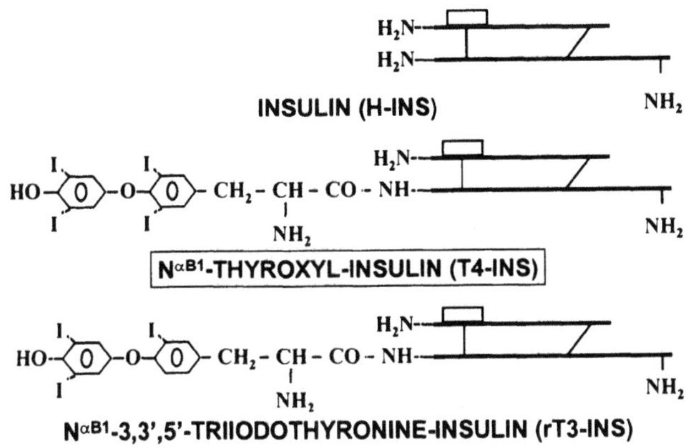

Figure 5. Schematic presentation of thyronine insulin analogues.

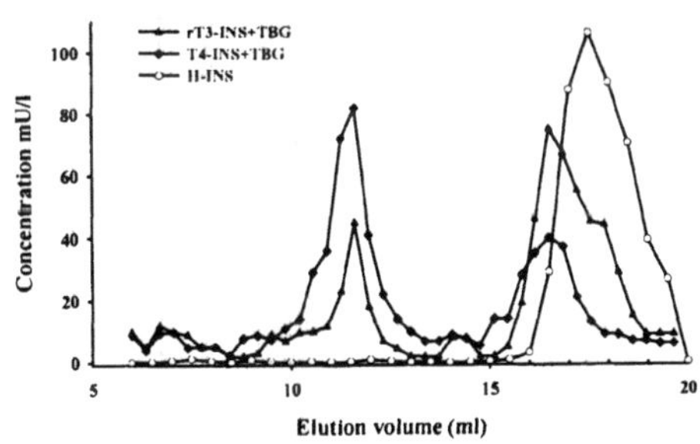

Figure 6. Elution profile of rT3-Ins, T4-Ins and human insulin after incubation with THBPs using a calibrated FPLC system. On the basis of their molecular weight the free and bound forms of these materials were separated on a Suprose 12 HR 10/30 column equilibrated with 0.1 mol/l phosphate buffered saline pH7.2. The eluate was collected and the concentration of the analogue in the fractions was measured by radioimmunoassay. Protein bound insulin analogue was eluted between 5-15 ml and free, unbound analogue or insulin was eluted between 15-20 ml of elution volume.

Normal human subjects received a subcutaneous bolus injection of either $N^{\alpha B1}$ thyroxyl-insulin (T4-Ins) (3.44 nmol/kg) or NPH insulin (2.06 nmol/kg; 0.3 U/kg) in random order (Shojaee-Moradie et al., 2000). Insulin kinetics, relative effects on hepatic glucose production and peripheral glucose uptake were studied using euglycaemic clamp and stable isotope (D-6, 6-$^2H_2$ glucose) dilution techniques.

T4-Ins was well tolerated, rapidly absorbed, had a long serum half-life and was highly protein bound (~86%).

Comparison of the effects of the test materials on glucose Ra and Rd revealed that T-4Ins displayed significant relative hepatoselectivity (p=0.025), with no effect on metabolic clearance rate of glucose. Further experiments with a lower dose of

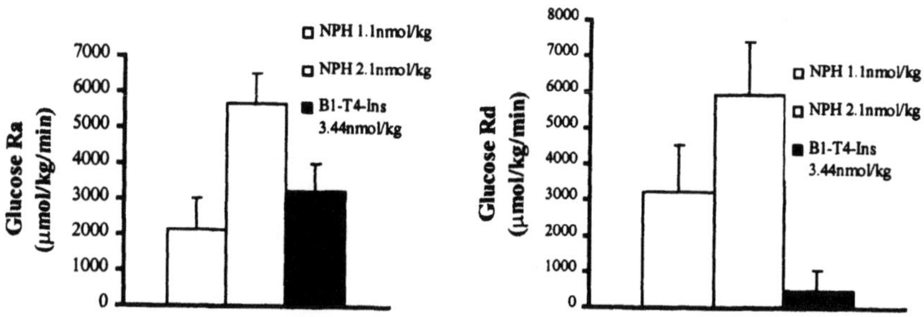

Figure 7. Area under the curve calculated for glucose Ra and Rd during euglycaemic clamp studies after a subcutaneous bolus injection of NPH insulin (1.1 or 2.1 nmol/kg) and B1-T4-Ins (3.44 nmol/kg) in normal subjects (n=5). Values are mean ±SEM.

NPH insulin (1.1 nmol/kg; 0.15 U/kg) in normal subjects confirmed that the analogue had less effect on glucose Rd than a dose of insulin with a similar effect on the liver (Figure 7).

Similarly comparison of the fall in concentration profiles of non-esterified free fatty acids confirms a reduced effect of T4-Ins on adipose tissue (Figure 8).

Hepatoselective insulin analogues may offer a more physiological balance of insulin action than currently available insulin preparations. Potential long-term benefits will be evaluated in clinical studies.

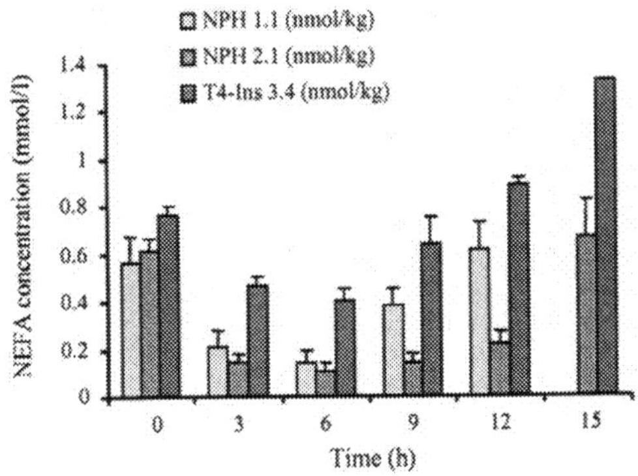

Figure 8. Non-esterified fatty acid concentrations (NEFA) after a subcutanenous injection of NPH insulin and B1-T4-Ins during clamp studies, n=5. Values are mean ±SEM.

6. REFERENCES

Amiel S.A., Sherwin R.S., Hintz R.L., Gertner J.M., Press C.M. and Tamborlane W.V. "Effects of diabetes and its control on insulin-like growth factors in young subjects with type I diabetes." *Diabetes* 33(1984):1175-1179.

Bolli G.B., Marchi R.D., Park G.D., Pramming S. and Koivisto V.A. "Insulin analogues and their potential in the management of diabetes mellitus." *Diabetologia* 42 (1999): 1151-1167.

Carpentier J.L., Gorden P., Freychet P., LeCam A. and Orci L. "Intracellular localisation of ^{125}I-labelled insulin in hepatocytes from intact rat liver." *Proc. Natl. Acad. Sci. USA* 76 (1979): 2803-2807.

Chap Z., Ishida T., Chou J., Hartley C.J., Entman M.L., Brandenburg D., Jones R.H. and Field J.B. "First pass hepatic extraction and metabolic effects of insulin and insulin analogues." *AM. J. Physiol.* 252 (1987): E209-E217.

Chap Z., Jones R.H., Chou J., Hartley C.J., Entman M.L. and Field J.B. "Effect of dexamethasone on hepatic glucose and insulin metabolism after oral glucose on conscious dogs." *J. Clin. Invest.* 78 (1986): 1355-1361.

Ciofetta M., Lalli C., Del Sindaco P., Torlone E., Pampanelli S., Mauro L., Chiara D.L., Brunetti P., Bolli G.B. "Contribution of postprandial versus interprandial blood glucose to HbA_{1C} in Type I diabetes on physiologic intensive therapy with lispro insulin at mealtime." *Diabetes Care* 22 (1999): 795-800.

De Meyts P. "The structural basis of insulin and insulin-like growth factor-I receptor binding and negative co-operativity, and its relevance to mitogenic versus metabolic signalling." *Diabetologia* 37 Suppl. 2 (1994): S135-S148.

Field J.B. "Extraction of insulin by liver." *Annu. Rev. Med.* 24 (1973):309-314.

Glauber H.S., Revers R.R., Henry R., Schmeiser L., Wallace P., Kolterman O.G., Cohen R.M., Rubenstein A.H., Galloway J.A., Frank B.H. and Olefsky J.M. *Diabetes* 35 (1986):311-317.

Hodgkin D.C. "The Banting Memorial Lecture: The structure of insulin." *Diabetes* 21 (1972): 1131-1150.

Horwitz D.L., Starr J.I., Mako M.E., Blackard W.G. and Rubenstein A.H. "Proinsulin, insulin and C-peptide concentrations in human portal and peripheral blood." *J. Clin. Invest.* 55 (1975): 1278-1283.

Jones, R.H. "Insulin action and metabolism." In: *Clinical Diabetes*: An Illustrated Text. (Besser G.M., Bodansky H.J., Cudworth A.G., eds.)Publisher: Mosby, pp. 5.1-5.16 (1988).

Jones R.H., Dron D.I., Ellis M.J., Sönksen P.H. and Brandenburg D. "Biological properties of chemically modified insulins. I Biological activity of proinsulin and insulin modified at A1 Glycine and B29-Lysine." *Diabetologia* 12 (1976): 601-608.

Jones R.H., Sönksen P.H., Boroujerdi M.A. and Carson E.R. "Number and affinity of insulin receptors in intact human subjects." *Diabetologia* 27 (1984): 207-211.

Ryle A.P., Sanger F., Smith L.F. and Kitai R. "The disulphide bonds of insulin." *Biochem. J.* 60 (1955): 541-556.

Shojaee-Moradie F., Jackson N.C., Brandenburg D., Sönksen P.H. and Jones R.H. "Demonstration of a relatively hepatoselective effect of covalent insulin dimers on glucose metabolism in dogs." *Diabetologia* 38 (1995): 1007-1013.

Shojaee-Moradie F., Powrie J.K., Sundermann E., Spring M.W., Schüttler A., Sönksen P.H., Brandenburg D. and Jones R.H. "Novel hepatoselective insulin analog: studies with a covalently linked thyroxyl-insulin complex in humans." *Diabetes Care* Aug; 23 (2000): 1124-1129.

Sönksen P.H., Russell-Jones D. and Jones R.H. "Growth hormone and diabetes mellitus." *Horm. Res.* 40 (1993): 68-79.

Sönksen P.H., Tompkins C.V., Srivastava M.C. and Nabarro J.D.N. "A comparison study on the metabolism of human insulin and porcine proinsulin in man." *Clin. Sci. Mol. Med.* 45 (1973): 633-654.

Telfer M.J., Shojaee-Moradie F., Sundermann E., Schüttler A., Brandenburg D. and Jones R.H. "The effects of thyroid hormone binding proteins on insulin receptor binding of thyroxyl-insulin analogues in vitro." *Diab. Med.* 15 (1998):Suppl. 2; 55.

The Diabetes Control and Complications Trial Research Group. "The effect of intensive treatment of diabetes on the development and progression of long-term complications in insulin-dependent diabetes mellitus." *N. Eng. J. Med.* 329 (1993): 977-986.

P. DE MEYTS AND J. WHITTAKER

STRUCTURE-FUNCTION RELATIONSHIPS OF INSULIN AND INSULIN-LIKE GROWTH FACTOR-I RECEPTOR BINDING

Abstract. In this chapter, we review the current knowledge of the structure of the ligsands and receptors of the insulin/insulin-like growth factors (IGFs) peptide family, their structure-function relationships and mapping of the binding sites on the ligands and receptors by site-directed mutagenesis. We discuss a plausible bivalent crosslinking binding model that explains the complex binding properties of insulin and the IGFs.

1. THE INSULIN PEPTIDE FAMILY

Insulin and the insulin-like growth factors IGF-I and II belong to a superfamily of structurally related peptide hormones essential for growth, development, metabolism and survival (Chan, 2000). In vertebrates, members of this family include insulin, the insulin-like growth factors IGF-I and II, relaxin, a relaxin-like factor (RLF), also called Leydig insulin-like peptide (Ley I-L), and placentin. Several family members have been identified in invertebrates as well, including bombyxin from the silkworm, *Bombyx Mori*, a locust insulin-related peptide (LIRP) from *Locusta migratoria* and seven different molluscan insulin-like peptides (MIP) from the pond snail, *Lymnea Stagnalis*. In addition, three new gene families of insulin-like peptides with atypical disulphide bond pattern (37 peptides in all) have been recently identified in the nematode *C. Elegans* that may be ligands for the *daf-2* insulin-like receptor. The amino acid sequences of insulins from over 50 species have been determined to date, and reveal a substantial degree of conservation. The three-dimensional structure of pig insulin has been determined by X-ray crystallography at a resolution of 1.5 Å (Baker et al., 1988) and structures of various insulins and analogues established by crystallographic or NMR methods are available. IGF structures have also been determined by NMR (Van den Brande, 1992) and X-ray crystallography (Vajdos et al., 2001). The structural aspects of the insulin molecule are reviewed in detail elsewhere in this volume, as well as in previous exhaustive reviews (Baker et al., 1988; Blundell et al., 1972). The evolutionary relationship of the members of this peptide family has also been recently reviewed (Chan, 2000).

Human insulin consists of 51 amino acids arranged in two polypeptide chains (A and B) linked by two disulphide bonds between A7 and B7, and A20 and B19. In addition, an intra-chain disulphide bond links A6 to A11. The hormone is synthesized as a large precursor, preproinsulin, which is processed to the mature insulin molecule by cleavage of a 24 residue hydrophobic N-terminal signal peptide

M. Federwisch, M. Leyck Dieken and P. De Meyts (eds.), Insulin & Related Proteins, 131–149.

and of a 30 amino acid connecting peptide that links A1 to B30. The IGFs I and II are similarly structured, except for a permanent C-region of respectively 10 and 8 amino acids linking the A and B chains that is not removed during biosynthesis, and the presence of an extension (D-region) of 8 (IGF-I) and 6 (IGF-II) amino acids at the C-terminus of the A chain.

This brief review will focus on the current status of understanding the molecular basis of insulin and IGF-I binding to their respective receptors, and is appropriately dedicated to Prof. Axel Wollmer on the occasion of his retirement. Indeed, much of the initial progress in the understanding of insulin structure-function relationships, and its convergence with the emerging knowledge of the receptor structure and properties, relied upon the fruitful collaboration between most groups involved in this research and the Aachen group, as exemplified by the various contributions to the Alcuin Symposium excerpted in this volume.

2. STRUCTURE OF THE INSULIN RECEPTOR FAMILY

2.1. Introduction

The primary event in the initiation of signalling by insulin and the insulin-like family of hormones and growth factors is the interaction of the peptide with a specific cell surface glycoprotein receptor. While the function of this class of receptors has been characterized in detail (Ullrich and Schlessinger, 1990), the elucidation of the molecular basis of the interaction of ligands with the receptors and the subsequent initiation of signalling has been slow, largely because of the low abundance and biochemical complexity of the protein. However the application of recombinant technology and molecular biological and biophysical techniques, now provides the beginning of an understanding of the molecular details of these events.

The existence of a specific insulin receptor was first identified in the early 1970s and over the next decade rapid progress was made in its functional characterization using conventional biochemical and cell biological techniques (Siddle, 1992; Lee and Pilch, 1994). These studies demonstrated that the insulin receptor and the related IGF-I receptor (Siddle, 1992; Adams et al., 2000) were heterotetramers composed of disulphide-linked heterodimers. Each heterodimer is in turn composed of an extracellular α subunit linked by disulphide bonds to a β subunit which is partly extracellular, traverses the plasma membrane and contains a large intracellular domain. Affinity labelling studies demonstrated that the α subunit contained the ligand binding site. Further it was shown that these receptors exhibited ligand regulated tyrosine kinase activity and thus belong to the superfamily of receptor tyrosine kinases (Ullrich and Schlessinger, 1990).

The cDNA for the human insulin receptor was cloned in 1985 (Ullrich et al., 1985; Ebina et al., 1985). It encodes a 1,370 amino acid single chain precursor protein. The α and β chains are generated by the cleavage of this precursor at a prototypical furin cleavage site. The α subunit contains a cysteine rich domain which is homologous to that in the extracellular domain of the EGF receptor family of proteins. The α chain and the N-terminal part of the β chain form the extracellular part of the protein which is connected by a single transmembrane domain to the cytoplasmic portion of the protein which has prototypical features of a protein tyrosine kinase. The IGF-I receptor (Ullrich et al., 1986) and a third member of IR family, the insulin receptor related receptor (Shier and Watt, 1989) whose ligand is unknown, have been cloned and their deduced amino acid sequences indicate that they exhibit the same structural organization as the insulin receptor. More recently insulin receptor family homologues have been identified in protochordates, insects, gastropods, nematodes and cnidarians and their amino acid sequences indicate that they also exhibit the same overall structural organization as their mammalian counterparts, although some of them have additional domains (Chan, 2000; Adams et al., 2000).

The descriptions of the structural organization of the receptors and of structural details of this family of receptors that follow, will be confined to the extracellular domains. Discussion of the structure and function of the cytoplasmic domains is beyond the scope of this review.

2.2. *Structural Organization*

Comparative sequence analysis of the insulin receptor and the related EGF receptor families of proteins has facilitated the formulation of a hypothetical tertiary structure of the extracellular part of the receptor heterodimer (Bajaj et al., 1987). The amino terminal half consists of two homologous globular domains (L1, amino acids 1-120 and L2, amino acids 311-428) separated by a cysteine rich region containing 24-26 cysteines. The crystal structure of this region of the IGF-I receptor has recently been determined and will be discussed in detail below. The carboxy terminal portion, amino acids 471-929, consists of 3 Fibronectin Type III (FnIII) repeats (Adams et al., 2000). FnIII repeats are common components of many proteins, particularly extracellular matrix protcins, ccll surface adhesion proteins and cell surface receptors. In the human insulin receptor, the first FnIII domain is composed of residues 471-593, the second extends from amino acid 594-808 and contains a 104-115 amino acid insertion, and the third is formed from amino acids 809-912. The folds of FnIII domains are similar in conformation to those of immunoglobulins but with a distinctive sequence motif. It consists of a 7 stranded β sandwich in a three on four topology (EBA:GFCC). The cysteine residues (C524) involved in one of the α–α disulphide bonds and the cysteines forming the α–β disulphide are located in equivalent positions in the domains. The first FnIII domain also contains the epitope to which many insulin receptor antibodies, both monoclonal and polyclonal, are directed.

The extracellular domain of the IGF-I receptor is similarly organized, see Adams et al. (2000) for a detailed description.

2.3. *Crystal Structure of the L1/Cys-rich/L2 Domains (L_1CL_2) of the IGF-I Receptor*

Recently a high resolution crystal structure of an amino-terminal fragment of the insulin-like growth factor I receptor (amino acids 1-460) has been reported (Garrett et al., 1998). The molecule is composed of an extended bilobed structure composed of the two globular L domains flanking the cysteine rich domain with dimensions of 40 x 48 x 105 Å. The amino terminal globular domain contacts the cysteine rich domain along its length. In contrast there is minimal contact between the C-terminal domain and the cysteine rich domain. In the crystal structure, L1 (amino acids 1-150) and L2 (amino acids 300 - 460) occupy very different positions relative to the cysteine rich domain. However this could be an artefact of crystal packing in this fragment and, as discussed below, the position of L2 may be very different in the native molecule. It is possible that it is rotated into a position similar to that of L1 in relation to the cysteine rich domain. However, irrespective of this, a cavity of approximately 24Å diameter occupies the centre of the molecule and probably represents a binding pocket.

Each L domain resembles a loaf of bread with dimensions 24 x 32 x 37 Å and is formed from a single right handed β helix capped at the ends by short α helices and disulphide bonds. The base of the loaf is formed from a six stranded β sheet five residues in length. Both sides are formed from β sheets 3 amino acids in length and the top is composed of irregular loops connecting the short β strands. As predicted from sequence comparisons, the cysteine rich domain is composed of repetitive modules resembling parts of laminin and the TNF receptor. These form a rod like structure connecting the two globular L domains from which a large mobile loop projects into the putative binding pocket.

2.4. *Structure of the Receptor Dimer – Electron Microscopic (EM) Studies*

While the studies described above provide reasonably accurate predictions of the structures of the modules forming the extracellular domain of the receptors, they do not provide any indication of their overall organization in the receptor dimer. Some clues to the answer to this question have been provided by single molecule electron microscopic imaging of either purified recombinant receptor extracellular domain (Schaefer et al., 1992; Tulloch et al., 1999) or of purified full length receptor using a variety of different techniques (Woldin et al., 1999; Christiansen et al., 1991; Luo et al. 1999; Tranum-Jensen et al., 1994). However the results of such studies have been somewhat variable.

The first such studies were performed on renatured holoreceptors purified by SDS/PAGE and the images of the holoreceptor suggested a T shaped structure (Christiansen et al., 1991). Subsequently studies of the purified ectodomain indicated a Y shaped molecule to be the consensus structure of the ectodomain (Schaefer et al., 1992). It was suggested that they might represent the upper half of the T shaped images observed with the whole receptor. More recently cryo electron microscopic studies of detergent solubilized and vesicle reconstituted receptor have also suggested a Y shaped structure (Woldin et al., 1999).

Tulloch et al. have reported the structure of EM images of the insulin receptor ectodomain (Figure 1) decorated with Fab fragments of conformation specific monoclonal anti-receptor antibodies (Tulloch et al., 1999). This approach has the advantages of providing both controls for the structural integrity of the molecules imaged and also, within the limits of the resolution of the technique, clues to the localization of subdomains. In contrast to the Y shaped structure deduced from the studies discussed above, the consensus image of the molecular envelope of the receptor ectodomain, in this study, was that of a U shaped prism of overall dimensions 90 x 80 x 120 Å; despite the differences in the overall image, the dimensions are similar to those found in the other studies. The prism has a central cleft 30 - 40 Å wide, adequate to accommodate a single molecule of insulin. Fab fragments directed towards epitopes in the first FnIII domain or in the insert region of the second FnIII domain appear to bind to the base of the U. A third Fab fragment directed towards an epitope in the distal part of the cysteine rich domain binds half way up the uprights of the U on different sides at each end. This suggests an arrangement of the receptor dimer in which the 2 L_1CL_2 fragments are arranged in an anti-parallel configuration and form the uprights and the base of the cavity of the U and the FnIII domains form the membrane proximal base, implying that the L2 domain occupies a similar position to the L1 domain relative to the cysteine rich domain, in contrast to what is observed in the crystal structure. The dimensions of the crystallized fragment can be accommodated within such a structural model.

More recently Luo and co-workers have reported a quarternary structure of the insulin-insulin receptor complex based on 3D reconstructions from low-dose, low temperature, dark-field scanning electron microscopic studies of the complex (Luo et al., 1999). The complexes were found to be approximately spherical with a diameter of approximately 150Å. The amino terminus of the extracellular domain was identified by covalent labeling with a photoreactive nanogold insulin analogue. Using the published structures of the IGF-I receptor N-terminal fragment, the insulin receptor tyrosine kinase catalytic domain, and of FnIII repeats, they produced a reconstruction of the domain structure within the protein envelope of the receptor dimer obtained from the EM images. In this reconstruction the L_1CL_2 structure occupies the central core of the extracellular domain dimer with the FnIII domains surrounding them just proximal to the membrane (in this model the authors ignored the recently published evidence for the existence of a third FnIII domain).

A number of assumptions have been made in this reconstruction and the details should thus be regarded with some scepticism. Principally it is assumed that the position of the L2 domain in relation to the cysteine rich domain is that observed in the crystal structure of the IGF-I receptor and as we have discussed, this does not appear to be compatible with the finding that the insulin minireceptor, in which the amino acids 704-719 have been fused to the C-terminus of the L_1CL_2 fragment, binds insulin with relatively high affinity (Kristensen et al., 1998). Further recent theoretical predictions regarding the receptor ligand binding site made by these

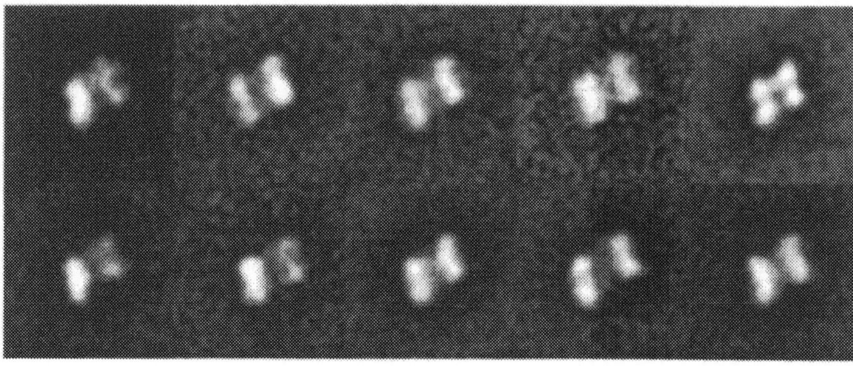

Figure 1. EM single-molecule imaging of human insulin receptor ectodomain. Figure kindly given by Dr Colin W. Ward. See Tulloch et al., 1999 for experimental details.

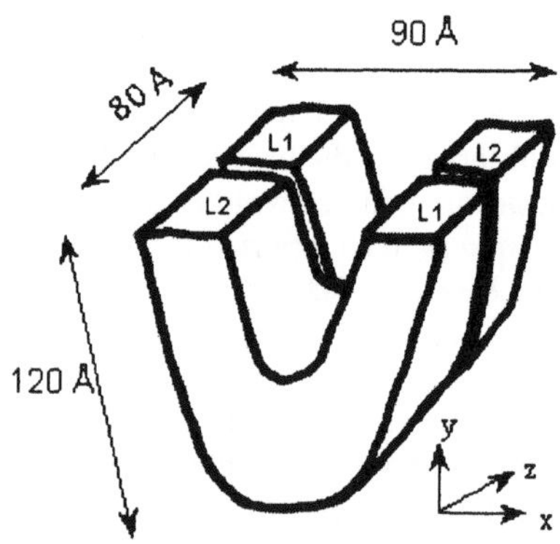

Figure 2. Proposed schematic structure of the insulin receptor ectodomain based on single molecule imaging using monoclonal antibodies Fab complexes. From Tulloch et al., 1999, used with permission.

authors from their model (Ottensmeyer et al., 2000) propose an insulin-receptor interface which is not entirely supported by mutational analysis of the insulin molecule. For example, the authors ignore the important free energy contributed by Tyr A19 (Kristensen et al., 1997; Jensen A.M. and De Meyts P., in preparation), as well as of Leu A13 (Schäffer, 1994; De Meyts, 1994), whose mutation to Ala causes a nearly 20-fold reduction in the rate of association to the receptor (Jensen A.M., Holst P.A. and De Meyts P., in preparation). In contrast, their proposed interface involves insulin residues whose mutation to alanine has no significant effect on binding such as Glu A4, Asn A15, Ser B9 or Lys B29. They also propose a strong electrostatic interaction between charged residues on the receptor and insulin's Glu B21 and Arg B22 while if anything mutation of these insulin residues to Ala increases binding affinity (Kristensen et al., 1997; Jensen A.-M., Holst P. A. and De Meyts P., manuscript in preparation). Thus, in our opinion, the authors have somewhat overextended the predictive capacity of the low resolution approach provided by electron microscopy.

In summary, the lack of consensus between these studies implies that only limited conclusions can be drawn at this point about the detailed domain organization of the receptor dimer. It is likely that it is a globular molecule with a diameter of 150Å. In the extracellular component the L_1CL_2 domains form the membrane distal core and the FnIII domains form the membrane proximal part.

3. MECHANISM AND KINETICS OF LIGAND BINDING

The equilibrium and kinetic properties of insulin and IGF-I binding can be summarized as follows. Scatchard plots of ligand binding are curvilinear, indicating non-uniformity of ligand binding affinity over the saturation range. This was proposed to result from negative cooperativity between insulin receptor binding sites, whereby binding of one insulin molecule reduces the affinity for further binding (De Meyts, 1994). Recent studies with purified receptors have established that the intact receptor heterotetramer binds only one insulin molecule with high affinity, but at saturation binds at least a second insulin molecule with lower affinity (Schäffer, 1994; De Meyts, 1994). In contrast, a purified secreted receptor ectodomain made of the two extracellular α subunits and the two extracellular portions of the β subunits, shows a linear Scatchard plot and binds two insulin molecules with an equal affinity, which is about twenty times lower than the affinity of the intact holoreceptor (Schäffer, 1994). Likewise, half insulin receptors (αβ) obtained by reduction of the holoreceptor bind insulin, but with low affinity (Sweet et al., 1987). These data suggest that each receptor half contains partial binding sites for insulin and have equivalent but low binding affinity, but that cooperation between the two receptor halves is required to create one high affinity binding site (Schäffer, 1994; De Meyts, 1994).

The negative cooperativity model was supported by the observation that the dissociation of a tracer amount of ^{125}I-insulin in an "infinite" dilution is markedly accelerated in the presence of unlabeled insulin in the dilution buffer (De Meyts. 1994). A dose-response curve for this acceleration of dissociation as a function of the concentration of unlabeled insulin in the medium resulted in a "bell-shaped" curve, often the hallmark of phenomena linked to ligand-induced receptor dimerization or cross-linking (De Meyts, 1994).

Taken altogether, the above data as well as detailed studies of the kinetics of binding of a number of insulin analogues, have led Schäffer (1994) and De Meyts (1994) to propose binding models whereby insulin acts as a bivalent ligand featuring two binding sites on the opposite faces of the molecule, that cross-links distinct sites on the two α subunits of the insulin receptor and thereby creates high affinity binding with a stoichiometry of one insulin bound per receptor heterotetramer (see Figure 1 for details of the proposed mechanism).

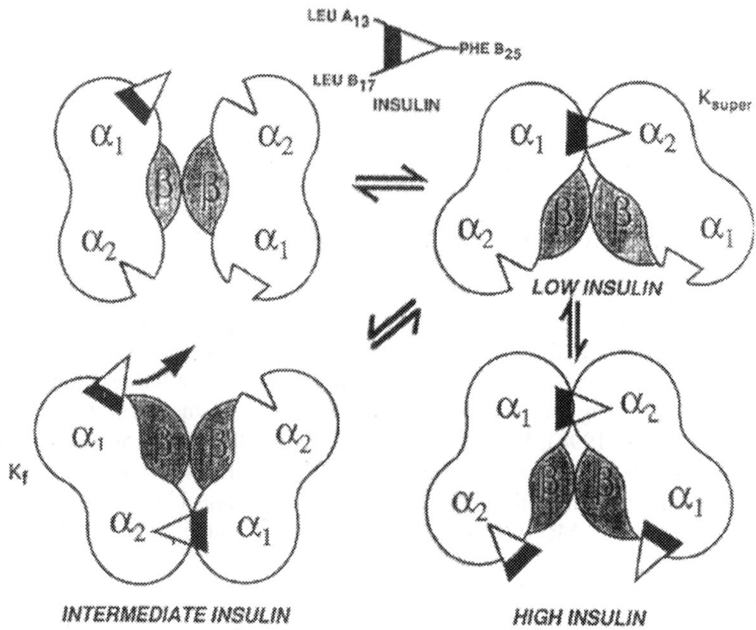

Figure 3. The symmetrical, alternative cross-linking model for insulin binding to its receptor. The receptor is viewed from the top. Each α subunit is represented as a pseudo-symmetrical dimer, as suggested by Bajaj et al. (1987), containing two binding subsites α1 and α2. The first insulin molecule binds through its hexamer-forming surface to α1, and then cross-links through its dimer-forming surface to α2 on the second α subunit. The resulting tight bivalent binding is referred to as the K_{super} state (De Meyts, 1994). If the concentration of insulin is increased, partial dissociation of the first bound insulin allows a second insulin molecule to cross-link the opposite α1 - α2 pair, which allows the first molecule to dissociate completely. At very high insulin concentrations, α1 and α2 opposite the first cross-link are both occupied, which prevents the second cross-linking and thereby maintains the first bound insulin molecule in the K_{super} state, explaining the bell-shaped dose-response curve of dissociation kinetics. This model is supported by recent EM studies of the insulin receptor ectodomain (Tulloch et al., 1999).
Modified from De Meyts (1994).

Our studies of the equilibrium and kinetics of IGF-I binding to the IGF-I receptor suggest that the binding mechanism is generally similar and exhibits the same degree of negative cooperativity, with the exception that the dose-response curve is not bell-shaped (Christoffersen et al., 1994; De Meyts et al., 1994). The implications of this difference are discussed in De Meyts (1994).

4. LOCATION OF THE LIGAND BINDING SITE ON THE INSULIN AND IGF-I RECEPTORS

The localization of the ligand binding sites of the insulin and IGF-I receptors has been the subject of intense study for many years. This problem has been approached experimentally both by affinity labelling, by chemical cross-linking and by mutational analysis. In the latter category, insights into this problem have been obtained from studies of the binding properties of chimeric receptors, naturally occuring and site directed point mutants and more recently by systematic alanine scanning mutagenesis.

Affinity labelling studies with either insulin and bifunctional crosslinkers or various photoreactive insulin analogues suggests that insulin contacts at least three different regions of the α subunit of the receptor. Waugh et al. (1989) and Wedekind et al. (1989), using methods that probably crosslink insulin to the receptor predominantly through lysine 29 of the B chain, isolated labelled receptor derived proteolytic fragments comprising amino acids 1 - 320 and 1 - 120 respectively. In contrast Fabry et al. (1992), using an insulin analogue with a photoreactive group in the position of Phe B1, i.e. crosslinking insulin to the receptor through Phe B1, provided evidence for a contact site between amino acids 430 – 488. More recently Kurose et al. (1994) using a photo-active insulin analogue that crosslinked to the receptor via the Phe B25 position found evidence for contact with a peptide between amino acids 704 and 718 at the C-terminus of the α subunit.

Functional studies of recombinant secreted chimeric insulin/IGF-I receptors indicate that the major determinant of insulin specificity resides between amino acids 1 and 68 of the receptor (Andersen et al., 1992; Schumacher et al., 1993; Schumacher et al., 1991), thus providing further support for the affinity labelling experiments of Waugh (1989) and Wedekind (1989). Further indications of the involvement of this region in ligand binding have been provided by the identification of a mutation in a patient with extreme insulin resistance, Asn 15 to Lys, which results in compromised insulin binding (Wedekind et al., 1990). The finding that site directed mutations of Phe 89 of the receptor to Leu (De Meyts et al., 1990) or Ala compromise binding suggests that the contact site is more extensive than the N-terminal 68 amino acids and may therefore possibly involve the whole of the L1 domain (Schumacher et al., 1991).

The most definitive evidence of the involvement of L1 in ligand binding has been provided by systematic alanine scanning mutagenesis of this region of the secreted recombinant receptor. The results of these studies indicate that 4 discontinuous peptide segments, amino acids 12-15, 34-44, 64-67 and 89-91, form a ligand binding site (Williams et al., 1995). When mapped on to the structure of this region of the homologous IGF-I receptor, these amino acids form a discrete footprint on the base of the domain (Garrett et al., 1998; Figure 2). Those amino acids, providing the majority of the free energy of binding, are located centrally within the footprint and those making more minor contributions are more peripheral, as has been observed for interactions of growth hormone with its receptor. However the L1 domain is unable on its own to bind insulin, although it can be expressed as a stable secreted protein suggesting that it folds normally.

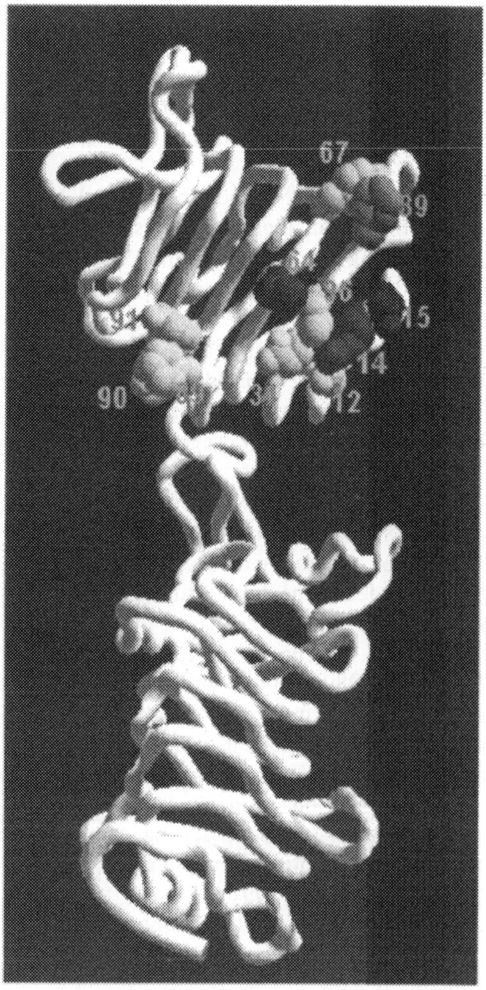

Figure 4. Alanine scanning mutagenesis of the insulin receptor L1 binding domain. The insulin receptor L1, cysteine rich and L2 domains (amino acids 4-470) were modeled from the published coordinates of the crystal structure of the N-terminal fragment of the IGF-I receptor (Garrett et al., 1998) using the program SWISS-MODEL (Peitsch et al., 1997). The Cα backbone of the the L1, cysteine rich domains and L2 is shown as a worm representation in white. The amino acids whose mutation affected binding are shown in space-filling representation. Alanine mutants of amino acids colored yellow produced a 2-10 fold reduction in affinity. Alanine mutants of those colored orange produced a 10-100 fold reduction in affinity and of those coloured red a > 100 fold reduction. This figure was prepared with the Swiss-PDB Viewer (Peitsch et al., 1997).

Alanine scanning mutagenesis of the secreted receptor also confirms that the C-terminus of the α subunit, amino acids 704-715, is also an essential component of this ligand binding site (Mynarcik et al., 1996). In this region, three discontinuous peptides appear to be essential for high affinity insulin binding and analogue binding studies (Mynarcik et al., 1997) reveal that the free energy contribution of this region to insulin binding is considerably greater than that of the L1 domain, consistent with the observation that both the L1 and L_1CL_2 domains are incapable of binding insulin when expressed on their own. It has recently been shown that fusion of the C-terminal peptide to the C-terminus of L_1CL_2 results in a mini-insulin receptor with an affinity for insulin similar to that of the full-length ectodomain (Kristensen et al., 1998). These findings also suggest that this peptide is situated close to L1 in the intact molecule and that some of the residues composing this peptide must interact with residues of L_1CL_2 to maintain the integrity of the ligand binding site.

At present there is no definitive functional confirmation that the contact site identified by Fabry et al. (1992) in the domain encoded by exon 6 plays a role in insulin binding, although the binding properties of insulin/IGF-I chimeric receptors in which regions of L2 of the insulin receptor, containing this site, have been substituted for the equivalent regions of the IGF-I receptor suggest that it may be a determinant of insulin specificity in the full length receptor molecule. Also naturally occurring mutations in this region, identified in patients with extreme insulin resistance, produce altered insulin binding properties which suggest that insulin binding determinants may reside within this region. Thus, mutation of Lys 460 to Glu (as found in a leprechaun patient) alters the pH dependence of insulin binding and nearly abolishes the loss of negative cooperativity seen at insulin concentrations greater than 0.1 μM (Kadowaki et al., 1990). Mutation of Lys 460 to Arg abolishes the acceleration of ^{125}I-insulin dissociation by unlabeled insulin and linearizes the Scatchard plot (Kadowaki et al., 1990; Wallach B. and De Meyts P., unpublished data). Interestingly, the corresponding amino acid in the IGF-I receptor – which displays accelerated ligand dissociation - is Arg, suggesting again that different amino acid side chains are involved in ligand binding properties of the two receptors. In addition, site-directed mutagenesis of residues Phe 400 and Tyr 401 in the insulin receptor also suggests a role for these residues in insulin binding or site-site interactions (Carbonnelle et al., 1992). Moreover, a number of antibodies directed against this region either inhibit insulin binding (residues 450-601) or activate the insulin receptor (residues 469-592). Finally, Christoffersen et al. (1994) have shown that inserting a domain encoded by exon 6 and part of exon 7 of the insulin receptor into the IGF-I receptor sequence restores a bell-shaped dose-response curve for negative cooperativity.

Participation of other regions of the L2 domain in the ligand binding site has been proposed on the basis of the finding that a mutation of Ser323 to leucine identified in a patient with extreme insulin resistance severely compromises insulin binding to the receptor without perturbing the structure of the full length receptor (Roach et al., 1994). We have also found that this mutation impairs binding both to the holoreceptor and to the recombinant secreted receptor (Whittaker and Mynarcik, 1998). However, while the posttranslational processing and structure of the mutant holoreceptor are normal this mutation causes a marked impairment of secretion of recombinant receptor extracellular domain suggesting that it perturbs the structure of the protein. In addition, alanine mutations in the same position are without significant effect on function of the holoreceptor or the receptor ectodomain, suggesting that the Ser323 side chain is not a direct participant in ligand binding. Thus it is likely that this mutation produces its effect on ligand binding by perturbing the structure of the ligand binding site.

Very recently, Brandt et al. (2001) have shown that by adding to the L_1CL_2 domain either the first FnIII domain encoded by exons 7 and 8 (which contains the Cys 524 involved in α–α disulphide bonding) or the full-length sequence of exon 10 (which contains the Cys 682, 683 and 685 also involved in α–α bonding, as well as the residues 704-717 involved in ligand binding as discussed above), dimeric receptor constructs were generated, but they did not have a higher affinity than monomeric mini-receptors. However, if the dimeric construct contained *both* the first FnIII domain and the exon 10-encoded domain, the dimeric receptor acquired a 1000 times higher affinity, similar to that of the intact holoreceptor, and furthermore showed negative cooperativity in ligand dissociation kinetics. These data strongly support the proposed ligand crosslinking model described above, and show that the minimal structure that contains all the binding epitopes as well as the architecture required for high affinity binding and negative cooperativity is made of two disulfide-linked α-subunits, lacking the domains encoded by exons 9, 11 and 12, and show that most of the second FnIII domain (involved in α–β contact) and the extracellular part of the β-subunits play no role in ligand binding (Brandt et al., 2001).

In summary, the studies discussed above lend support to the concept that insulin binds asymmetrically to two discrete sites in the receptor dimer, crosslinking the constituent monomers. One of these sites, the binding site of the secreted recombinant receptor, has been characterized in detail. It is partly located in the L1 domain of the extracellular portion of the receptor and also contains a peptide from the C-terminus of the α subunit. The second site, in contrast, is considerably less well studied but is located in L2 domain. These findings are consistent with the topology of the extracellular portion of the receptor that has been proposed on the basis of crystallographic and electron microscopic studies.

5. LOCATION OF THE RECEPTOR BINDING SITES ON THE INSULIN AND IGF-I MOLECULES

At increasing insulin concentrations, the insulin monomer assembles first into dimers and, in the presence of Zn, into hexamers (Baker et al., 1988). The insulin molecule has one obvious flat surface studded with aromatic and aliphatic residues, which aggregates through primarily non-polar forces. The residues of this dimer-forming surface are mostly B-chain amino acids: GlyB8, SerB9, ValB12, TyrB16, GlyB23, PheB24, PheB25, TyrB26, ThrB27, ProB28, AsnA21. In the hexamer, the dimer surfaces buried comprise both polar and non-polar residues from the A and B chains: LeuA13, TyrA14, GluA17, PheB1, ValB2, GlnB4, GluB13, AlaB14, LeuB17, ValB18, CysB19, GlyB20.

Not surprisingly given the relatively small and compact size of insulin, the surfaces involved in receptor binding have considerable overlap with the surfaces involved in self-aggregation. It has been recognized for many years that a number of surface residues that have been widely conserved during the evolution of vertebrates are probably involved in receptor binding: GlyA1, GlnA5, TyrA19, AsnA21, ValB12, TyrB16, GlyB23, PheB24, PheB25, TyrB26 ("classical binding surface") (Baker et al., 1988; Kristensen et al., 1998). The involvement of these residues was recently confirmed by alanine scanning mutagenesis of the insulin molecule, using a soluble secreted insulin receptor ectodomain as binding partner (Kristensen et al., 1997). De Meyts et al. (1978) showed that a patch of residues composed of B23-26 and A21 ("cooperative site") was essential for the negative cooperativity in binding. In addition, ValA3 (a residue mutated in a diabetic patient), which is not on the molecule surface, probably becomes exposed and interacts with the receptor upon displacement of the B-chain C-terminus during the receptor binding process (Hua et al., 1991). The positive effect of positively charged substitutions at Thr A8 (as in chicken and turkey insulins) suggests that this residue may also be part of this binding domain (Ivanov-Piron A. and De Meyts P., unpublished).

The concept that the "classical binding surface" was the only one involved in receptor binding was challenged by studies of the binding properties of the phylogenetically ancient insulin from hagfish (and more recently lamprey insulin) (Baker et al., 1988; De Meyts et al., 1978; Muggeo et al., 1979; Holst P.A., Kristensen C., Larsø L., Larsen U.D., Shymko R.M., Conlon J.M., Whittaker J., De Meyts P., submitted for publication). These insulins displayed low affinity for the insulin holoreceptor (but higher affinity for the secreted ectodomain), low biological activity (e.g. lipogenesis in rat adipocytes) and slow association kinetics, despite a fully conserved classical binding surface. Moreover, the dose-response curve for negative cooperativity obtained with hagfish insulin is not bell-shaped (De Meyts et al., 1978). These properties were shared by hystricomorph insulins (De Meyts, 1994; De Meyts et al., 1978; Ivanov-Piron M.A. and De Meyts P., unpublished), which have in common with hagfish insulin a number of mutations in the hexamer-forming surface. Studies of insulin analogues (designed for decreased aggregation and fast subcutaneous absorption) bearing single mutations at the hexamer-forming surface (specifically Leu A13 and Leu B17) reproduced the alterations in binding kinetics

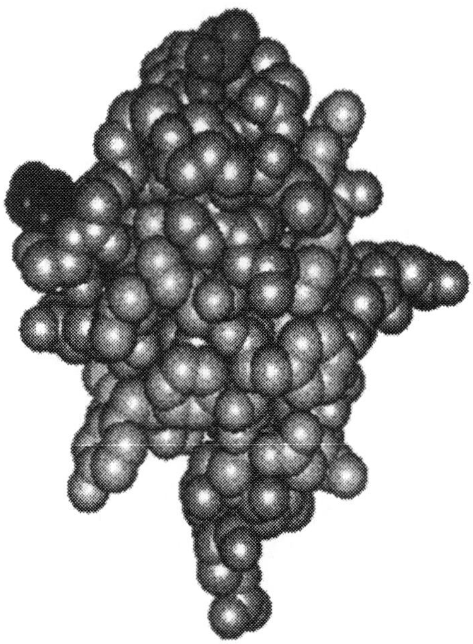

*Figure 5a. Binding site 2 of the insulin molecule.
The "classical binding surface" is shown in orange and ThrA8 in red.
This figure was prepared with WebLab ViewerLite .*

observed in hagfish and hystricomorph insulins, supporting the concept of an additional receptor binding surface coinciding partially with the hexamer-forming surface (Schäffer, 1994; De Meyts, 1994; Figure 3). Using the IM-9 human lymphocyte as a receptor binding partner, we have recently mapped by alanine scanning mutagenesis the amino acids that constitute this new binding surface, with Leu A13 and Leu B17 making the major contributions (Jensen et al., 2000; Jensen A.-M. and De Meyts P., manuscript in preparation). The positive effect of negatively charged substitutions (Asp, Glu) at His B10 suggest that this residue is also part of this binding surface.

The structure-function relationships of the IGF-I molecule have not been subjected to the same amount of scrutiny as insulin (see the exhaustive review by Van den Brande, 1992). Obtaining properly folded IGF-I analogues, which have a much greater propensity than insulin to misfold through mismatching of disulphide bonds, has been a major problem. It is however clear that the permanent C-peptide (non-existent in insulin) plays an essential role in IGF-I binding to its receptor (e.g. Tyr 31), and that the B23 (Tyr in insulin and Phe in IGF-I) and B24 (Phe in insulin

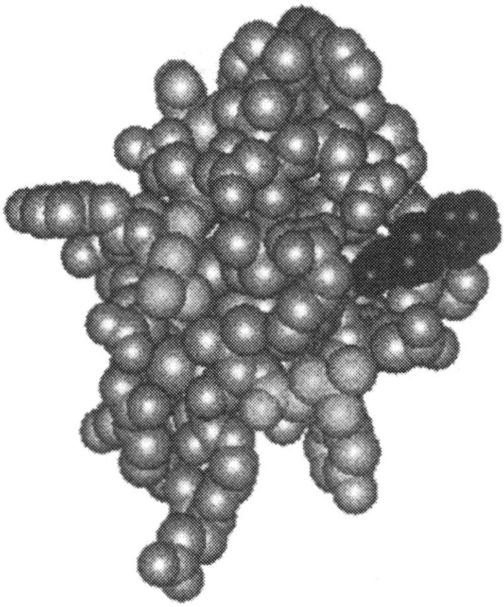

Figure 5b. Binding site 1 of the insulin molecule.
The residues from the hexamer forming surface whose mutation to alanine is most deleterious to insulin binding (Leu A13 and Leu B17) are shown in green; His B10 is shown in dark blue. This figure was prepared with WebLab ViewerLite .

and Tyr in IGF-I) residues play an equally important role in each ligand binding to their cognate receptor, as well as to the heterologous receptor. In contrast, mutation of ValB12 in insulin dramatically decreases receptor binding while the equivalent mutation in IGF-I has little effect (Chakravarty A. and De Meyts P., unpublished). Furthermore, the C-terminal end of the B-chain (B27-30, which plays no role in insulin binding) and initial C-peptide residues, especially positively charged Arg31 and 32, may be important for IGF-I receptor binding. Mutations in this region may enhance the mitogenic potency of analogues through enhanced IGF-I receptor binding (Slieker et al., 1997; Kurtzhals et al., 2000). See also Van den Brande (1992) and Adams et al. (2000) for more detailed description of the structure-function relationships of the IGFs.

These data suggest, like the abovementioned receptor mutagenesis data, that the insulin and IGF-I receptor interactions share a general binding pattern, but that within the overall contact zone, different patches of side chains provide the major contributions to binding energy.

In conclusion, although we do not have yet a definitive picture of the ligand–receptor interaction for insulin and IGF-I, substantial progress has recently been made in the definition of the functional epitopes of both ligand and receptors by molecular biological techniques, as well as in the three dimensional characterisation of structural modules of the receptors. The recent successful design of minimized receptor constructs that bind insulin with high affinity and properties comparable to the intact receptor represents a critical step towards solving the structure of the complex.

6. REFERENCES

Adams T.E., Epa V.C., Garrett T.P. and Ward C.W. "Structure and function of the type 1 insulin-like growth factor receptor." *Cell Mol. Life Sci.* 57 (2000): 1050-1093.

Andersen A.S., Kjeldsen T., Wiberg F.C., Vissing H., Schäffer L., Rasmussen J.S., De Meyts P. and Moller N.P. "Identification of determinants that confer ligand specificity on the insulin receptor." *J. Biol. Chem.* 267 (1992): 13681-13686.

Bajaj M., Waterfield M.D., Schlessinger J., Taylor W.R. and Blundell T. "On the tertiary structure of the extracellular domains of the epidermal growth factor and insulin receptors." *Biochim. Biophys. Acta* 916 (1987): 220-226.

Baker E.N., Blundell T.L., Cutfield J.F., Cutfield S.M., Dodson E.J., Dodson G.G., Crowfoot Hodgkin D.M., Hubbard R.E., Isaacs N.W., Reynolds C.D., Sakabe K., Sakabe N. and Vijayan N.M. "The structure of 2Zn Pig insulin crystals at 1.5 Å resolution." *Phil. Trans. R. Soc. Lond. B* 319 (1988): 369-456.

Blundell T.L., Dodson G.G., Hodgkin D.C. and Mercola D.A. "Insulin: the structure in the crystal and its reflection in chemistry and biology." *Adv. Prot. Chem.* 26 (1972): 279-402.

Brandt J., Andersen A.S. and Kristensen C. "Dimeric fragment of the insulin receptor a-subunit binds insulin with full holoreceptor affinity." *J. Biol. Chem.* (2001), in press (M009402200).

Carbonnelle C., Gu J.L., Vissing H., Shymko R.M. and De Meyts P. "Mutagenesis of the insulin receptor domain encoded by exon 6 alters ligand binding properties, pH and temperature dependence and receptor recycling." The Endocrine Society 74th Annual Meeting, 326 (Abstract 1098) (1992).

Chan S.J. "Insulin through the ages: phylogeny of a growth promoting hormone and metabolic regulatory hormone." *American Zoologist* 40 (2000): 213-222.

Christiansen K., Tranum-Jensen J., Carlsen J. and Vinten J. "A model for the quaternary structure of human placental insulin receptor deduced from electron microscopy." *Proc Natl Acad Sci USA* 88 (1991): 249-252.

Christoffersen C.T., Bornfeldt K.E., Rotella C.M., Gonzales N., Vissing H., Shymko R.M., ten Hoeve J., Groffen J., Heisterkamp N. and De Meyts P. "Negative cooperativity in the insulin-like growth factor-I receptor and a chimeric IGF-I /insulin receptor." *Endocrinology* 135 (1994): 472-475.

De Meyts P. "The structural basis of insulin and insulin-like growth factor-I receptor binding and negative cooperativity, and its relevance to mitogenic versus metabolic signalling." *Diabetologia* 37, Suppl. 2 (1994): 135-148.

De Meyts P., Gu J.L., Shymko R.M., Kaplan B., Bell G.I. and Whittaker J. "Identification of a ligand-binding region of the human insulin receptor encoded by the second exon of the gene." *Mol. Endo.* 4 (1990): 409-416.

De Meyts P., Van Obberghen E., Roth J., Brandenburg D. and Wollmer A. "Mapping of the residues of the receptor binding region of insulin responsible for the negative cooperativity." *Nature* 273 (1978): 504-509.

De Meyts P., Wallach B., Christoffersen C.T., Ursø B., Grønskov K., Latus L.J., Yakushiji F., Ilondo M.M., Shymko R.M. "The insulin-like growth factor-I receptor." *Horm. Res.* 42 (1994): 152-169.

Ebina Y., Ellis L., Jarnagin K., Edery M., Graf L., Clauser E., Ou J., Masiarz F., Kan Y.W., Goldfine I.D., Roth R.A. and Rutter W.J. "The human insulin receptor cDNA: the structural basis for hormone-activated transmembrane signalling." *Cell* 40 (1985): 747-748.

Fabry M., Schaefer E., Ellis L., Kojro E., Fahrenholz F. and Brandenburg D. "Detection of a new hormone contact site within the insulin receptor ectodomain by the use of a novel photoreactive insulin." J. Biol. Chem. 267 (1992): 8950-8956.

Garrett T.P., McKern N.M., Lou M., Frenkel M.J., Bentley J.D., Lovrecz G.O., Elleman T.C., Cosgrove L.J. and Ward C.W. "Crystal structure of the first three domains of the type-1 insulin-like growth factor receptor." *Nature* 394 (1998): 395-399.

Hua X.H., Shoelson S.E., Kochoyan M. and Weiss M.A. "Receptor binding redefined by a structural switch in a mutant human insulin." *Nature* 354 (1991): 238-241.

Jensen A.M., Kristensen C., Larsen U.D., Havelund S., Shymko R.M. and De Meyts P. "Mapping on the high affinity receptor binding surface of the insulin molecule by alanine scanning mutagenesis." 11th International Congress of Endocrinology (ICE 2000), Sydney, Australia, Abstract.

Kadowaki H., Kadowaki T., Cama A., Marcus-Samuels B., Rovira A., Bevins C.L. and Taylor S.I. "Mutagenesis of lysine 460 in the human insulin receptor. Effect upon receptor recycling and cooperative interactions among binding sites." *J. Biol. Chem.* 265 (1990): 21285-21296.

Kadowaki T., Kadowaki H., Accili D. and Taylor S.I. "Substitution of lysine for asparagine at position 15 in the alpha-subunit of the human insulin receptor. A mutation that impairs transport of receptors to the cell surface and decreases the affinity of insulin binding." *J. Biol. Chem.* 265 (1990): 19143-19150.

Kristensen C., Kjeldsen T., Wiberg F.C., Schäffer L., Hach M., Havelund S., Bass J., Steiner D.F. and Andersen S.A. "Alanine scanning mutagenesis of insulin." *J. Biol. Chem.* 272 (1997): 12978-12983.

Kristensen C., Wiberg F.C., Schäffer L. and Andersen A.S. "Expression and characterization of a 70-kda fragment of the insulin-receptor that binds insulin - minimizing ligand- binding domain of the insulin-receptor." *J. Biol. Chem.* 273 (1998): 17780-17786.

Kurose T., Pashmforoush M., Yoshimasa Y., Carroll R., Schwartz G.P., Burke G.T., Katsoyannis P.G. and Steiner D.F. "Cross-linking of a B25 azidophenylalanine insulin derivative to the carboxyl-terminal region of the alpha-subunit of the insulin receptor. Identification of a new insulin-binding domain in the insulin receptor." J. Biol. Chem. 269 (1994): 29190-29197.

Kurtzhals P., Schäffer L., Sørensen A., Kristensen C., Jonassen I., Schmid C. and Trüb T. "Correlations of receptor binding and metabolic and mitogenic potencies of insulin analogs designed for clinical use." *Diabetes* 49 (2000): 999–1005.

Lee J. and Pilch P. "The insulin receptor: structure, function and signaling." *Am. J. Physiol.* 266 (*Cell Physiol.* 35) (1994): C319-C334.

Luo R.Z.T., Beniac D.R., Fernandes A., Yip C.C. and Ottensmeyer F.P. "Quaternary structure of the insulin-insulin receptor complex." *Science* 285 (1999): 1077-1080.

Muggeo M., Van Obberghen E., Kahn C.R., Roth J., Ginsberg B.H., De Meyts P., Emdin S.O. and Falkmer S. "The insulin receptor and insulin of the Atlantic hagfish: extraordinary conservation of binding specificity and negative cooperativity in the most primitive vertebrate." *Diabetes* 28 (1979): 175-181.

Mynarcik D.C., Williams P.F., Schäffer L., Yu G.Q. and Whittaker J. "Analog binding-properties of insulin-receptor mutants - identification of amino-acids interacting with the COOH terminus of the B-chain of the insulin molecule." *J. Biol. Chem.* 272: 2077-2081.

Mynarcik D.C., Yu G.Q. and Whittaker J. "Alanine-scanning mutagenesis of a C-terminal ligand binding domain of the insulin receptor alpha subunit." *J. Biol. Chem.* 271 (1996): 2439-2442. Published erratum in *J. Biol. Chem.* 271 (1996): 30297.

Ottensmeyer F.P., Beniac D.R., Luo R.Z.T. and Yip C.C. "Mechanism of transmembrane signaling: insulin binding and the insulin receptor." *Biochemistry* 39 (2000): 12103-12112.

Peitsch M.C., Wilkins M.R., Tonella L., Sanchez J.-C., Appel R.D. and Hochstrasser D.F. "Large scale protein modelling and integration with the SWISS-PROT and SWISS-2DPAGE databases: the example of *Escherichia coli.*" *Electrophoresis* 18 (1997): 498-501.

Roach P., Zick Y., Formisano P., Accili D., Taylor S.I. and Gorden P. "A novel human insulin receptor gene mutation uniquely inhibits insulin binding without impairing posttranslational processing." *Diabetes* 43 (1994): 1096-1102.

Schaefer E.M., Erickson H.P., Federwisch M., Wollmer A. and Ellis L. "Structural organization of the human insulin receptor ectodomain." *J. Biol. Chem.* 267 (1992): 23393-23402.

Schäffer L. "A model for insulin binding to the insulin receptor." *Eur. J. Biochem.* 221 (1994): 1127-1132.

Schumacher R., Mosthaf L., Schlessinger J., Brandenburg D. and Ullrich A. "Insulin and insulin-like growth factor-1 binding specificity is determined by distinct regions of their cognate receptors." *J. Biol. Chem.* 266 (1991): 19288-19295.

Schumacher R., Soos M.A., Schlessinger J., Brandenburg D., Siddle K. and Ullrich A. "Signaling-competent receptor chimeras allow mapping of major insulin receptor binding domain determinants." *J. Biol. Chem.* 268 (1993): 1087-1094.

Shier P. and Watt V.M. "Primary structure of a putative receptor for a ligand of the insulin family." *J. Biol. Chem.* 264 (1989): 14605-14608.

Siddle K. "The insulin receptor and type I IGF receptor: comparison of structure and function." *Progress in Growth Factor Research* 4 (1992): 301-320.

Slieker L.J., Brooke G.S., DiMarchi R.D., Flora D.B., Green L.K., Hoffman J.A., Long H.B., Fan L., Shields J.E., Sundell K.L., Surface P.L. and Chance R.E. "Modifications in the B10 and B26-30 regions of the B chain of human insulin alter affinity for the human IGF-I receptor more than for the insulin receptor." *Diabetologia* 40 (1997): 54-61.

Sweet L.J., Morrison B.D. and Pessin J. "Isolation of functional a-b heterodimers from the purified a2-b2 heterotetrameric insulin receptor complex. A structural basis for insulin binding heterogeneity." *J. Biol. Chem.* 262 (1987): 6939-6942.

Tranum-Jensen J., Christiansen K., Carlsen J., Brenzel G. and Vinten J. "Membrane topology of insulin receptors reconstituted into lipid vesicles." *J. Membr. Biol.* 140 (1994): 215-223.

Tulloch P.A., Lawrence L.J., McKern N.M., Robinson C.P., Bentley J.D., Cosgrove L., Ivancic N., Lovrecz G.O., Siddle K. and Ward C.W. "Single-molecule imaging of human insulin receptor ectodomain and its Fab complexes." *J. Struct. Biol.* 125 (1999): 11-18.

Ullrich A. and Schlessinger J. "Signal transduction by receptors with tyrosine kinase activity." *Cell* 61 (1990): 203-212.

Ullrich A., Bell J.R., Chen E.Y., Herrera R., Petruzelli L.M., Dull T.J., Gray A., Coussens L., Liao Y.C., Tsubokawa M., Mason A., Seeburg P.H., Grunfeld C., Rosen O.M. and Ramachandran J. "Human insulin receptor and its relationship to the tyrosine kinase family of oncogenes." *Nature* 313 (1985): 756-761.

Ullrich A., Gray A., Tam A.W., Yang-Feng T., Tsubokawa M., Collins C., Henzel W., Le Bon T., Kathuria S., Chen E., Jacobs S., Francke U., Ramachandran J. and Fujita-Yamaguchi Y. "Insulin-like growth factor-I receptor primary structure: comparison with insulin receptor suggests structural determinants that define functional specificity." *EMBO J.* 5 (1986): 2503-2512.

Vajdos F.F., Ultsch M., Schaffer M.L., Deshayes K.D., Liu J., Skelton N.J. and de Vos A.M. "Crystal structure of human insulin-like growth factor-I: detergent binding inhibits binding protein interactions." *Biochemistry* 40 (2001): 11022-11029.

Van den Brande J.L. "Structure of the insulin-like growth factors: relationship to function." In: *The insulin-like growth factors. Structure and biological functions.* Schofield P, ed. Oxford University Press (1992): pp 12-44.

Waugh S.M., DiBella E.E. and Pilch P.F. "Isolation of a proteolytically derived domain of the insulin receptor containing the major site of cross-linking/binding." *Biochemistry* 28 (1989): 3448-3455.

Wedekind F., Baer-Pontzen K., Bala-Mohan S., Choli D., Zahn H. and Brandenburg D. "Hormone binding site of the insulin receptor: analysis using photoaffinity-mediated avidin complexing." *Biol. Chem. Hoppe Seyler* 370 (1989): 251-258.

Whittaker J. and Mynarcik D.C. "Phenotype of the Ser-323 to Leu mutation of the insulin- receptor is isoform dependent." *Diabetes* 47 (Suppl 1) (1998): 264.

Williams P.F., Mynarcik D.C., Yu G.Q. and Whittaker J. "Mapping of an NH2-terminal ligand binding site of the insulin receptor by alanine scanning mutagenesis." *J. Biol. Chem.* 270 (1995): 3012-3016.

Woldin C.N., Hing F.S., Lee J., Pilch P.F. and Shipley G.G. "Structural studies of the detergent-solubilized and vesicle- reconstituted insulin receptor." *J. Biol. Chem.* 274 (1999): 34981-34992.

7. ACKNOWLEDGEMENTS

The work of the first author (P.D.M.) has heavily relied for many years on the collaboration with the groups of Dietrich Brandenburg and Axel Wollmer in Aachen, as well as with other colleagues, many of whom attended this Alcuin Symposium.

Our enjoyable collaboration with Jan Markussen's Insulin Research group at Novo Nordisk over the last 10 years is gratefully acknowledged, especially with Claus Kristensen, Asser Sloth Andersen, Thomas Kjeldsen, Svend Havelund and Lauge Schäffer, as well as with Ulla Dahl Larsen in Isotope Chemistry and Ronald M. Shymko in Scientific Computing. This work is supported by the Danish Medical Research Council (through the Danish Center for Growth and Regeneration) and the Juvenile Diabetes Foundation International (Grant # 1-2000-198).

C.W. WARD, T.P.J. GARRETT, N.M. McKERN, L.G. SPARROW, M.J. FRENKEL

STRUCTURAL RELATIONSHIPS BETWEEN MEMBERS OF THE INSULIN RECEPTOR FAMILY

Abstract. Comparative sequence analyses have revealed that many proteins are composed of a number of different, sometimes repeated, structural units. In the case of the insulin receptor subfamily, 11 distinct regions have been identified in each monomer. The N-terminal half of the ectodomain contains two homologous domains (L1 and L2), separated by a cys-rich region containing 24-26 cysteines. The C-terminal half of the ectodomain consists of three fibronectin type III domains, the second of which contains a large insert domain. Experimental structural information has been obtained from crystals of a fragment (residues 1-462) comprising the L1-cysteine-rich-L2 region of the IGF-IR ectodomain. The molecule adopts an extended structure with a central space, bounded by all three domains, of sufficient size to accommodate ligand. Two regions of the receptor which are involved in hormone binding, map to this central site. Each L domains adopts a compact shape consisting of a single stranded right-handed ß-helix. The cys-rich region is composed of eight disulfide-bonded modules, seven of which form a rod-shaped domain with modules associated in a novel manner. While they resemble the modules found in the TNF receptor and laminin repeats, this novel arrangement suggests that EGF repeats found in many proteins are composed of two smaller modules containing two or one disulfide bond. Single-molecule electron microscope imaging has revealed that the IR ectodomain resembles a compact U-shaped prism, where the L1-cys rich-L2 domains occupy the membrane distal region and the fibronectin type III and insert domains are located in the membrane-proximal third.

1. INTRODUCTION

For several years now we have been interested in determining the structure of the insulin receptor and its complex with insulin to understand the molecular details of insulin-induced receptor activation. This paper will discuss the developments in our knowledge of the structure of members of the IR family but will not deal with the issues of ligand binding or specificity. These will be covered by other presentations at this symposium and have been recently reviewed (Adams et al., 2000).

The first evidence for the presence of an insulin receptor (IR) came in 1971 from the pioneering studies of Cuatrecasas using ^{125}I-insulin to label a protein in the plasma membrane of insulin-responsive cells (Cuatrecasas et al., 1971). The IR could be solubilized with non-ionic detergents and was subsequently shown by SDS gel electrophoresis to be a homodimer composed of two α– and two β–chains held together by disulfide bonds (Jacobs et al., 1979). The next key discovery was the demonstration by Kasuga and coworkers that the IR was a tyrosine kinase which was activated following insulin binding (Kasuga et al., 1982). Similar data have been established for the insulin-like growth factor receptor (IGF-IR) and its activation by IGF-I and IGF-II (see Adams et al., 2000, for refs).

M. Federwisch, M. Leyck Dieken and P. De Meyts (eds.), Insulin & Related Proteins, 151–164.

The next landmark discovery was the cloning and sequencing of the cDNA for human IR in 1985 (Ullrich et al., 1985; Ebina et al., 1985). It coded for a 1,370 amino acid precursor which is cleaved by furin into an α- and β-chain. The α-chain and 194 residues of β-chain comprise the extracellular portion of the IR. There is a single transmembrane sequence and a 403 residue cytoplasmic domain containing the tyrosine kinase. The cDNA for the IGF-IR and a third member of the IR subfamily, the insulin receptor-related receptor (IRR) have been cloned and sequenced and are similarly organized (Ullrich et al., 1986; Shier and Watt 1989). These receptors are heavily glycosylated. The ligands, insulin and the two IGFs share a common 3D architecture and can bind to each other's receptor in a competitive manner. The IRR ligand is unknown.

The IR subfamily is related to another class of cell surface receptors, the epidermal growth factor receptor (EGFR) subfamily. Like the IR subfamily, EGFR members have a single transmembrane region and a cytoplasmic region containing tyrosine kinase catalytic domain flanked by regulatory juxtamembrane and C-tail sequences (Ullrich et al., 1984). The major feature which separates members of the IR subfamily from most other receptor families is that they exists on the cell surface as disulfide-linked dimers and require domain rearrangement on binding ligand, rather than receptor oligomerization, to initiate signal transdiction. The EGFRs do not form disulfide-linked dimers and the ligands (EGF and TGFα) are structurally unrelated to insulin and the IGFs.

Representatives of both of these receptors are found in some of the simplest multicellular animals, including cnidarians (polyps and jellyfish), nematodes, trematodes, gastropods and insects. Some of the primitive receptors have additional domains when compared with the mammalian counterparts (see Adams et al., 2000 for refs). For example the *Caenorhabditis elegans* (nematode) and *Drosophila melanogaster* (insect) IRs have additional sequences at their N- and C-termini and the EGFRs from *D. melanogaster, C. elegans* and *Schistosoma mansoni* (trematode) have further duplications of the cys-rich sequences at the end of their ectodomains. Despite these differences functional activity is retained within a receptor class as illustrated by the ability of porcine insulin to activate the IR from the mosquito, *Aedes aegypti* and stimulate the secretion of ecdysteroids from mosquito ovaries (Graf et al.,1997).

2. SEQUENCE ANALYSES AND STRUCTURAL PREDICTIONS

There are three approaches to gaining information about the 3D structure of proteins: i) homology searching for protein domains whose structures and functions are known, ii) *de novo* structural predictions using a variety of programs and iii) direct experimentation. All three have been applied to the insulin receptor family. The first significant analysis was the observation that the first 120 or so residues of the IR (and EGFR) ectodomain upstream of the cys-rich region showed sequence similarity with the 120 residues immediately downstream of the cys-rich region (Bajaj et al.,

1987). These domains were termed L1 and L2 (where the L stands for large) and were further shown to contain at four internal repeats in the region equivalent to 1-119 in hIR (Bajaj et al., 1987). The repeats ranged in size from 20-42 residues and included a 13 residue repeating motif X--X-(x)X-G-X(x)X where X is hydrophobic and (x) is often hydrophobic. In addition they applied several structural prediction methods to conclude that the repeats were composed of α-helix/β-strand/turn/β-strand secondary structure with the conserved G residue being part of the turn (Bajaj et al., 1987).

These authors also suggested, on the basis of sequence alignments, that the cys-rich regions of IR and EGFR (approximately 150 amino acid residues with 24-26 Cs) consisted of three repeats of eight C residues. The spacing between the homologous C residues and the nature of the amino acid residues in these loops were variable (Bajaj et al., 1987). The report on the 3D structure of the tumor necrosis factor receptor p55 by Banner et al. (1993) prompted us to investigate whether this structure might represent the first view of the modules that make up the cys-rich regions of IR and EGFR (Ward et al., 1995). The TNFR ectodomain is of similar size and cysteine content as the IR and EGFR cys-rich regions and consists of four repeating modules each containing six C residues in C1-C2 (loop 1), C3-C5, C4-C6 (loop 2) disulfide linkages. The C3-C5 pair are occasionally missing and additional disulfides are found in some repeats (Banner et al., 1993; Ward et al., 1995). When the SwissProt database was searched with profiles generated from multiple sequence alignments of TNFR repeats the first 14 hits were known members of the TNFR family. Five of the next seven highest ranking alignments were members of the IR family suggesting that the cys-rich region of IR might be structurally related to the TNFR. The data allowed the disulfide bonds in the cys-rich regions of IR and EGFR to be predicted and indicated that these cys-rich regions were not simple repeats of eight C residues but were more complicated combinations of loop 1-type and loop 2-type disulfide-linked modules (Ward et al., 1995). The validity of these alignments took on greater significance with the publication of the 3D structure of three consecutive repeats of laminin (Stetefeld et al., 1996). Laminins A (18th and 21st) and B (27th and 36th) were very highly ranked in the SwissProt searches with the TNFR profiles (Ward et al., 1995) and the 3D structure was consistent with the predictions of disulfide bonded modules obtained from the profile analyses.

Sequence analyses have also been used to show that the C-terminal half of the IR ectodomain contains three fibronectin type III (FnIII) modules, one of the most common structural modules found in many proteins (see Dickinson et al., 1992; Campbell and Spitzfaden, 1994). The first report was by O'Bryan et al. (1991), who described the last two fibronectin type III (FnIII) repeats at residues 594-801 and 811-907 (IR numbers) in detail, following their cloning and characterization of the tyrosine kinase *axl* which appeared closely related to the insulin receptor family. The first of these FnIII modules was shown to contain a large insert from residues 661 to 768 (O'Bryan et al., 1991). Subsequently Schaefer et al. (1992) made some structural assignments by aligning these two FnIII with the two FnIII domains of human growth hormone receptor and the third domain of tenascin for which 3D structures were available (De Vos et al., 1992; Leahy et al., 1992). Recently, three independent reports have described the identification of a third FnIII module in the region (residues 471-593) previously referred to as the connecting domain (Mulhern et al., 1998; Marino-Buslje et al., 1998; Ward, 1999). These authors differ in their assignments of residues to the seven β-strands in each of these FnIII domains with the regions of greatest disagreement being the locations of the C' and E strands in all three modules (see Adams et al., 2000). Cys524 which is involved in one of the dimer disulfide bonds is predicted to be in the C-C' loop (Marino-Buslje et al., 1998; Mulhern et al., 1998) or the C'-E loop (Ward, 1999). Cys647, which is involved in the α-β disulfide bond is predicted to lie in the C'-E loop (Mulhern et al., 1998; Ward, 1999) or the E-strand (Schaefer et al., 1992). Cys860 the other partner in the α-β disulfide bond, is predicted to be located in the C'-E loop (Ward, 1999) or the C' strand (Schaefer et al., 1992; Mulhern et al., 1998). Residues forming part of the epitope for the monoclonal antibody 83-14 (i.e. those differing between mouse IR and human IR) are predicted to be in the C'-E loop (Marino-Buslje et al., 1998) at the top of the first FnIII domain (Figure 7) or the E-F loop at the bottom of this domain (Mulhern et al., 1998, Ward, 1999). This antibody has been used to locate the FnIII domain in the hIR ectodomain dimer (Tulloch et al., 1999).

Given the extensive scrutiny of the insulin receptor family sequences, it is surprising that this third FnIII domain had been missed by us (Sparrow et al., 1997) and others (Schaefer et al., 1992) including the detailed "pastiche" model of the IR developed (McDonald et al., 1995) after the publication of the 3D structure of the IR kinase domain (Hubbard et al., 1994). Although it was not discussed by O'Bryan et al (1991), they appear to have identified this additional FnIII domain as it is illustrated unequivocally in their Figure 4. However in that paper the main focus was on comparisons with *axl* (which contains only two FnIII domains) and other tyrosine kinases and their sequence alignments were confined to the two C-terminal FnIII repeats (O'Bryan et al., 1991). A cartoon summarizing the domain arrangements and inter-chain disulfide bonds is shown in Figure 1.

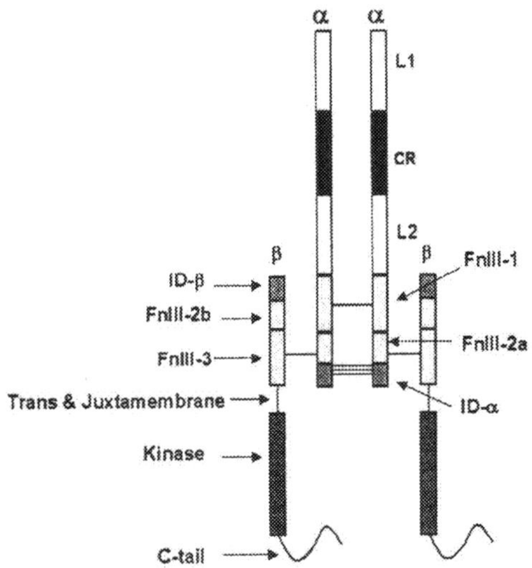

Figure 1. Cartoon of the IR dimer showing the distribution of domains across the α- and β-chains and the approximate location of the α-β disulfides and the α-α dimer disulfide bonds. Based on Sparrow et al., (1997).

3. SECONDARY STRUCTURE

The ectodomain of IR contains 41 cysteine residues per monomer, 37 in the α-chain and 4 in the β-chain. Some of the disulfide bonds have been established by chemical analysis. C^{8} is linked to C^{26} in the L1 domain (Sparrow et al., 1997) and C^{435} is linked to C^{468} in the L2 domain (Schäffer and Ljungqvist, 1992). Sequence alignments suggest the homologous pairs of C residues at the end of the L1 domain (C^{126} and C^{155}) and the start of the L2 domain (C^{312} and C^{333}) are also disulfide bonded (Ward et al., 1995) as confirmed in the 3D structure (Garrett et al., 1998). None of the disulfide bonds in the Cys-rich region have been established chemically except for the additional disulfide bond C^{266}-C^{274}, in the large loop of sixth cys-rich module of hIR (Schäffer and Hansen, 1996).

C^{524}, in the first FnIII domain, was shown to form a dimer disulfide bond with C^{524} in the second monomer (Schäffer and Ljungqvist 1992). However, mutation of C^{524} to A (Macaulay et al., 1994) or S (Bilan and Yip, 1994; Lu and Guidotti, 1996) resulted in an IR that was still dimeric, indicating that more than one disulfide bond is involved in dimer formation. The additional dimer disulfides of IR were shown, by chemical analysis, to involve the triplet of C residues at positions 682, 683 and 685 in the insert domain of IR (Sparrow et al., 1997). It was not possible to determine which one of the three, or whether all three of these C residues are involved in dimer disulfides. The sequence around this triplet resembles those found in the hinge region of antibodies (Kabat et al., 1991) where multiple disulfide bonds occur. It is interesting to note that the IGF-I receptor has an additional C, seven residues upstream of the C triplet in the insert domain and lacks the C residue equivalent to 884 in hIR. *Drosophila* IR (Ruan et al., 1995; Fernandez et al., 1995) also lacks the cysteine residue equivalent to C^{884} in hIR as well as the equivalent of one of C^{682} or C^{683}.

There is only a single α-β disulfide link between C^{647} in the first FnIII repeat and C^{872} (exon 11+ isoform) in the second FnIII repeat of human IR (Sparrow et al., 1997), which is consistent with the mutagenesis data (Cheatham and Kahn, 1992; Lu and Guidotti 1996) and the predictions of Ward *et al.* (1995). It differs from the suggestions made by Schaefer *et al.* (1992), that the disulfide links between the four β-chain cysteines, 798, 807, 872 and 884 (exon 11+ numbering) were similar to those in the growth hormone receptor (De Vos et al., 1992). Such an arrangement would confine the two disulfide bonds to the β-chain and would not allow for the presence of the known α-β disulfide linkage. The structural implications of the disulfide bond between C^{647} and C^{872} are that the two FnIII domains are aligned side by side (Ward et al., 1995), not end to end, as is the more common configuration (Campbell and Spitzfaden, 1994). Finally, there is an intra-chain disulfide linkage between the β-chain residues C^{798} and C^{807} (exon 11+ numbering) in the predicted F and G strands of the third FnIII domain (Sparrow et al., 1997). The remaining Cys residue in human IR, C^{884} (exon 11+ isoform), which has no counterpart in human IGF-IR or human IRR, has been shown to exist as a buried free thiol (Sparrow et al., 1997).

4. THREE-DIMENSIONAL STRUCTURE OF THE L1/CYS-RICH/L2 DOMAINS OF THE IGF-IR

Recently we solved the 3D structure of the L1/cys-rich/L2 domain fragment of the IGF-IR (Garrett et al., 1998). As shown in Figure 2, the molecule adopts an extended bilobal structure (approximately 40 x 48 x 105 Å) with the L domains at either end. The cys-rich region runs two-thirds the length of the molecule, making contact along the length of the L1 domain but having very little contact with the L2 domain. This leaves a space at the center of the molecule of approximately 24 Å diameter and of sufficient size to accommodate the ligands, IGF-I or IGF-II. The space is bounded on three sides by the regions of IGF-IR which are known to contribute to ligand binding based on studies of chemical cross-linking, receptor chimeras and natural or site-specific mutants (see Garrett et al., 1998, and Adams et al., 2000, for references).

4.1. The L Domains

Each L domain of human IGF-IR (residues 1-150 and 300-460) adopts a compact shape (approx. 24 x 32 x 37 Å) being formed from a single-stranded right handed β-helix, capped on the ends by short α-helices and disulfide bonds. The body of each domain looks like a loaf of bread with three flat sides and an irregular top (Figures 2 and 3). The two domains are superimposable with an rms deviation in position of 1.6 Å for 109 C^{α} atoms (Garrett et al., 1998). The repetitive nature of the β-helix is reflected in the sequence and, as mentioned in Section 2 above, a five-fold repeat, centered on a conserved glycine, had been identified by sequence analyses (Bajaj et al., 1987). The structure, however, revealed that the L domains comprised six helical turns and a fold that was quite unexpected (Garrett et al., 1998).

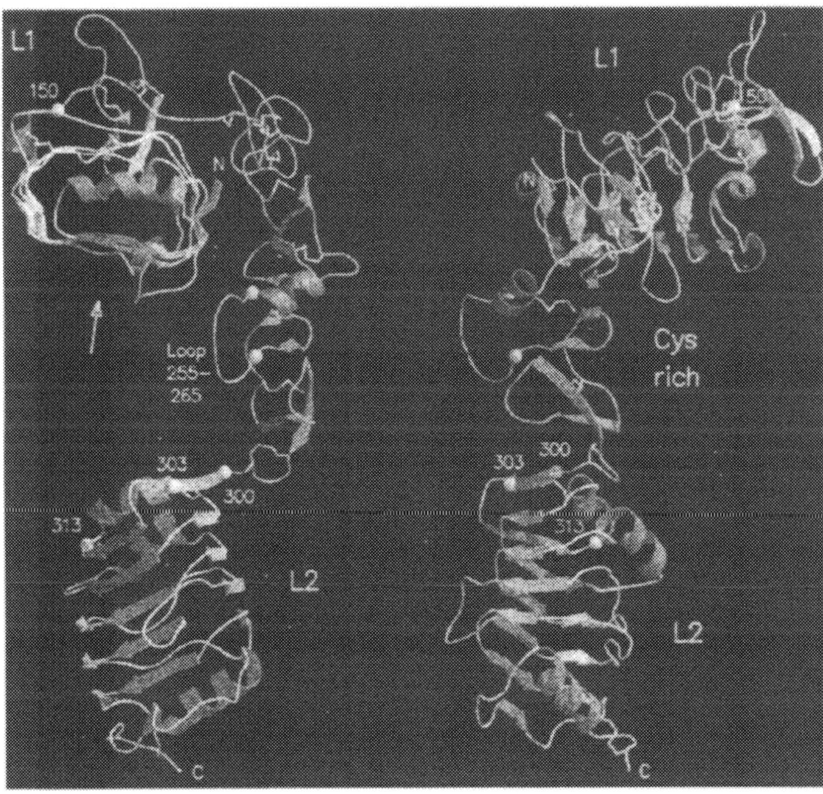

Figure 2. Polypeptide fold for residues 1-459 of the human IGF-I receptor. In the left hand panel, the L1 domain is at the top viewed from the N-terminal end. In the right hand view, the model has been rotated clockwise 90°. Helices are depicted as curled ribbons and β-strands as broad arrows. Based on Garrett et al., (1998).

A notable difference between the two domains is found at the C-terminal end. For L1 the indole ring of W^{176} from the cys-rich region is inserted between the last two turns of β-helix into the hydrophobic core of the domain and the C-terminal helix of L1 becomes vestigial. The sequence motif of residues which form the Trp pocket in L1 does not occur in L2 of the IR family (Garrett et al., 1998). However in EGFR, which has an additional cys-rich region after the L2 domain, the motif can be found in both L domains and the W is conserved in both cys-rich regions (Garrett et al., 1998).

Upon comparing the L domains with other right-handed β-helix structures such as pectate lyase (Yoder et al., 1993), ribonuclease inhibitor (Kobe and Deisenhofer 1995) and the p22 tailspike protein (Steinbacher et al., 1994) there are some striking similarities as well as differences. The other β-helix domains are considerably larger (up to 16 β-helical turns) and often contain many, and sometimes quite substantial, insertions. Although the sizes of the helix repeats are similar (here 24-25 residues *vs* 22-23 for pectate lyase) the type of helix is best characterized by its cross-section. The L domains have a slightly skewed rectangular cross-section like the ribonuclease inhibitor while the other β-helix proteins have a triangular or V-shaped cross-section (Figure 3). In the hydrophobic core a common feature is the stacking of aliphatic residues from successive turns of the β-helix. Near the C-terminus of each L domain there is a short ladder of asparagines on the third sheet, reminiscent of the long Asn ladder observed in pectate lyase (Yoder et al., 1993). On the opposite side of the L domains the Gly bend (between the green and blue sheets) and the blue sheet itself correspond to the sequence motif identified previously (Bajaj et al., 1987) but this feature has no counterpart in the other β-helix domains.

In most cases only the N-terminal end of the other β-helix domains is capped by an α-helix but L domains are capped at both ends by α-helices and have a disulfide bond at each end anchoring the termini.

4.2. The Cys-rich Domain

As anticipated from the TNFR profile analyses (Ward et al., 1995), the cys-rich domain is composed of modules with disulfide bond connectivities resembling parts of the TNF receptor (Banner et al., 1993) and laminin (Stetefeld et al., 1996) repeats (Figure 4). The first module sits at the end of L1 while the remaining seven form a curved rod running diagonally across L1 and reaching to L2 (Figure 2). The strands in modules 2-7 run roughly perpendicular to the axis of the rod in a manner more akin to laminin than to the TNF receptor, where the strands run parallel to the axis (Figure 4). The modular arrangement of the IGF-IR cys-rich domain is different to other cys-rich proteins for which structures are known (Figure 4). The first 3 modules of IGF-IR have a common core, containing a pair of disulfide bonds, but show considerable variation in the loops. These modules are referred to here as C2 (two disulfide bonds). The connectivity of the cysteines is the same as the first part of an EGF motif (Cys 1-3 and 2-4) but their structures do not appear to be closely

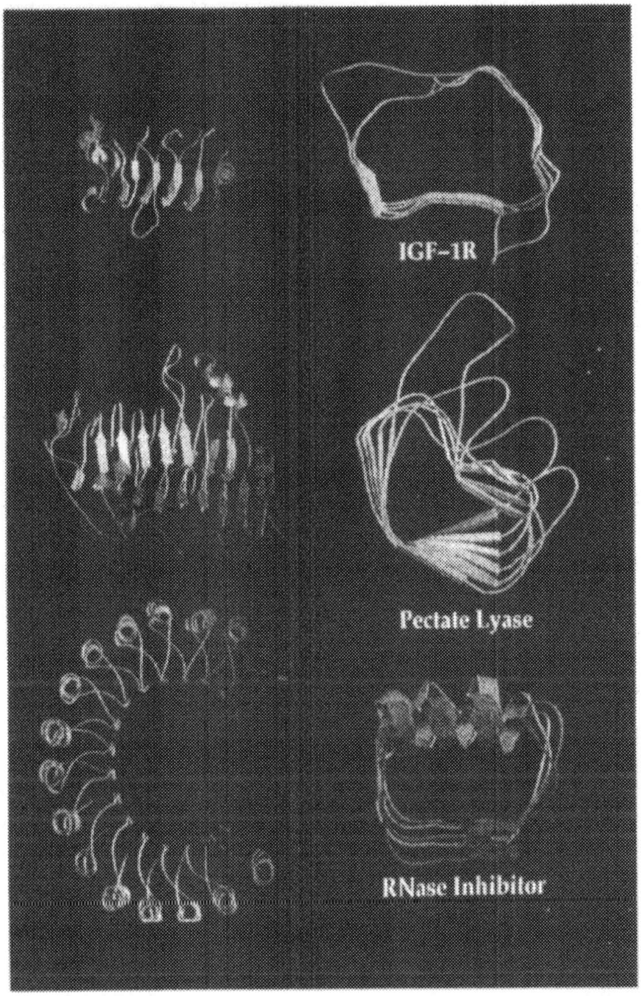

Figure 3. Comparison of the polypeptide folds of the L1 domain of the IGF-IR (Garrett et al., 1998) with those of pectate lyase (Yoder et al., 1993) and porcine ribonuclease inhibitor (Kobe and Deisenhofer 1995).

related to any member of the EGF family. Modules 4 to 7 have a different motif, a β-finger, seen previously in residues 2152-2168 of fibrillin (Downing et al., 1996).Each is composed of three polypeptide strands, the first and third being disulfide bonded and the second and third forming a β-ribbon. These are referred to here as C1, because of the single disulfide bond. The β-ribbon of each β-finger (or C1) module lines up antiparallel to form a tightly twisted 8-stranded β-sheet (Figure

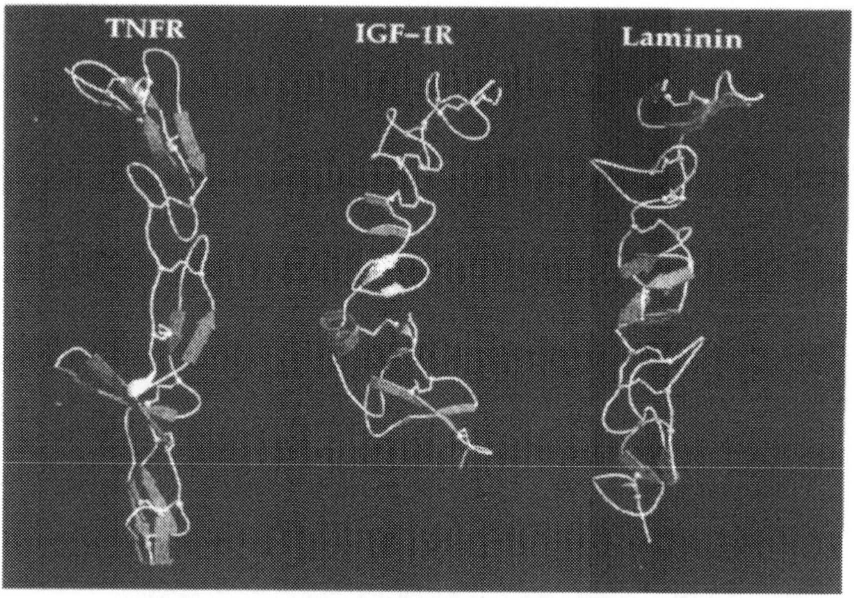

Figure 4. Comparison of the folds of modules 2-8 in the cys-rich region of the IGF-IR (Garrett et al., 1998) with the four repeats (8 modules) in the extracellular domain of the TNFR (Banner et al., 1993) and the first three repeats (9 modules) of laminin (Stetefeld et al., 1996).

4). Module 6 deviates from the common pattern with the first segment being replaced by an α-helix followed by a large loop that is implicated in ligand binding (see Garrett et al., 1998). As modules 4-7 are similar it is possible that they arose from a series of gene duplications. The final module is a disulfide linked bend of five residues.

The first cys-rich region in the EGFR ectodomain has the same eight modules in the same order, C2-C2-C2-C1-C1-C1-C1-C1', as the IR family (Ward et al., 1995, Abe et al., 1998). In contrast, the second cys-rich region of EGFR has the C1 and C2 modules arranged in a different combination $[C2\text{-}C1\text{-}C1]_2\text{-}C2$ (Ward et al., 1995, Abe et al., 1998) where the C2-C1-C1 repeat is similar to that seen in laminin (Stetefeld et al., 1996) and the furin-like proteases (Ward et al., 1995). The fact that the C1 and C2 cys-rich modules are grouped according to type implies that these are the minimal building blocks of the EGF-like cys-rich domains found in many proteins. Although it can be as short as 16 residues, the β-finger (C1) motif is clearly distinct and capable of forming a regular extended structure. Thus cys-rich domains such as laminin (Stetefeld et al., 1996) and fibrillin (Downing et al., 1996) can be considered not as modified EGF repeats but as a series of repeat units each composed of a small number of C1 and C2 modules. The EGF repeats in proteins such as fibrillin can be viewed as a [C2-C1] repeats while the laminin repeat is [C2-C1-C1].

This is further supported by the observation that EGFRs from insects (*D. melanogaster*), nematodes (*C. elegans, C. vulgaris*) and trematodes (*S. mansoni*) contain a third cys-rich region, with a similar repeat pattern of C1 and C2 modules. In *S. mansoni* there are large insertions between two of the modules. The combination of the second and third cys-rich repeats in these primitive organisms could also be viewed as five repeats of the C2-C1-C1 set of modules (see Adams et al., 2000 for refs).

5. THE INSULIN RECEPTOR ECTODOMAIN DIMER

A remaining issue is how are these various building blocks (domains) organized in the dimeric, native receptor. The first clues have come from single molecule images of the IR ectodomain and its complexes with three Fabs obtained by electron microscopy (Tulloch et al., 1999). These images show that the IR ectodomain dimer resembles a U-shaped prism of approximate dimensions 90 x 80 x 120 Å. The images show clearly the dimeric structure of the IR ectodomain. Measurement of the images yields a length of approximately 80 Å along each bar, and a width of approximately 90 Å across the two bars. The width of the cleft (assumed membrane-distal) between the two side arms is approximately 30 Å, sufficient to accommodate ligand (Tulloch et al., 1999).

Fab molecules from the monoclonal antibody 83-7, which recognizes the cys-rich region of IR, bound half-way up one end of each side arm in a diametrically opposite manner, indicating a two-fold axis of symmetry normal to the membrane surface (Tulloch et al., 1999). Fabs 83-14 and 18-44, which have been mapped respectively to the first FnIII repeat (residues 469-592) and residues 765-770 in the insert domain, bound near the base of the prism at opposite corners (Tulloch et al., 1999).

The single molecule images, together with the 3D structure of the first three domains of the insulin-like growth factor type I receptor (Garrett et al., 1998), suggest that the IR dimer is organized into two layers with the L1/cys-rich/L2 domains occupying the upper (membrane distal) region of the U-shaped prism and the fibronectin type III domains, the insert domains (and the disulfide bonds involved in dimer formation) located predominantly in the membrane-proximal region (Tulloch et al., 1999).

High resolution data is required to establish the precise arrangement of the fourteen modules that make up the ectodomain dimer and the way they interact with the respective ligands insulin and the IGFs to generate signal transduction. Recently whole receptors solubilized from human placental membranes have been examined by electron cryomicroscopy and 3D reconstruction using a library of 700 images (Luo et al., 1999). Gold-labeled insulin was used to locate the insulin-binding domain. The images seen were compact and globular, measuring 150 Å diameter. Some domain-like features became evident at intermediate-density thresholds, which indicated a strong two-fold vertical rotational symmetry. When this symmetry was applied to the reconstruction some structural features became evident. Their overall model showed the L1/cys-rich/L2 domains arranged in an antiparallel manner and at an angle to each other when viewed from the side. The six FnIII domains and the two L2 domains are placed in a central band with the two tyrosine kinase domains at the base of the model (Luo et al., 1999). The images described in these studies (Tulloch et al., 1999; Luo et al., 1999) are substantially different from the "T"-, "X"- or "Y"-shaped objects reported for recombinant ectodomain (Schaefer et al., 1992), detergent solubilized or vesicle-reconstituted whole receptors (Christiansen et al., 1991; Tranum-Jensen et al., 1994; Woldin et al., 1999).

6. REFERENCES

Abe Y., Odaka M., Inagaki F., Lax I., Schlessinger J. and Kohda D. "Disulfide bond structure of human epidermal growth factor receptor." *Journal of Biological Chemistry* 273 (1998): 11150-11157.

Adams T.E., Epa V.C., Garrett T.P.J. and Ward C.W. "Structure of the type I insulin-like growth factor receptor." *Cellular and Molecular Life Sciences* 57 (2000):1050-1093.

Bajaj M., Waterfield M.D., Schlessinger J., Taylor W.R. and Blundell T. "On the tertiary structure of the extracellular domains of the epidermal growth factor and insulin receptors." *Biochimica et Biophysica Acta* 916 (1987): 220-226.

Banner D.W., D'Arcy A., Janes W., Gentz R., Schoenfeld H.-J., Broger C., Loetscher H. and Lessiauer W. "Crystal structure of the soluble human 55 kd TNF receptor-human TNF-β complex: implications for TNF receptor activation." *Cell* 73 (1993): 431-445.

Bilan P.J. and Yip C.C.Q. "Unusual insulin binding to cells expressing an insulin receptor mutated at cysteine 524." *Biochemical and Biophysical Research Communications* 205 (1994): 1891-1898.

Campbell I.D. and Spitzfaden C. "Building proteins with fibronectin type III modules." *Structure* 2 (1994): 333-337.

Cheatham B. and Kahn C.R. "Cysteine 647 in the insulin receptor is required for normal covalent interaction between ?- and ?-subunits and signal transduction." *Journal of Biological Chemistry* 267 (1992): 7108-7115.

Christiansen K., Tranum-Jensen J., Carlsen J. and Vinten J. "A model for the quaternary structure of human placental insulin receptor deduced from electron microscopy." *Proceedings of the National Academy of Sciences, USA* 88 (1991): 249-252.

Cuatrecasas P. "Insulin-receptor interactions in adipose tissue cells: direct measurement and properties." *Proceedings of the National Academy of Sciences, USA* 68 (1971):1264-1268.

De Vos A.M., Ultsch M. and Kossiakoff A.A. "Crystals of the complex between human growth hormone and the extracellular domain of its receptor." *Science* 255 (1992): 306-312.

Dickinson C.D., Veerapandian B., Dai X-P., Hamlin R.C., Xuong N-H., Ruoslahti E. and Ely K.R. "Crystal structure of the tenth type III cell adhesion module of human fibronectin." *Journal of Molecular Biology* 236 (1992): 1079-1092.

Downing A.K., Knott V., Werner J.M., Cardy C.M., Campbell I.D. and Handford P.A. "Solution structure of a pair of calcium-binding epidermal growth factor-like domains: implications for the Marfan syndrome and other genetic disorders." *Cell.* 85 (1996): 597-605.

Ebina Y., Ellis L., Jarnagin K., Edery M., Graf L., Clauser E., Ou J-H; Masiarz F; Kan Y.W., Goldfine I.D, Roth R.A. and Rutter W.J. "The human insulin receptor cDNA: the structural basis for hormone-activated transmembrane signalling." *Cell* 40 (1985): 747-758.

Fernandez R., Tabarini D., Azpiazu N., Frasch M. and Schlessinger J. "The *Drosophila* insulin receptor homolog - a gene essential for embryonic development encodes two receptor isoforms with different sigaling potential." *EMBO Journal* 14 (1995): 3373-3384.

Garrett T.P.J., Mckern N.M., Lou M.Z., Frenkel M.J., Bentley J.D., Lovrecz G.O., Elleman T.C., Cosgrove L.J. and Ward C.W. "Crystal structure of the first three domains of the type-1 insulin-like growth factor receptor." *Nature* 394 (1998): 395-399.

Graf R., Neuenschwander S., Brown M.R. and Ackermann U. "Insulin-mediated secretion of ecdysteroids from mosquito ovaries and molecular cloning of the insulin receptor homologue from ovaries of bloodfed *Aedes aegypti*." *Insect Molecular Biology* 6 (1997): 151-163.

Hubbard S.R., Wei L., Elis L. and Hendrickson W.A. "Crystal structure of the tyrosine kinase domain of the human insulin receptor." *Nature* 372 (1994): 746-754.

Jacobs S., Hazum, E., Shechter, Y. and Cuatrecasas, P. "Insulin receptor: covalent labeling and identification of subunits." *Proceedings of the National Academy of Sciences, USA*. 76 (1979): 4918-4921.

Kabat E.A., Wu T.T., Perry H.M., Gottesman K.S. and Foeller C. *Sequences of Proteins of Immunological Interest*. 5th edition, (1991). US Department of Health and Human Services, Bethesda, MD.

Kasuga M., Karlsson F.A. and Kahn C.R. "Insulin stimulates the phosphorylation of the 95,000-dalton subunit of its own receptor." *Science* 215 (1982): 185-187.

Kobe B. and Deisenhofer J. "Proteins with leucine-rich repeats." *Current Opinion in Structural Biology* 5 (1995): 409-416.

Leahy D.J., Hendrickson W.A., Aukhil I. and Erickson H.P. "Structure of a fibronectin type III domain from tenascin phased by MAD analysis of the selenomethionyl protein." *Science* 258 (1992): 987-991.

Lu K. and Guidotti G. "Identification of the cysteine residues involved in the class I disulfide bondsof the human insulin receptor: properties of insulin receptor monomers." *Molecular Biology of the Cell*. 7 (1996): 679-691.

Luo R.Z.T., Beniac D.R., Fernandes A., Yip C.C. and Ottensmeyer F.P. "Quaternary structure of the insulin-insulin receptor complex." *Science* 285 (1999): 1077-1080.

Macaulay S.L., Polites M., Hewish D.R. and Ward C.W. "Cysteine-524 is not the only residue involved in the formation of disulfide-bonded dimers of the insulin receptor." *Biochemical Journal* 303 (1994): 575-581.

Marino-Buslje C., Mizuguchi K., Siddle K. and Blundell T.L. "A third fibronectin type III domain in the extracellular region of the insulin receptor family." *FEBS Letters* 441 (1998): 331-336.

McDonald N.Q., Murray-Rust J., and Blundell, T.L. "The first structure of a receptor tyrosine kinase domain - a further step in understanding the molecular basis of insulin action." *Structure* 3 (1995): 1-6.

Mulhern T.D., Booker G.W., and Cosgrove L.: "A third fibronectin type-III domain in the insulin-family receptors." *Trends in Biochemical Sciences* 23 (1998): 465-466.

O'Bryan J.P., Frye R.A., Cogswell P.C., Neubauer Z., Kitch B., Prokop C., Espinosa III R.; Le Beau M.M., Earp H.S., and Liu E.T. "*axl*, a transforming gene isolated from primary human myeloid leukemia cells, encodes a novel receptor tyrosine kinase." *Molecular and Cellular Biology* 11 (1991): 5016-5031.

Ruan Y.M., Chen C., Cao Y.X. and Garofalo R.S. "The *Drosophila* insulin receptor contains a novel carboxy-terminal extension likelt to play an important role in signal transduction." *Journal of Biological Chemistry* 270 (1995): 4236-4243.

Schaefer E.M., Erickson H.P., Federwisch M., Wollmer A., and Ellis L. "Structural organization of the human insulin receptor ectodomain." *Journal of Biological Chemistry* 267 (1992): 23393-23402.

Schaffer L. and Ljungqvist L. "Identification of a disulfide bridge connecting the alpha-subunits of the extracellular domain of the insulin receptor." *Biochemical and Biophysical Research Communications* 189 (1992): 650-653.

Schäffer L. and Hansen P.H. "Partial characterization of the disulphide bridges of the soluble insulin receptor." *Experimental and Clinical Endocrinology and Diabetes*. 104 (Suppl. 2, 1996): 89.

Shier P. and Watt V.M. "Primary structure of a putative receptor for a ligand of the insulin family." *Journal of Biological Chemistry* 264 (1989): 14605-14608.

Sparrow L.G., Mckern N.M., Gorman J.J., Strike P.M., Robinson C.P., Bentley J.D., and Ward C.W. "The disulfide bonds in the C-terminal domains of the human insulin receptor ectodomain." *Journal of Biological Chemistry* 272 (1997): 29460-29467.

Steinbacher S., Seckler R., Miller S., Steipe B., Huber R., and Reinemer P. "Crystal structure of P22 tailspike protein: interdigitated subunits in a thermostable trimer." *Science* 265 (1994): 383-386.

Stetefeld J., Mayer U., Timpl R., and Huber R. "Crystal structure of three consecutive laminin-type epidermal growth factor-like (LE) modules of laminin gamma-1 chain harboring the nidogen binding site." *Journal of Molecular Biology* 257 (1996): 644-657.

Tranum-Jensen J., Christiansen K., Carlsen J., Brenzel G., and Vinten J. "Membrane topology of insulin receptors reconstituted into lipid vesicles." *Journal of Membrane Biology* 140 (1994): 215-223.

Tulloch P.A., Lawrence L.J., McKern N.M., Robinson C.P., Bentley J.D., Cosgrove L. et al. "Single-molecule imaging of human insulin receptor ectodomain and its Fab complexes." *Journal of Structural Biology* 125 (1999): 11-18.

Ullrich A., Coussens L., Hayflick J.S., Dull T.J., Gray A., Tam A.W., Lee J., Yarden Y., Libermann T.A., Schlessinger J., Downard J., Mayes E.L.V., Whittle N., Waterfield M.D., and Seeburg P.H. "Human epidermal growth factor receptor cDNA sequence and aberrant expression of the amplified gene in A431 epidermoid carcinoma cells." *Nature* 309 (1984): 418-425.

Ullrich A.; Bell J.R., Chen E.Y., Herrara R., Petruzelli L.M., Dull T.J., Gray A., Coussens L., Liao Y-C., Tsubokawa M., Mason A., Seeburg P.H., Grunfeld C., Rosen O.M., and Ramachandran J. "Human insulin receptor and its relationship to the tyrosine kinase family of oncogenes." *Nature* 313 (1985): 756-761.

Ullrich A., Gray A., Tam A.W., Yang-Feng T., Tsubokawa M., Collins C., Henzel W., Le Bon T., Kathuria S., Chen E., Jacobs S., Francke U., Ramachandran J., and Fujita-Yamaguchi Y. "Insulin-like growth factor 1 receptor primary structure: comparison with insulin receptor suggests structural determinants that define functional specificity." *EMBO Journal* 5 (1986): 2503-2512.

Ward C.W. "Members of the insulin receptor family contain three fibronectin type III domains." *Growth Factors* 16 (1999): 315-322.

Ward C.W.; Hoyne P.A. and Flegg R.H. "Insulin and epidermal growth factor receptors contain the cysteine repeat motif found in the tumor necrosis factor receptor." *Proteins: Structure Function and Genetics*. 22 (1995): 141-153.

Woldin C.N., Hing F.S., Lee J., Pilch P.F., and Shipley G.G. "Structural studies of the detergent-solubilized and vesicle-reconstituted insulin receptor." *Journal of Biological Chemistry* 274 (1999): 34981-34992.

Yoder M.D., Lietzke S.E., and Jurnak F. "Unusual structural features in the parallel beta-helix in pectate lyases." *Structure* 1 (1993): 241-251.

C. KRISTENSEN AND A.S. ANDERSEN

INSULIN INTERACTION WITH MINIMIZED RECEPTORS AND BINDING PROTEINS

Abstract. In order to characterize regions of the insulin receptor that are essential for ligand binding we have minimized the ligand-binding domain of IR by site directed mutagenesis. The smallest receptor fragment identified comprised the first three domains of IR (1-468) fused to 16 amino acids from the C-terminal of the α-subunit. The mass of this receptor fragment was 70 kD, and the affinity for insulin was 5 nM, which is similar to what was found for the full length soluble ectodomain of the insulin receptor. A similarly minimized IGF-I receptor construct (mIGF-IR) was shown to bind IGF-I with an affinity of 1.5 nM.
The role of the C-terminal peptide was further investigated by characterizing chimeric mini-receptor constructs. The carboxy terminal domain of the insulin receptor related receptor (IRRR) was found to abolish binding in IR and IGF-IR context, whereas swapping the carboxy terminal domains in either mIR or mIGF-IR context only cause minor changes in affinities, demonstrating that the carboxy terminal of IR and IGF-IR α-subunits are interchangable suggesting that this domain is part of the common binding site. The last part of this chapter describes a novel high affinity insulin binding protein (IBP) secreted from cells from at least three insects. This IBP is composed of two Ig-like C2-domains, has a molecular weight of 27 kDa, binds human insulin and related peptides with high affinity (10-200 pM) and inhibits insulin signaling through the insulin receptor. The ligand binding profile suggests that IBP recognizes a region that is highly conserved in the insulin superfamily but distinct from the classical insulin receptor-binding site.

1. INTRODUCTION

The insulin superfamily is an ancient group of structurally homologous proteins, including IGF-I, IGF-II, relaxin, bombyxin and related peptides (Steiner and Chan, 1988; Murray-Rust et al., 1992). Insulin related peptides have been identified in both vertebrates and several invertebrates, including insects (Ishizaki and Suzuki, 1988; Gregoire et al., 1998; Duret et al., 1998; Smit et al., 1993). The evolution of the insulin hormone family has been reviewed recently by Chan and Steiner (2000). Many of these hormones have been shown to act through receptors structurally similar to the insulin receptor, being heterotetrameric, membrane-spanning tyrosine kinases (De Meyts et al., 1994; Fullbright et al., 1997).

The amino acid sequences of the three human proteins belonging to the insulin receptor family, IR, IGF-IR and IRRR, are approximately 50% identical (Ullrich et al., 1985; Ebina et al., 1985; Shier and Watt, 1989) and they are thought to have the same overall structure and conformation. The structure of the insulin and IGF-I receptors has been investigated extensively and reviewed (Lee and Pilch, 1994; Adams et al., 2000). Recently, a crystal structure of the first three domains of the IGF-I receptor (residues 1-462) has been solved (Garrett et al., 1998) and this confirmed the L1-Cys-L2 domain structure predicted from alignment with epidermal

M. Federwisch, M. Leyck Dieken and P. De Meyts (eds.), Insulin & Related Proteins, 165–176.

growth factor receptor sequences (Bajaj et al., 1987; Ward et al., 1995). However, this IGF-IR fragment does not bind ligand, so only indirect information can be used to identify important structural features necessary for ligand binding.

Modulators of these hormone/receptor systems have been identified, as the IGFs are known to associate with high affinity to a family of circulating binding proteins (IGFBPs) that influence IGF availability and activity. In addition there have been reports of proteins with homology to the IGFBP family that supposedly bind and possibly modulate the activity of insulin. The IGFBP family and related proteins have been reviewed by Hwa et al. (1999).

2. MINIMIZING THE INSULIN RECEPTOR

2.1. Mini-IR Constructs and Binding Affinities.

To characterize the domains of the IR that are essential for ligand binding, we took an indirect approach by investigating which domains could be deleted from the IR without compromising insulin binding. Using site directed mutagenesis, a number of deletions were made in the IR region 469-703 (Figure 1, Table 1). The constructs were stably expressed in baby hamster kidney (BHK) cells as fusion protein of IR ectodomain and Fc, the constant region of IgG heavy chain (Bass et al., 1996).

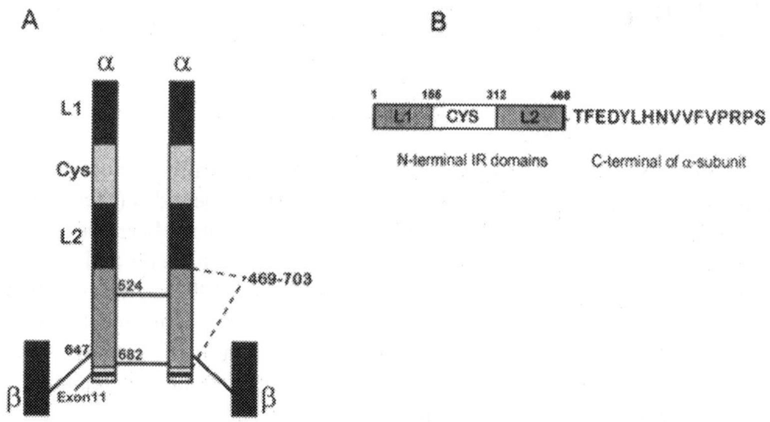

Figure 1. Panel A. ***Major structural features of the insulin receptor α-subunit****. Shown is the insulin receptor ectodomain.* ***L1*** *and* ***L2*** *are homologous domains that contain EGF like regions (see Bajaj et al. (Bajaj et al., 1987));* ***Cys*** *is the cysteine rich domain; and* ***Exon11*** *is the alternatively spliced region (Yamaguchi et al., 1993). The 469-703 region deleted in the present study is indicated to the right. Panel B. The minimized IR. The IRΔ703 protein is 70 kD and consists of the first three domains of the IR (****L1****,* ***Cys****, and* ***L2****) and 16 aminoacids from the C-terminal of the α-subunit*

The full-length fusion protein IRwt has been shown to bind insulin with biphasic binding curves (Kristensen et al., 1997), comprising a pM component and a nanomolar component. In contrast to the biphasic binding curves obtained with the intact IRwt receptor all deletion constructs yielded one-site binding curves with nM affinity (Table 1), similar to what is found for the intact IR ectodomain secreted from BHK cells (sIR), indicating that the minimized IR binds insulin in a 'sIR mode', nM affinity.

Table 1. Insulin affinity of receptor deletion constructs

Construct	*Deletion (amino acids)*	*Affinity (K_d) (nM)*
IRwt	-	3.1 ± 0.9
IRΔ599	487-599	2.3 ± 0.6
IRΔ613	487-613	2.5 ± 0.3
IRΔ629	487-629	1.8 ± 0.1
IRΔ649	469-649	1.9 ± 1.0
IRΔ673	469-673	1.1 ± 0.2
IRΔ685	469-685	6.4 ± 2.4
IRΔ685*	469-685,718-729	7.3 ± 1.6
IRΔ703	469-703,718-729	4.4 ± 0.8
sIR	-	5.0 ± 0.2

2.2. Mini-IR. - Discussion

The receptor region investigated here (amino acids 469-685) contains the cysteines that are responsible for α-α and α-β disulfides in the β-α-α-β subunit composition (Figure 1) (Schäffer and Ljungqvist, 1992; Sparrow et al., 1997). The importance of this region for subunit interactions is clearly evident from the stepwise decomposition of the receptor that was observed using the deletion strategy (Kristensen et al., 1998) and this supports the disulfide contacts suggested by Sparrow at al. (Sparrow et al., 1997) (Figure 1).

In the mini-IR we have deleted 234 amino acids in the central part of the IR α-subunit. The fact that this large domain can be deleted without compromising insulin binding indicates that the N-terminal and C-terminals that are known to be important for binding of insulin (reviewed by Adams et al., 2000) must remain in close contact in the IR structure. The deleted region (469-703) is apparently not essential for the insulin binding region and what actually keeps the C-terminal 16 amino acids of the 70 kD receptor fragment (Figure 6) in close contact to the N-terminal 468 amino acids is not clear. Using alanine scanning mutagenesis Mynarcik et al. (1996) found that 7 alanine mutations in the 704-716 region were deleterious to insulin binding (loss of more than 100 fold on K_d). Clearly all of these effects cannot be direct effects on ligand-receptor interaction, it may be that some of these residues are essential for docking the C-terminal of the α-subunit in close contact to regions N-terminal of amino acid 468 i.e. maintaining the insulin binding pocket intact.

Attempts to obtain smaller IR fragments that bind insulin have not been met with much success, only the full length ectodomain has been expressed and purified in substantial quantities (Andersen et al., 1990; Markussen et al., 1991). Given the size and complexity of the IR ectodomain protein, the identification of a minimal ligand-receptor complex is expected to facilitate crystallization and detailed structural analysis by X-ray crystallography. The mini-IR demonstrates that we can minimize the ligand binding domain of the insulin receptor by deleting regions of the IR α-subunit, and thus obtain a 70 kD monomeric receptor fragment that binds insulin with the same affinity as the soluble IR ectodomain.

3. THE MINI-IGF-IR AND CHIMERIC MINI-RECEPTORS

3.1. Chimeric Mini-receptor Constructs and Binding of Ligands

To elucidate the role of the C-terminal domain of the α-subunit for ligand binding specificity in the insulin receptor family we decided to make a mini-IGF-IR (mIGF-IR) as well as chimera between mIR and mIGF-IR also including constructs with the C-terminal region of IRRR. Except for the mini-IR (IRΔ703) described above all constructs were expressed as FLAG peptide fusions allowing detection and purification of all constructs using anti-FLAG antibody. A schematic representation is shown in Figure 2 and all constructs are described in Table 2.

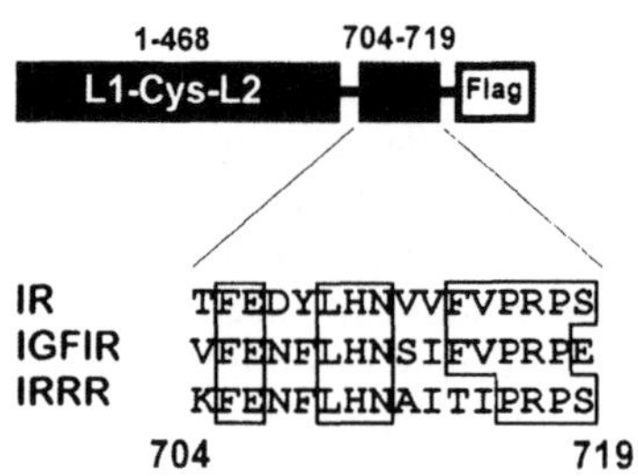

Figure 2. ***Schematic representation of the mini-receptor.*** *Below is shown an alignment of the three C-terminal domains that have been used. For more details see Kristensen et al. (Kristensen et al., 1999)*

All receptor constructs were purified using an anti-FLAG column. After purification, immunoblotting showed that all constructs were efficiently expressed and purified (Kristensen et al., 1999). When setting up binding assays with the constructs a first conclusion was that there is no detectable binding with any of the constructs with the C-terminal domain from IRRR whereas all other constructs bound ligands (Table 2).

The binding data show that the minimized IGF-IR construct bound IGF-I with high affinity (1.5 nM), which is approximately 4 fold lower affinity than found for the soluble ectodomain of IGF-IR (sIGF-IR) (0.4 nM) (Andersen et al., 1990). In terms of binding cognate ligand, the minimized IR more faithfully reflected the binding characteristics of the ectodomain; mIR and IRΔ703 have retained the full binding activity of the IR ectodomain (sIR) all having affinities of 5-6 nM for insulin. Although mIR (and IRΔ703) have unchanged affinity for insulin they have higher affinity for IGF-I compared to sIR, resulting in the mini-IR's having almost same affinities for insulin and IGF-I, the affinity for IGF-I being only 3-6 fold lower than affinity for insulin (Table 2). In contrast the mIGF-IR remained highly specific for IGF-I compared to insulin (>1000 fold).

Table 2. Chimeric constructs - ligand binding affinities

	Domain structure		Binding affinities		
Construct	Amino terminal 1-468	Carboxy terminal 704-719	Insulin (nM)	IGF-I (nM)	ICP (nM)
mIR	IR	IR	5.8	15	3.6
IRΔ703	IR	IR	5.9	35	5.3
mIR.IGF-IR	IR	IGF-IR	12	7.2	8.1
mIR.IRRR	IR	IRRR	-	-	-
mIGF-IR	IGF-IR	IGF-IR	3100	1.5	9.8
mIGF-IR.IR	IGF-IR	IR	3800	3.4	6.2
mIGF-IR.IRRR	IGF-IR	IRRR	-	-	-

3.2. Chimeric Mini-receptor Constructs. - Discussion.

The minimized IGF-IR construct is particularly interesting because the X-ray structure of the first three domains of IGF-IR was solved recently (Garrett et al., 1998). In the X-ray structure the C-terminal of the L2 domain does not face the putative ligand binding lobe whereas our data show that only 16 amino-acid fused to this C-terminal induces binding of IGF-I. One possible explanation may be that the L2 domain is rotated so that the C-terminal of L2 is closer to L1 when ligand is bound.

The low specificity of the minimized insulin receptors is puzzling, we speculate that since the sIR have all the epitopes required for binding IGF-I with 20 nM affinity (like mIR and IRΔ703) there must be something inhibiting receptor-IGF-I interaction in sIR. Possibly some of the regions deleted in the minimized IR's interfere with IGF-I binding in sIR or maybe the dimeric conformation of sIR interferes with IGF-I attaining the higher affinity.

Despite that the primary sequence of IRRR is as similar to IR and IGF-IR as the IR and IGF-IR are to each other (Shier and Watt, 1989) no binding of insulin, IGF-I or any other ligand has been observed with the IRRR. In the chimeric receptor study we investigated the role of the short C-terminal domain of IRRR in mIR and mIGF-IR context. The clear conclusion from these chimeric receptors is that the C-terminal of IRRR abolishes binding of insulin and IGF-I. This is somewhat surprising because only four of the residues in the C-terminal region (IR residues 704-719) are different from corresponding positions of both IR and IGF-IR (Figure 2). One of these four residues, alanine 694, has been shown by alanine scanning mutagenesis to be acceptable at the corresponding position 712 in IR (Mynarcik et al., 1996) and because alanine mutations at two of the other positions (Mynarcik et al., 1997) did not influence binding of IGF-I to IGF-IR, it seems that Phe714 in IR is the most important difference between IR and IRRR in this region.

The chimeric mini-receptors show that replacing the C-terminal domain of IR with corresponding domain of IGF-IR only has minor effect on ligand binding. This is also true for mIGF-IR with C-terminal of IR, so within IR and IGF-IR the C-terminal domains are interchangable in terms of ligand binding suggesting that this domain is part of the common binding site of these receptors.

We have previously characterized the hybrid hormone ICP, which consists of insulin with IGF-I C-domain, against various IR and IGF-IR. The conclusion was that ICP binds with high affinity to IR (55-113% of insulin) and with 19-28% of IGF-I affinity to IGF-IR (Kristensen et al., 1995). Here we get similar results for relative binding affinities for mini-receptors, the affinity of ICP for mIR and IRΔ703 was not different from affinity of insulin. For mIGF-IR the affinity of ICP was 16% of IGF-I, and thus previous conclusions that the C-domain of IGF-I is a major determinant of IGF-IR specificity for IGF-I is true also for the minimized IGF-IR.

4. THE NOVEL INSULIN BINDING PROTEINS.

4.1. The Discovery of Insulin Binding Protein (IBP) from S. frugiperda

The *Sf9* insect cell expression system has been employed extensively for production of recombinant proteins, and has proved to be a good tool for expressing various insulin receptor constructs (Sissom and Ellis, 1989; Paul et al., 1990). When trying to express insulin receptor fragments in *Sf9* insect cells using a baculovirus system we discovered insulin-binding activity in medium from untransfected *Sf9* cells. Chemical crosslinking of labeled insulin to the proteins from the supernatant of *Sf9* cells showed that a protein of approximately 27 kD specifically crosslinked the labeled insulin. We were able to track down this binding activity and it turned out to be a novel secreted 27 kDa protein that binds proteins belonging to the insulin superfamily and inhibits insulin activation of the insulin receptor (Andersen et al.,

2000). These binding proteins are not structurally related to the IGF binding proteins and we called them IBP's (for Insulin related peptide Binding Protein).

4.2. Cloning of IBP's from Insects

For cloning the IBP we first purified the protein from SF9 cell culture medium. Due to the high affinity for insulin the protein was efficiently purified on an insulin-mini-leak column. The purified protein was digested proteolytically and amino acid sequencing of resulting fragments gave short amino-acid sequences that allowed us

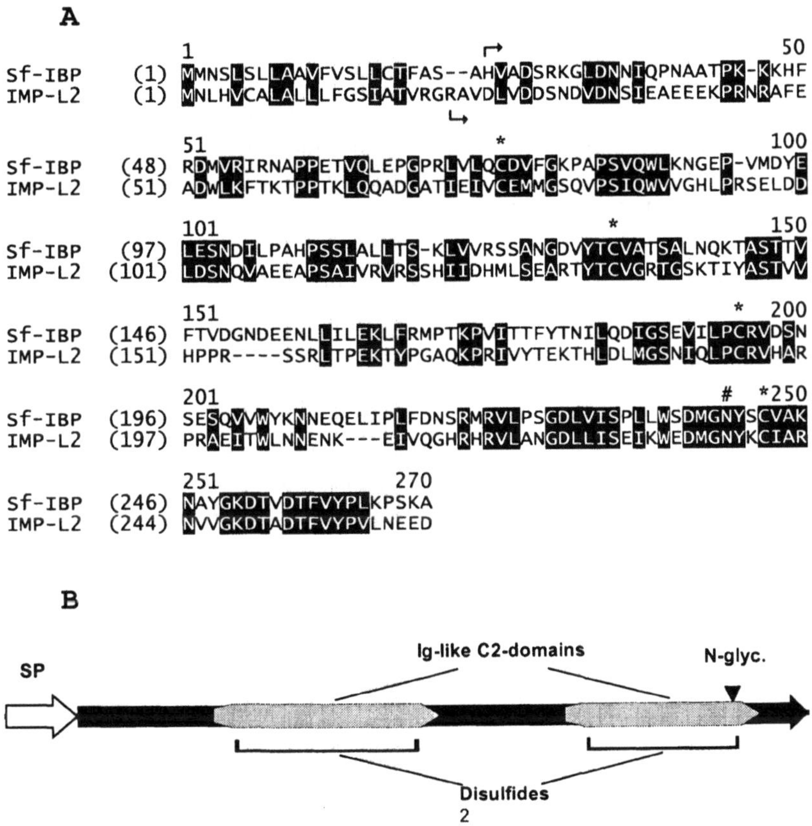

Figure 3. Amino acid sequence comparison of Sf-IBP and IMP-L2.
A: Conserved residues and conservative substitutions are in reversed contrast. The signal peptide cleavage sites are marked with arrows and the four cysteines of the Ig domains are indicated by asterisk, the N-linked glycosylation site is denoted by #.
B: Schematic representation of the domain structure of Sf-IBP and IMP-L2.

to PCR clone the full length cDNA sequence. The protein was named *Sf*-IBP (*S*podoptera *f*rugiperda *I*nsulin related peptide *B*inding *P*rotein), and is composed of two IgG-like C2 domains (Figure 3). Searching for similar proteins in the Non-Redundant GenBank^{TM} database revealed that IMP-L2, from *drosophila* had the highest homology (Figure 3). The IMP-L2 was expressed in BHK cells and was found to bind insulin. Many other homologous proteins are found in the data bank including a number of cell surface receptors, but these are all probably only structurally related due to the Ig-like C2-domains, and not necessarily functionally related.

In addition to Spodoptera frugiperda and Drosophila we also found insulin-binding activity in medium from the HI5 cell line, which is derived from the cabbage looper, Trichoplusia Ni. The initial characterization clearly suggests that the binding activity from the Trichoplusia Ni cells involves a binding protein homologous to Sf-IBP and IMP-L2. We have called the Tricoplusia Ni binding protein Tn-IBP. With the identification of homologous insulin binding proteins from three different insect organisms, we suggest that insulin binding proteins containing Ig-like C2 domains may be present in all insects.

4.3. Characterization of the IBP's. – Binding of Insulin and Related Peptides

The binding data show that Sf-IBP and Tn-IBP have very high affinity for insulin (lower pM range), whereas IMP-L2 has an affinity of 80 nM for insulin. When looking at the other insulin-like peptides including the naturally occurring human peptides IGF-I, IGF-II and proinsulin as well as the synthetic insulin analogs, the overall picture is that these ligands all have affinities close to insulin affinity for the respective binding protein. But there are a few dramatic deviations from this general pattern, most strikingly is it that Tn-IBP has approximately 4000 fold poorer affinity for IGF-I and IGF-II than for insulin. The other exception is that insulin analogs with the B10Asp mutation (X92 and X10) have lost 10-100 fold in binding affinity.

The synthetic insulin analogs selected for initial characterization of the new insulin binding protein family are characterized by aberrant behavior in binding to the human insulin receptor. The X92 analog is a super potent receptor binder (Schäffer, 1994), but for the IBP's the X92 had lower affinity than insulin and the X10 analog demonstrates that the poor affinity is due to the B10Asp mutation.

The deleterious effect of the B10 Asp mutation on Sf-IBP binding and the fact that des-octapeptide insulin (X145) bind with even higher affinity than insulin clearly suggests that Sf-IBP recognize a region of insulin that is highly conserved in the insulin superfamily but distinct from the classical receptor binding site (which involves among others the C-terminal B23-B26 region of the insulin B-chain) (Kristensen et al., 1997).

When allowed to compete for insulin in an insulin receptor autophosphorylation assay, Sf-IBP acted as an inhibitor of insulin action. This observation points to a function of Sf-IBP and related proteins in inhibiting or modulating the effect of a putative insect insulin related peptide.

The drosophila homolog IMP-L2 has previously been studied by Fristrom and co-workers (Osterbur et al., 1988; Garbe et al., 1993). IMP-L2 is being induced by the molting hormone 20-hydroxyecdysone and implicated in epithelial spreading and fusion. Garbe et al. found expression of IMP-L2 in neuronal cells, and genetic analysis demonstrated that IMP-L2 is an essential gene in Drosophila as disruptions of this gene results in embryonic lethality (Garbe et al., 1993). However no morphological abnormalities were evident in the embryos and the authors concluded that they had not identified the cause of lethality in IMP-L2 null Drosophila. Our characterization of IMP-L2 suggests that this protein may be involved in regulating the availability or activity of a *Drosophila* insulin-like peptide, and that lack of IMP-L2 causes a defective signaling through the *Drosophila* insulin receptor that in turn leads to the observed lethal phenotype.

Table 3. Ligand binding data for Sf-IBP, IMP-L2 and Tn-IBP.

Ligand	**Substitutions**	***S. frugiperda*** ***Sf*-IBP** **(nM)**	***Drosophila*** **IMP-L2** **(nM)**	*T. ni* ***Tn*-IBP** **(nM)**
Insulin	-	**0.07**	**81**	**0.02**
Proinsulin	-	**0.02**	**87**	**<0.01**
IGF-I	-	**0.17**	**17**	**45**
IGF-II	-	**0.37**	**42**	**194**
Mini-proinsulin	B29-A1 peptide link	**0.04**	**218**	**0.03**
X92	A8H,B10D,B25Yamide	**1.7**	**1270**	**0.15**
X137	B25Y-amide	**0.01**	**94**	**0.01**
X10	B10D	**11**	**2650**	**1.1**
X145	des(B23-B30)	**<0.01**	**17**	**<0.01**
ICP	Insulin w/ IGF-I C-domain	**0.04**	**77**	**0.01**

We have searched the NCBI database for other *Sf*-IBP and IMP-L2 homologues but so far we have only identified three open reading frames from C. elegans that encode secreted proteins composed of two Ig-like C2 domains and a predicted molecular weight around 27 kDa. We have expressed one of these proteins but have not been able to detect insulin binding to this (data not shown). Thus, the possibility remains that this family of IBPs is limited to insects, and serve a function that is specific to this class. If a human homologue should exist that is involved in binding proteins belonging to the insulin superfamily, such a protein could be involved in certain disease states (e.g. diabetes or cancer) and as such would represent an obvious and attractive target for drug development.

6. CONCLUDING REMARKS

These are exiting years in terms of the structure function of interaction between the hormones and receptors in the insulin superfamily. The first crystal structure of part of the extracellular domain of the IGF-IR is available, and here we present minimized ligand binding domains that may be a first step towards a crystal structure of insulin in complex with a receptor fragment.

Another small protein that binds insulin with high affinity is the binding protein found in insects. The initial structure function studies of the interaction between insulin and these binding proteins indicate that the binding mechanisms are completely different from those known from the interaction with the receptor, studies are ongoing to further elucidate these interactions.

7. REFERENCES

Andersen A.S., Hansen P.H., Schaffer L. and Kristensen C. "Structure and function of the type 1 insulin-like growth factor receptor." *Cell. Mol. Life Sci.* 57 (2000): 1050-1093.

Andersen A.S., Hansen P.H., Schaffer L. and Kristensen C. "A new secreted insect protein belonging to the immunoglobulin superfamily binds insulin and related peptides and inhibits their activities." *J. Biol. Chem.* 275 (2000): 16948-16953.

Andersen A.S., Kjeldsen T., Wiberg F.C., Christensen P.M., Rasmussen J.S., Norris, ., Møller K.B. and Møller N.P.H. "Changing the Insulin Receptor To Posses Insulin-like Growth Factor I Ligand Specificity." *Biochemistry* 29 (1990): 7363-7366.

Bajaj M., Waterfield M.D., Schlessinger J., Taylor W.R. and Blundell T. "On the tertiary structure of the extracellular domains of the epidermal growth factor and insulin receptors." *Biochem. Biophys. Acta* 916 (1987): 220-226.

Bass J., Kurose T., Pashmforoush M. and Steiner D.F. "Fusion of Insulin Receptor Ectodomains to immunoglobin Constant Domains Reproduces High-affinity Insulin Binding *in vitro*." *J. Biol. Chem.* 271 (1996): 19367-19375.

Chan S.J. and Steiner D.F. "Insulin through the ages: Phylogeny of a growth promoting and metabolic regulatory hormone." *American Zoologist* 40 (2000): 213-222.

De Meyts P., Wallach B., Christoffersen C.T., Ursø B., Grønskov K., Latus L.J., Yakushiji F., Ilondo M.M. and Shymko R.M. "The Insulin-Like Growth Factor-I Receptor." *Horm. Res.* 42 (1994): 152-169.

Duret L., Guex N., Peitsch M.C. and Bairoch A. "New insulin-like proteins with atypical disulfide bond pattern characterized in Caenorhabditis elegans by comparative sequence analysis and homology modeling." *Genome Res.* 8 (1998): 348-353.

Ebina Y., Ellis L., Jarnagin K., Edery M., Graf L., Clauser E., Ou J.-H., Masiarz F., Kann Y.W., Goldfine I.D., Roth R.A. and Rutter W.J. "The human Insulin Receptor cDNA: The Structural Basis for Hormone-Activated Transmembrane Signalling." *Cell* 40 (1985): 747-758.

Fullbright G., Lacy E.R. and Bullesbach E.E. "The prothoracicotropic hormone bombyxin has specific receptors on insect ovarian cells." *Eur. J. Biochem.* 245 (1997): 774-780.

Garbe J.C., Yang E. and Fristrom J.W. "Imp-l2 - an essential secreted immunoglobulin family member implicated in neural and ectodermal development in drosophila." *Development* 119 (1993): 1237-1250.

Garrett T.P., McKern N.M., Lou M.Z., Frenkel M.J., Bentley J.D., Lovrecz G.O., Elleman T.C., Cosgrove L.J. and Ward C.W. "Crystal-structure of the first 3 domains of the type-1 insulin-like-growth-factor receptor." *Nature* 394 (1998): 395-399.

Gregoire F.M., Chomiki N., Kachinskas D. and Warden C.H. "Cloning and developmental regulation of a novel member of the insulin-like gene family in Caenorhabditis elegans." *Biochem. Biophys. Res. Commun.* 249 (1998): 385-390.

Hwa V., Oh Y. and Rosenfeld R.G. "The insulin-like growth factor-binding protein (IGFBP) superfamily." *Endocr. Rev.* 20 (1999): 761-787.

Ishizaki H. and Suzuki A. "An insect brain peptide as a member of insulin family." *Horm. Metab. Res.* 20 (1988): 426-429.

Kristensen C., Andersen A.S., Hach M., Wiberg F.C., Schäffer L. and Kjeldsen T. "A single-chain insulin-like growth factor I/insulin hybrid binds with high affinity to the insulin receptor." *Biochem. J.* 305 (1995): 981-986.

Kristensen C., Kjeldsen T., Wiberg F.C., Schäffer L., Hach M., Havelund S., Bass J., Steiner D.F. and Andersen A.S. "Alanine Scanning Mutagenesis of Insulin." *J. Biol. Chem.* 272 (1997): 12978-12983.

Kristensen C., Wiberg F.C. and Andersen A.S. "Specificity of insulin and insulin-like growth factor I receptors investigated using chimeric mini-receptors. Role of C-terminal of receptor alpha subunit." *J. Biol. Chem.* 274 (1999): 37351-37356.

Kristensen C., Wiberg F.C., Schäffer L. and Andersen A.S. "Expression and Characterization of a 70 kD Fragment of the Insulin Receptor that Binds Insulin. Minimizing ligand binding domain of the insulin receptor." *J. Biol. Chem.* 273 (1998): 17780-17786.

Lee J. and Pilch P.F. "The insulin receptor: structure, function, and signaling." *Am. J. Physiol.* 266 (1994): C319-C334.

Markussen J., Halstrøm J., Wiberg F.C. and Schäffer L. "Immobilized Insulin for High Capacity Chromatography of Insulin Receptors." *J. Biol. Chem.* 266 (1991): 18814-18818.

Murray-Rust J., McLeod A.N., Blundell T.L. and Wood S.P. "Structure and evolution of insulins: implications for receptor binding." *Bioessays* 14 (1992): 325-331.

Mynarcik D.C., Williams P.F., Schäffer L., Yu G.Q. and Whittaker J. "Identification of Common Ligand Binding Determinants of the Insulin and Insulin-like Growth Factor 1 Receptors." *J. Biol. Chem.* 272 (1997): 18650-18655.

Mynarcik D.C., Yu G.Q. and Whittaker J. "Alanine-scanning mutagenesis of a c-terminal ligand-binding domain of the insulin-receptor alpha-subunit." *J. Biol. Chem.* 271 (1996): 2439-2442.

Osterbur D.L., Fristrom D.K., Natzle J.E., Tojo S.J. and Fristrom J.W. "Genes expressed during imaginal discs morphogenesis: IMP-L2, a gene expressed during imaginal disc and imaginal histoblast morphogenesis." *Dev. Biol.* 129 (1988): 439-448.

Paul J.I., Tavare J., Denton R.M. and Steiner D.F. "Baculovirus-directed expression of the human insulin receptor and an insulin-binding ectodomain." published erratum appears in *J. Biol. Chem.* 1990 Nov 15;265(32):20051. *J. Biol. Chem.* 265 (1990): 13074-13083.

Schäffer L. "A model for insulin binding to the insulin-receptor." *Eur. J. Biochem.* 221 (1994): 1127-1132.

Schäffer L. and Ljungqvist L. "Identification of a Disulfide Bridge Connecting the a-Subunits of the Extracellular Domain of the Insulin Receptor." *Biochem. Biophys. Res. Comm.* 189 (1992): 650-653.

Shier P. and Watt V.M. "Primary structure of a putative receptor for a ligand of the insulin family." *J. Biol. Chem.* 264 (1989): 14605-14608.

Sissom J. and Ellis L. "Secretion of the extracellular domain of the human insulin receptor from insect cells by use of a baculovirus vector." *Biochem. J.* 261 (1989): 119-126.

Smit A.B., van Marle A., van Elk R., Bogerd J., van Heerikhuizen H. and Geraerts W.P. "Evolutionary conservation of the insulin gene structure in invertebrates: cloning of the gene encoding molluscan insulin- related peptide III from Lymnaea stagnalis." *J. Mol. Endocrinol.* 11 (1993): 103-113.

Sparrow L.G., McKern N.M., Gorman J.J., Strike P.M., Robinson C.P., Bentley J.D. and Ward C.W. "The Disulfide Bonds in the C-terminal Domains of the Human Insulin Receptor Ectodomain." *J. Biol. Chem.* 272 (1997): 29460-29467.

Steiner D.F. and Chan S.J. "An overview of insulin evolution." *Horm. Metab. Res.* 20 (1988): 443-444.

Ullrich A., Bell J.R., Chen E.Y., Herrera R., Petruzzelli L.M., Dull T.J., Gray A., Coussens L., Liao Y.C., Tsubokawa M., Mason A., Seeburg T.J., Grunfeld C., Rosen O.M. and Ramachandran J. "Human insulin receptor and its relationsship to the tyrosine kinase family of oncogenes." *Nature* 313 (1985): 756-761.

Ward C.W., Hoyne P.A. and Flegg R.H. "Insulin and Epidermal Growth Factor Receptors Contain the Cysteine Repeat Motif Found in the Tumor Necrosis Factor Receptor." *Proteins: Structure, Function and Genetics* 22 (1995): 141-153.

Yamaguchi Y., Flier J.S., Benecke H., Ransil B.J., and Moller D.E. "Ligand-Binding Properties of the Two Isoforms of the Human Insulin Receptor." *Endocrinol.* 132 (1993): 1132-1138.

L. PLUM, K. GIESEN, R. KLUGE, H.-G. JOOST

IDENTIFICATION OF SUSCEPTIBILITY LOCI FOR OBESITY, INSULIN RESISTANCE, AND HYPERGLYCEMIA IN A BACKCROSS MODEL OF NEW ZEALAND OBESE (NZO) AND LEAN SJL MICE

Abstract: New Zealand Obese (NZO) mice exhibit several characteristics that resemble the human metabolic syndrome or syndrome X. A backcross model of New Zealand Obese (NZO) mice with the lean SJL strain was established in order to locate genes responsible for obesity, insulin resistance, and type 2 diabetes-like hyperglycemia. In male NZOxF1 backcross mice, a major susceptibility locus for the development of severe hyperglycemia (*Nidd/SJL*) was identified on chromosome 4. The diabetogenic allele appeared to be contributed by the SJL genome and was responsible for approximately 60% of the total prevalence of diabetes in the total male backcross population. In female NZOxF1 animals, a significant quantitative trait locus for parameters of obesity (body weight, body mass index, total body fat) and insulin resistance (hyperinsulinemia) (*Nob1*) was detected on chromosome 5. The aberrant allele was presumably contributed by the NZO genome. Phenotypic effects of this locus were also observed in the male population. *Nidd/SJL* and *Nob1* showed additional effects on the development of insulin resistance and diabetes and were responsible for a major part of obesity, hyperinsulinemia and hyperglycemia in the backcross population.

1. INTRODUCTION

New Zealand Obese (NZO) mice exhibit a polygenic syndrome of obesity, insulin resistance, and hypertension (Bielschowsky and Bielschowsky, 1953; Herberg and Coleman, 1977; Hunt et al., 1976; Igel et al., 1997; Ortlepp et al., 2000) that resembles the human syndrome X (Reaven, 1993). Thus, the NZO strain is an ideal animal model for the identification of genes that are responsible for impaired body weight regulation and insulin action. It has previously been suggested that leptin resistance may be a primary cause of the obesity in NZO mice, and a leptin receptor variant with 4 amino acid exchanges including two non-conservative substitutions (A720T, T1044I; $Lepr^{A720T/T1044I}$) was found in the NZO strain (Figure 1) (Igel et al., 1997; Halaas et al., 1997). However, the contribution of this leptin receptor variant to the insulin resistance syndrome of the NZO mouse remains unclear, since the allele is also present in the related New Zealand Black (NZB) strain that shows neither obesity nor insulin resistance.

M. Federwisch, M. Leyck Dieken and P. De Meyts (eds.), Insulin & Related Proteins, 177–188.

The NZO strain has previously been used for the characterization of interactions of diabetes and obesity genes ('diabesity') in mice (Leiter et al. 1998). However, the genes predisposing for insulin resistance, islet cell failure and consecutive hyperglycemia are for the most part still unknown. In order to assess a possible contribution of the $Lepr^{A720T/T1044I}$ variant to the insulin resistance syndrome of the NZO mouse, and in order to identify other susceptibility loci for morbid obesity and hyperglycemia, a backcross model of NZO mice with the lean SJL strain was established. This review summarizes the data obtained with this backcross that lead to the identification of two major susceptibility loci (Plum et al., 2000; Kluge et al., 2000).

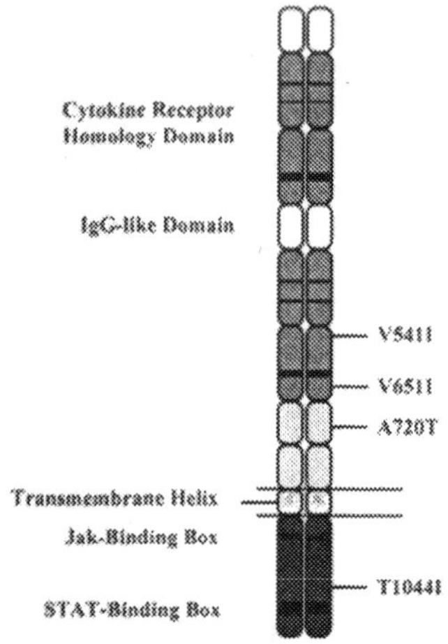

Figure 1. The leptin receptor of the NZO strain.

2. RESULTS

2.1. A Backcross Model of NZO and SJL Mice

SJL mice were chosen as the control strain because they show low percentages of carcass lipid and are resistant to the development of atherosclerosis, even when fed a high fat diet (Paigen, 1995; Festing, 1997).

Female SJL mice and male NZO mice were mated in order to produce an F1 generation. NZO-backcrosses were generated by mating NZO females with F1 males and also by mating NZO males with F1 females. After weaning at three weeks of age, backcross mice were fed a high fat diet, containing 16% fat, 46.8% carbohydrates and 17.1% protein (15.4kJ/g ME), and were kept under standardized climatic and circadian conditions Plum et al., 2000). Parental strains and F1 were treated in the same way in order to characterize the parental phenotypes. Body weight was determined once a week during the study period. All other phenotypic parameters (body length, total body fat, blood glucose, serum parameters) were measured when the mice were killed at the age of 22 weeks.

Characterization of the parental strains and the F1 progeny with regard to body weights, blood glucose, and serum parameters showed that NZO mice developed a more severe form of the syndrome than F1 animals (Table 1A). However, it is striking that male and female NZO mice exhibited considerably different abnormalities of glucose homeostasis. By the age of 22 weeks, male NZO mice showed severe hyperglycemia correlated with low serum insulin levels (low insulin secretion). Therefore, the inability to compensate for the increasing insulin resistance appears to be due to a loss of beta cell function. Note that male NZO mice did not gain weight between weeks 12 and 22 because of secondary weight loss due to their diabetes. In contrast, NZO females maintained a state of compensatory hyperinsulinemia, and decompensation of glucose tolerance was not observed in NZO females by the age of 22 weeks (Table 1A).

NZOxF1 backcross mice suffered from the syndrome almost to the same extent as parental NZO mice. Phenotypic parameters of this backcross population are given in Table 1B. Male backcross mice developed decompensated hyperglycemia and secondary weight loss almost as early as NZO males. Interestingly, some male backcross mice maintained normoglycemia, reflecting the great heterogeneity of the phenotype in the backcross population. These data also suggested that susceptibility loci for diabetes genes can be detected in the male NZOxF1 backcross mice, whereas genes responsible for obesity and insulin resistance (hyperinsulinemia) can be identified in the female backcross population.

Table 1. Body weights and serum parameters in parental (SJL, NZO), F1, and NZOxF1 backcross mice.

Males	A SJL	F1 (SJLxNZO)	NZO	B NZOxF1
Body weight wk.12 (g)	23.3 ± 1.5	32.7 ± 2.7***	50.3 ± 2.9***	51.1 ± 6.7
Body weight wk.22 (g)	27.4 ± 2.2	45.8 ± 3.3***	50.2 ± 9.4***	61.8 ± 11.5
BMI wk.22 (g/cm²) (1)	0.290 ± 0.01	0.377 ± 0.02***	0.391 ± 0.06***	0.449 ± 0.07
Total body fat (g)	2.6 ± 0.8	18.9 ± 7.5***	13.1 ± 4.9***	19.4 ± 7.3
Blood glucose (mM)	9.5 ± 1.1	12.5 ± 2.0***	22.7 ± 7.1***	21.6 ± 12.0
Serum insulin (ng/ml)	1.4 ± 1.2	11.9 ± 7.6***	6.8 ± 8.4**	14.1 ± 19.9
Insulin secretion (2)	0.157 ± 0.06	0.937 ± 0.49***	0.491 ± 0.73	1.097 ± 1.59
Cholesterol (mM)	3.14 ± 0.3	4.19 ± 0.2***	4.85 ± 0.9***	4.57 ± 0.9
Triglycerides (mM)	1.48 ± 0.1	1.58 ± 0.3	2.47 ± 0.9**	2.89 ± 2.75

Females	SJL	F1 (SJLxNZO)	NZO	NZOxF1
Body weight wk.12 (g)	19.8 ± 1.2	26.8 ± 1.9***	43.4 ± 4.9***	40.6 ± 5.7
Body weight wk.22 (g)	21.6 ± 0.8	33.6 ± 2.9***	58.6 ± 8.3***	54.1 ± 8.3
BMI wk.22 (g/cm²)(1)	0.242 ± 0.01	0.293 ± 0.02***	0.463 ± 0.06***	0.411 ± 0.05
Total body fat (g)	2.1 ± 0.3	9.2 ± 2.3***	24.5 ± 4.2***	20.7 ± 4.4
Blood glucose (mM)	8.5 ± 1.7	11.6 ± .18*	11.6 ± 3.4*	10.7 ± 4.1
Serum insulin (ng/ml)	1.1 ± 0.3	1.6 ± 0.7	39.5 ± 36.5***	21.3 ± 32.8
Insulin secretion (2)	0.133 ± 0.06	0.141 ± 0.07	3.053 ± 3.0*	1.957 ± 2.77
Cholesterol (mM)	1.71 ± 0.4	2.65 ± 0.2***	4.37 ± 0.4***	3.84 ± 0.6
Triglycerides (mM)	0.78 ± 0.2	1.10 ± 0.2**	1.78 ± 0.4***	1.58 ± 0.5

***A** Means ± S.D.; n=10; differences to SJL were tested for statistical significance by unpaired t-test (*, p<0.05; **, p>0.01, ***, p<0.001).*
***B** Means ± S.D.; n_{males}=111; $n_{females}$=98.*
BMI (body mass index, g/cm²) = Body weight wk.22 (g)/[Body length wk. 22]² (cm²)
Insulin secretion = Serum insulin (ng/ml)/Blood glucose (mM)

2.2. Identification of A Susceptibility Locus (Nidd/SJL) for Type 2 Diabetes-like hyperglycemia

Backcross animals were genotyped for 110 polymorphic microsatellite markers, spanning the whole genome in an average distance of 10 cM. Linkage analysis was performed in order to detect possible quantitative trait loci (QTL). After the initial genome wide scan, chromosomes with suggestive QTL were covered with additional polymorphic markers. Linkage maps and further QTL analyses of these chromosomes were calculated with the program Mapmaker/EXP 3.0 and Mapmaker/QTL 1.1 (Lincoln et al., 1992).

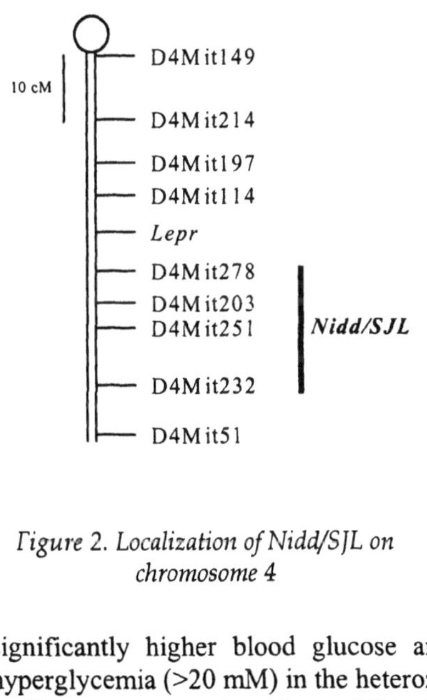

Figure 2. Localization of Nidd/SJL on chromosome 4

The genome wide scan revealed significant linkage disequilibrium on chromosomes 4 and 5. On chromosome 4, a major susceptibility locus for type 2 diabetes-like hyperglycemia was detected. The QTL on chromosome 4 was identified in male NZOxF1 backcross mice with maximal linkage disequilibrium between the markers D4Mit278 and D4Mit232 (Figure 2). LOD score maxima were detected for blood glucose (3.6), serum insulin (3.2), insulin secretion (3.8), serum triglycerides (3.6), and body weight at week 22 (3.0) in the proximity of the marker D4Mit278. Figure 3 summarizes the decisive phenotypic parameters of NZOxF1 males separated by their genotype at D4Mit278. The comparison of the two groups of genotypes showed a significantly higher blood glucose and a 2.5-fold higher prevalence of severe hyperglycemia (>20 mM) in the heterozygous group (genotype S/N).

Heterozygous mice also exhibited low serum insulin levels with secondary hypertriglyceridemia; extreme hypoinsulinemia (<0.1 ng/ml) was present in 17% of the heterozygous mice, but only in 3.5% of the homozygotes. As a consequence of the higher prevalence of severe hyperglycemia (diabetes), body weight, BMI, and total body fat content at week 22 were significantly lower in the heterozygous group.

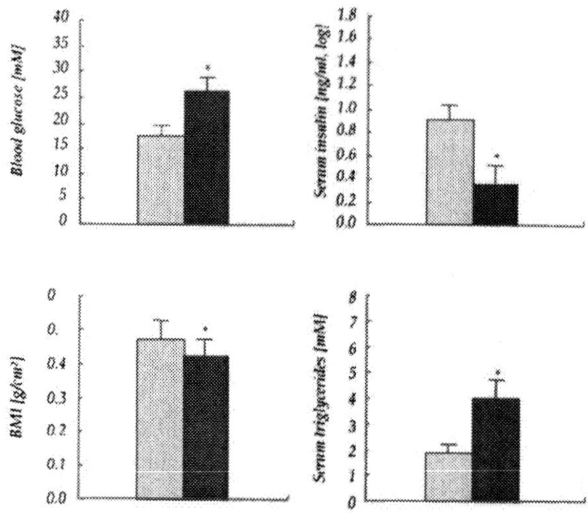

Figure 3. Characterization of phenotypic effects of Nidd/SJL in NZOxF1 males.
▨ D4Mit278N/N (n=58); ■ D4Mit278 S/N (n=53).
Means + S.E.M.; statistical significance was tested by unpaired t-test;
**, p≤0.001; investigated parameters were determined at the age of 22 weeks.*

In contrast, there was no difference between the two groups in early body weight (body weight at the age of 12 weeks). In female NZOxF1 backcross mice, an effect of different genotypes at D4Mit278 could not be detected with regard to any of the investigated parameters. This susceptibility locus for hyperglycemia and hypoinsulinemia was designated *Nidd/SJL*, because the diabetogenic allele appeared to be contributed from the SJL genome. In the male backcross population, *Nidd/SJL* was responsible for approximately 60% of the prevalence of diabetes.

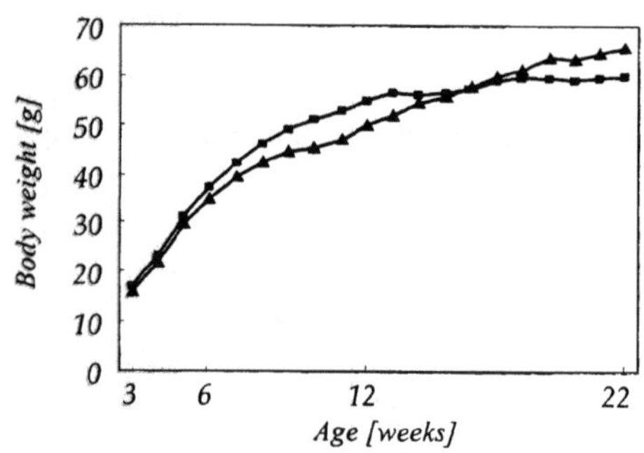

Figure 4. Development of body weight in diabetic and non-diabetic NZOxF1 males.
—■— Diabetics (n=53);
—▲— Non-diabetics (n=57).

2.3. Interaction of Obesity and Nidd/SJL

It has previously been shown that the development of type 2 diabetes in humans as well as the development of the type 2 diabetes-like syndrome in mice is substantially facilitated by obesity (Leiter, 1993; Leiter and Herberg, 1997; Vague and Raccah, 1992; Abate et al., 1996). Figure 4 indicates that early body weight increase is a predictor of hyperglycemia; male NZOxF1 backcross mice developing severe hyperglycemia (blood glucose ≥20mM) by week 22 exhibited significantly higher body weights between weeks 6 and 12 ($p<0.0001$).

An additional analysis of the relationship between body weight at week 12 and diabetes at week 22 is presented by Figure 5. Only 5% of mice weighing less than 45g at 12 weeks of age, but 66% of mice weighing more than 55g at that point of time developed severe hyperglycemia by the age of 22 weeks. Thus, body weight at week 12 appears to be a predictor for the prevalence of diabetes at week 22. Furthermore, the presence of the SJL allele at D4Mit278 (*Nidd/SJL*) seems to shift the critical degree of obesity predisposing to the development of type 2 diabetes-like hyperglycemia to lower body weights: animals with 12-week body weights of less than 45g did not develop diabetes when they were homozygous at D4Mit 278

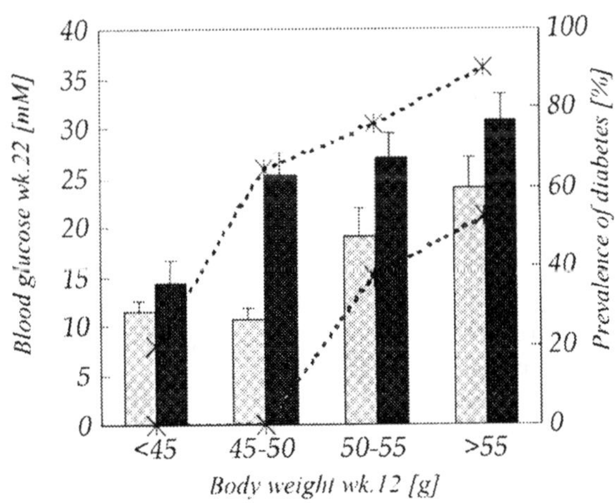

Figure 5: Influence of early body weight on the phenotypic effects of Nidd/SJL in male NZOxF1 mice.
Blood glucose at week 22, D4Mit278 N/N (mean+S.E.M
Blood glucose at week 22, D4Mit278 S/N (mean+S.E.M.);
Prevalence of diabetes, D4Mit278 N/N;
Prevalence of diabetes, D4Mit278 S/N;
Correlation between early obesity and blood glucose and between early obesity and decompensated hyperglycemia (diabetes), respectively. Mice were assigned to 4 different groups according to the weight category of their 12-week body weight (I.<45g, n=19; II.45-50g, n=26; III.50-55g, n=37; IV.>55g, n=29). The indicated parameters were calculated for the different genotypes.

(genotype N/N), whereas 42% of the heterozygous mice became diabetic. In the more obese group (12-week body weight >55g), the presence of *Nidd/SJL* intensified the contribution of obesity to the development of severe hyperglycemia; in that group, 90% of the heterozygotes developed diabetes. Thus, the data indicate that *Nidd/SJL* lowers the obesity threshold required for the development of diabetes; nevertheless a certain degree of obesity seems to be required for its detrimental effects.

2.4. Identification of A Susceptibility Locus (Nob1) for Obesity and Insulin Resistance

A susceptibility locus for obesity and insulin resistance was identified in the female NZOxF1 backcross mice. The QTL was located in the vicinity of the marker D5Mit392 on chromosome 5 (Figure 6). A summary of the characteristics of the female NZOxF1 backcross mice separated by their genotype at D5Mit392 is given in Table 2. Animals homozygous for D5Mit392 exhibited significantly higher body weights and BMI, a greater increase of body weight between weeks 12 and 22, and a greater quantity of total body fat. Furthermore, homozygotes showed 2 times higher serum insulin levels than heterozygotes. These data suggest that the responsible allele is contributed by the NZO genome; thus the QTL was designated *Nob1* (New Zealand obese 1). Highest LOD scores were obtained for body weight and BMI at week 22 (3.4, 3.8) and total body fat (3.0).

There was only suggestive linkage for body weight at 12 weeks in female backcross mice (LOD 1.8); in male mice, a significant linkage for said trait (LOD 3.1) could be detected at D5Mit392. A LOD score for the total backcross population (4.4) was calculated after correction for the difference between 12 week weights of males and females. A considerably lower LOD score (1.8) was obtained for serum insulin; this LOD score was markedly reduced (0.5), when serum insulin was corrected for the amount of body fat with the aid of the correlation between serum

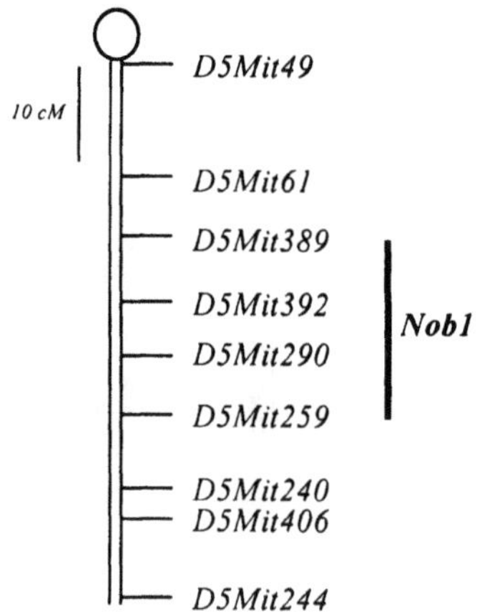

Figure 6. Localization of Nob1 on chromosome 5

insulin levels and total body fat (Ortlepp et al., 2000). Thus, the effect of *Nob1* on serum insulin levels is a function of its effect on body weight and is not independent of obesity.

Table 2. Phenotypic parameters in female NZOxF1 backcross mice separated by their genotype at D5Mit392.

Trait	D5Mit392 N/N	D5Mit392 S/N
Body weight wk.12 (g)	43.2 ± 5.6	39.7 ± 5.9**
Body weight wk.22 (g)	57.8 ± 8.7	51.7 ± 7.9***
Weight gain wk.12-22 (g)	14.7 ± 5.1	12.1 ± 4.4**
BMI wk.22 (g/cm²)	0.433 ± 0.05	0.394 ± 0.04***
Total body fat (g)	22.9 ± 4.6	19.7 ± 4.6***
Blood glucose (mM)	12.0 ± 4.9	10.7 ± 3.8
Serum insulin (ng/ml)(+)	24.18 ± 34.8	12.81 ± 21.7*
Insulin secretion	2.08 ± 2.9	1.26 ± 1.9
Serum cholesterol (mM)	3.86 ± 0.6	3.84 ± 0.5
Triglycerides (mM)	1.61 ± 0.6	1.50 ± 0.4

Means ± S.D.; $n_{N/N}$=60, $n_{S/N}$=71; differences were tested for statistical significance by unpaired t-test (, p<0.05; **, p>0.01, ***, p<0.001).(+) Because of the logarithmic distribution of serum insulin values, a conversion into log(serum insulin) was performed in order to calculate p-values/LOD scores.*

2.5. Combined Effects of Nob1 with Nidd/SJL and Lepr

In order to investigate the contribution of *Nob1* to the development of severe hyperglycemia in the male backcross population, and in order to detect possible synergistic effects of *Nob1* and *Nidd/SJL*, male backcross animals were analyzed with regard to their genotypes at both loci (Figure 7). Apparently, animals with 12 week weights of less than 50g and homozygosity at D4Mit278 (*Nidd/SJL*) did not develop decompensated hyperglycemia (blood glucose >20mM), independently of their genotype at D5Mit392. It is remarkable, however, that more than 75% of animals with homozygosity at D5Mit392 (*Nob1*) exhibited 12-week body weights of more than 50g. Thus, *Nob1* seems to lead to early obesity, which in return favors the development of insulin resistance and diabetes in male NZOxF1 backcross mice. Figure 7 also suggests the existence of at least one additional major genetic factor which could not be detected in the genome wide scan; a certain portion of animals lacking either one or both of the harmful alleles became morbidly obese and/or diabetic. This indicates that a multigenic phenotype may require the coinheritance of minor genetic elements for the effect of a major locus to be seen.

In order to assess the contribution of *Lepr* to the obesity syndrome of the NZO strain, female backcross animals were genotyped for the *Lepr* locus. Although the variant allele failed to show an adipogenic effect in the female backcross, it appeared to enhance the effect of *Nob1* on body weight. Females with homozygosity at both loci (*Lepr* N/N and D5Mit392 N/N, n=26) exhibited an average 22-week body weight of 60.6g, whereas animals heterozygous at both loci (*Lepr* S/N and D5Mit392 S/N, n=40) showed an average 22-week body weight of 51.9g (p=0.0002). Accordingly, homozygotes exhibited 2.2-fold higher serum insulin levels (35.7 ng/ml) than heterozygotes (16.0 ng/ml).

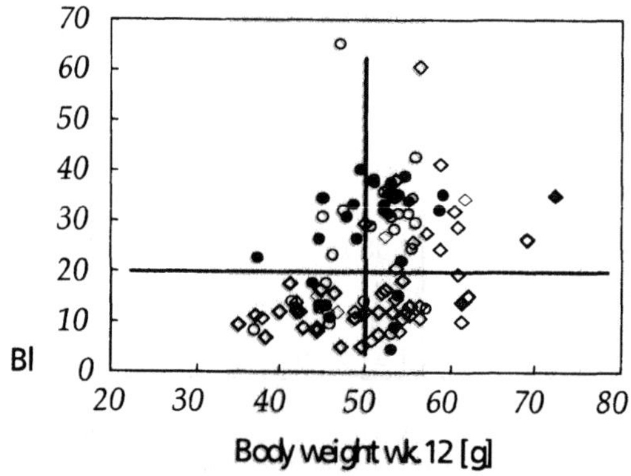

Figure 7: Phenotypic effects of Nidd/SJL and Nob1 in male NZOxF1 mice.
- ◇ *D4Mit278 N/N, D5Mit392 S/N (n=33);*
- ◆ *D4Mit278 N/N, D5Mit392 N/N (n=25);*
- ○ *D4Mit278 S/N, D5Mit392 S/N (n=26);*
- ● *D4Mit278 S/N, D5Mit293 N/N (n=26);*

The horizontal dotted line marks the cutoff (Blood glucose ≥ 20mM) for severe hyperglycemia. The vertical dividing line indicates a body weight of 50g.

3. CONCLUSIONS AND FUTURE PROSPECTS

The NZO mouse exhibits several characteristics that resemble the human metabolic syndrome or syndrome X. In our recent studies, phenotypic characterization and a comprehensive genetic analysis were performed in a backcross model of NZO and SJL mice. These studies resulted in the significant association of regions on chromosomes 4 and 5 with sensitivity to diabetes (Nidd/SJL) and obesity (*Nob1*), respectively. The two susceptibility loci *Nidd/SJL* and *Nob1* apparently make a combined contribution to the observed syndrome of severe hyperglycemia. Although obesity evidently aggravates insulin resistance in the described backcross model, diabetes fails to develop by the mere increase of body weight. The development of decompensated hyperglycemia appears to require the combination of genetic alterations affecting islet cell function together with gene variants producing a marked increase in body weight.

The finding that the lean and apparently healthy SJL strain contributed a diabetogenic allele is surprising and requires some comment. According to our analysis of the backcross population, *Nidd/SJL* requires a certain degree of obesity for its penetrance. This explains the finding that SJL mice do not develop hyperglycemia. Furthermore, it is also surprising that we failed to detect a major diabetogenic gene contributed by the NZO genome, although close to 100% of the parental mice had developed severe hyperglycemia by week 22. Thus, the diabetes in the NZO strain seems to be the consequence of several synergistically interacting genes which cannot individually cause diabetes in the backcross population. Alternatively, it is conceivable that *Nidd/SJL* substitutes (and aggravates) the effect(s) of the diabetogenic gene(s) in NZO. The latter possibility is supported by the identification of an NZO-derived diabetogenic allele (*Nidd3*; Leiter et al., 1998) in a different outcross.

The resolution of linkage analysis is limited by the number of genetic recombinations represented in the population examined. Thus, the regions comprising the susceptibility loci on chromosomes 4 and 5 are still large and can be estimated to contain at least 200 genes. In order to further define the susceptibility locus, a genetic locus can be isolated from other genes controlling the trait by transferring small chromosomal segments onto the genome of a second mouse strain by selective breeding to create congenic strains (Yui et al., 1996). This technique allows the study of the effects of single genetic intervals derived from the donor strain, isolated from the effects of other donor strain genes. However, this approach can only be successful, if the investigated gene variant does not require synergistic genes to exert its effects.

4. REFERENCES

Abate N., Garg A., Peshock R.M., Stray-Gundersen J., Adams-Huet B. and Grundy S.M. "Relationship of generalized and regional adiposity to insulin sensitivity in men with NIDDM." *Diabetes* 45 (1996): 1684-1693.

Bielschowsky M. and Bielschowsky F. "A new strain of mice with hereditary obesity." *Proc. Univ. Otago Med. School* 31 (1953): 29-31.

Festing M.F.W. "Inbred strains of mice." *Mouse genome* 95 (1997): 631-633.

Halaas J.L., Boozer C., Blai-West J., Fidahusein N., Denton D.A. and Friedman J.M. "Physiological response to long-term peripheral and central leptin infusion in lean and obese mice." *Proc. Natl. Acad. Sci. USA* 94 (1997): 8878-8883.

Herberg L. and Coleman D.L. "Laboratory animals exhibiting obesity and diabetes syndromes." *Metabolism* 26 (1977): 59-98.

Hunt C.E., Lindsey J.R. and Walkley S.U. "Animal models of diabetes and obesity, including the PBB/Ld mouse." *Fed. Proc.* 35(5) (1976): 1206-1217.

Igel M., Becker W., Herberg L. and Joost H.-G. "Hyperleptinemia, leptin resistance and polymorphic leptin receptor in the New Zealand Obese (NZO) mouse." *Endocrinology* 138(1997): 4234-4239.

Kluge R., Plum L., Bahrenberg G., Giesen K., Ortlepp J.R. and Joost H.-G. "Two quantitative trait loci for obesity and insulin resistance (*Nob1, Nob2*) in New Zealand Obese (NZO) mice and their interaction with the leptin receptor allele (*Lepr*$^{A720T/T1044I}$) of NZO". *Diabetologia* 43 (2000): 1565-1572.

Leiter E.H. "Obesity genes and diabetes induction in the mouse." *Crit. Rev. Food Sci. Nutr.* 33 (1993): 333-338.

Leiter E.H. and Herberg L. "The polygenetics of diabesity in mice." *Diabetes Rev.* 5(1997): 131-141.

Leiter E.H., Reifsnyder P.C., Flurkey K., Partke H.-J., Junger E. and Herberg L. "NIDDM genes in mice. Deleterious synergism by both parental genomes contributes to diabetogenic thresholds." *Diabetes* 47 (1998): 1287-1295.

Lincoln S., Daly M. and Lander E. "Constructing genetic maps with Mapmaker/EXP 3.0." *Whitehead Institute Technical Report*, 3rd edition, 1992.

Lincoln S., Daly M. and Lander E. "Mapping genes controlling quantitative traits with Mapmaker/QTL 1.1." *Whitehead Institute Technical Report*, 3rd edition, 1992.

Ortlepp J.R., Kluge R., Giesen K., Plum L., Radke P., Hanrath P. and Joost H.-G. "A metabolic syndrome of hypertension, hyperinsulinemia, and hypercholesterolemia in the New Zealand Obese (NZO) mouse." *Europ. J. Clin. Invest.* 30 (2000): 195-202.

Paigen B. "Genetics of responsiveness to high fat and high cholesterol diets in mice." *Am. J. Clin. Nutr.* 62 (1995): 458S-462S.

Plum L., Kluge R., Giesen K., Altmüller J., Ortlepp J.R. and Joost H.-G. "Type-2-diabetes-like hyperglycemia in a backcross model of New Zealand Obese (NZO) and SJL mice: Characterization of a susceptibility locus on chromosome 4 and its relation with obesity." *Diabetes* 49 (2000): 1590-1596.

Reaven G.M. "Role of insulin resistance in human disease (syndrome X): an expanded definition." *Annu. Rev. Med.* 44 (1993): 121-131.

Vague P. and Raccah D. "The syndrome of insulin resistance." *Horm. Res.* 38 (1992): 28-32.

Yui M.A., Muralidharan K., Moreno-Altamirano B., Perrin G., Chestnut K. and Wakeland E.K. "Production of congenic mouse strains carrying NOD-derived diabetogenic genetic intervals: an approach for the genetic dissection of complex traits." *Mamm. Genome* 7 (1996): 331-334.

L. PELLEGRINI, D.F. BURKE AND T.L. BLUNDELL

ACTIVATION MECHANISM OF FIBROBLAST GROWTH FACTOR RECEPTOR TYROSINE KINASE REVEALED BY CRYSTAL STRUCTURE OF FIBROBLAST GROWTH FACTOR RECEPTOR ECTODOMAIN BOUND TO FIBROBLAST GROWTH FACTOR AND HEPARIN.

Abstract. Fibroblast growth factors (FGF) are a family of polypeptides with a pleiotropic range of physiological and pathological activities. Association of fibroblast growth factor with its receptor tyrosine kinase (FGFR) and heparin-like heparan sulphate proteoglycan in a specific complex on the cell surface is essential for signal transduction. Here we report the crystal structure of the FGFR2 ectodomain in a dimeric form, induced by simultaneous binding to FGF1 and a heparin decasaccharide. The complex is assembled around a central heparin molecule linking two FGF1 ligands into a dimer that bridges between two receptor chains. Heparin binding is remarkably asymmetric and involves extensive contacts with both FGF1 molecules and one receptor chain. The structure of the FGF1-FGFR2-heparin ternary complex provides a structural basis for the essential role of heparan sulphate in FGF signalling. It suggests possible mechanisms for receptor activation, based on a heparin-induced conformational change in the receptor, and for higher order FGFR clustering.

1. INTRODUCTION

Fibroblast growth factors (FGFs) are a family of over 20 structurally related proteins that were initially identified by their strong mitogenic activity, and are now known to have essential biological roles in cell development, proliferation and differentiation (for a review see Galzie et al., 1997; Klein et al., 1997; Naski and Ornitz, 1998). FGF signalling requires binding and oligomerisation of the FGF receptor (FGFR), which leads to transphosphorylation of the intracellular kinase domains of the receptor and activation of the cytoplasmic signalling pathway. Due to their potent angiogenic activity, FGFs and their receptors play a crucial part in vascularisation of healthy and diseased tissues and have therefore become the focus of much attention in angiogenic therapy (Schumacher et al., 1998). Naturally occurring mutations of FGFRs are responsible for a family of phenotypically related syndromes that cause defects in skeletal development with different degrees of severity (for a review see Burke et al., 1998). Frequent mutations of FGFR3 are also found in human bladder and cervix carcinomas (Cappellen et al., 1999) and in multiple myeloma (Chesi et al., 1997).

M. Federwisch, M. Leyck Dieken and P. De Meyts (eds.), Insulin & Related Proteins, 189–200.

FGFRs belong to the family of receptor tyrosine kinases, with extracellular regions comprising three immunoglobulin (Ig)-like domains (D1-D3) (for a review see Johnson and Williams, 1993; Figure 1). A portion of FGFR spanning Ig-like domains D2 and D3 and the interdomain sequence is sufficient for high affinity binding of the ligand. Four closely related FGFRs (FGFR1-4) have been identified, but alternative splicing mechanisms generate a large number of isoforms. Two isoforms, IIIb and IIIc, incorporate different sequences in the C-terminal half of domain D3 that are major determinants of ligand binding specificity. Shorter isoforms lacking the D1 domain have higher affinity than full length forms and the change of splice form may be correlated with progression of prostate cancers towards malignancy (Wang et al., 1995). Biological activity is modulated by tissue-specific expression of distinct FGFs and FGFR isoforms that differ in their reciprocal affinities.

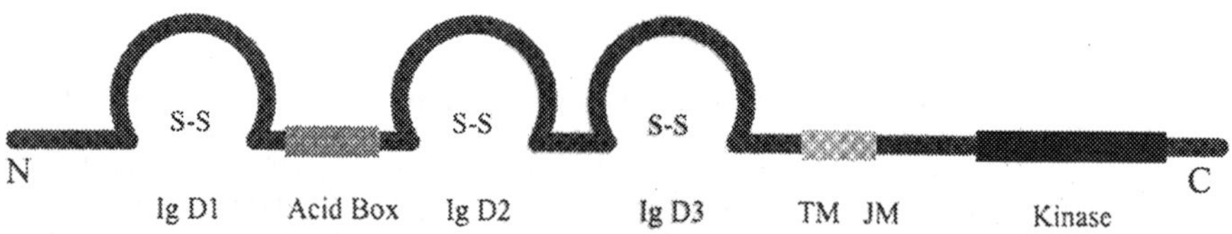

Figure 1. Schematic diagram of arrangement of domains in FGFR

FGFs have a high affinity for soluble heparin glycosaminoglycan and heparan sulphate proteoglycans present on the cell surface and in the extracellular matrix (for a review see Faham et al., 1998; Bernfield et al., 1999; Ornitz, 2000). The direct involvement of heparin-like heparan sulphate proteoglycans in the molecular association between FGF and its receptor is essential for biological activity. FGF fails to bind to FGFR and support cell proliferation in cell cultures exposed to heparinase or sodium chlorate (Rapraeger et al., 1991); cell lines lacking endogenous heparan sulphate require exogenous heparin for FGFR dimerisation, tyrosine kinase activation and cell proliferation (Spivak-Kroizman et al., 1994); heparan sulfate proteoglycans are essential for FGFR signaling during Drosophila embryonic development (Lin et al., 1999). An 18 aminoacid sequence mapping to domain D2 of FGFR, able to bind heparin independently of FGF and required for FGFR activity, has also been identified (Kann et al., 1993).

Although structural information about binary complexes of FGF-heparin (Faham et al., 1996; DiGabriele et al., 1998) and FGF-FGFR (Plotnikov et al., 1999; Stauber et al., 2000) has been obtained, and various models for the ternary association of FGF, FGFR and heparin have been proposed (Pantoliano et al., 1994; Huhtala et al., 1999; Kann et al., 1996; Pellegrini et al., 1998), no three-dimensional information about the mode of ligand-induced FGFR oligomerisation in the presence of heparin was available until recently (Pellegrini et al., 2000). Here we present the crystal structure of a ternary complex of FGFR2, FGF1 and heparin that shows how the ligand and heparin cooperate to promote receptor dimerisation.

2. STRUCTURE DETERMINATION

A ternary complex of human FGF1, the ligand-binding region of human FGFR2 (splice variant IIIc) spanning Ig-like domains D2 and D3, and a heparin decasaccharide was reconstituted by mixing FGF1, FGFR2 and heparin in 2:2:1 stoichiometric ratios and isolated by gel filtration as a single peak of the expected molecular weight (approx. 84 kDa). Crystals of the complex were grown from Li_2SO_4 solutions at mildly alkaline pH. They belong to space group $P6_2$ with a=b=195.9Å and c=68.9Å. The X-ray structure of the complex was determined by the multiple anomalous wavelength method using selenomethionine protein and refined to an R-factor of 26.7% (R-free of 30.9%) at 2.8Å resolution.

3. OVERALL ARCHITECTURE OF THE FGF-FGFR-HEPARIN COMPLEX

The FGF1-FGFR2-heparin complex is a heteropentamer formed by two receptors, two ligands and one heparin molecule (Figure 2). The ligand-binding region of the FGFR2 folds into two Ig-like domains, D2 (residues 152 to 249) and D3 (residues 254 to 361), separated by a short linker (residues 250 to 253), as identified in previous studies (Plotnikov et al., 1999; Stauber et al., 2000). D2 belongs to the I class as predicted (Harpaz and Chothia, 1994), while D3 approximates to a class C2 fold. FGF1 is composed of three antiparallel ß-sheets of four strands each that adopt the well known ß-trefoil fold (Zhu et al., 1991).

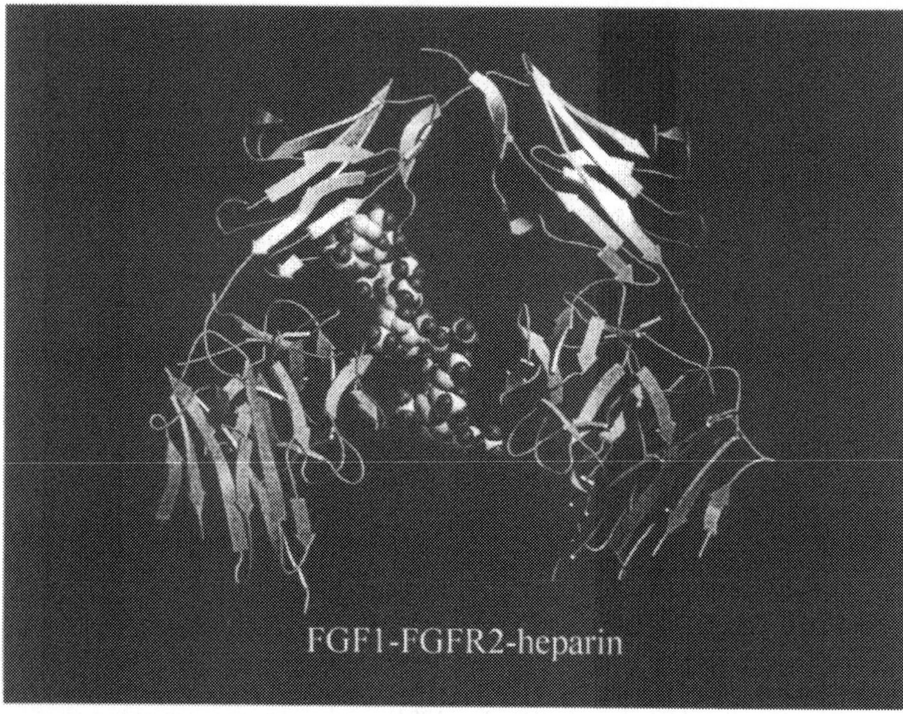

Figure 2. The FGF1-FGFR2-heparin heteropentameric complex. Ig-like domains 2 (D2) and 3 (D3) of the FGFR are shown in cyan and magenta, respectively and the FGF ligands in green. The heparin molecule is in CPK representation. The N- and C-termini of the receptor ectodomain are indicated. View perpendicular to the approximate molecular dyad.

The structure of the complex reveals that the mechanism of receptor dimerisation can be viewed as an heparin-linked FGF1 dimer that binds to and brings together two receptor molecules, or alternatively, as two 1:1 FGF1-FGFR2 complexes that associate via interaction with the heparin. The two protein halves of the complex are related by an approximate two-fold symmetry axis perpendicular to the cell surface. The pseudo two-fold symmetry is broken by the heparin oligosaccharide that lies across the two FGF1 molecules, tilted away from the dyad, and extends past the ligand dimer to interact with one receptor D2 domain.

The two D2 domains form the only receptor-receptor interface through their N-termini, that join at the top of the complex. The D3 domains are splayed away in opposite directions relative to a plane containing the two D2 and FGF1 molecules. The root mean square deviation (rmsd) for the C_α superposition of the two halves of the complex is 2.80Å, while the rmsd for the 229 C_α atoms of FGF1 and D2 only is

1.0Å. The two receptor chains in the ternary complex show marked bends in the linker region, with angles of 106° and 87° between D2 and D3, that orient the two domains towards the ligand.

We refer to the two 1:1 FGF-FGFR complexes within the heteropentamer as binary complexes A and B. In A both ligand and receptor interact with heparin whereas in B only the ligand is in contact with heparin.

4. PROTEIN-PROTEIN INTERACTIONS IN THE TERNARY COMPLEX

Protein-protein interactions in the ternary complex are dominated by ligand binding of each receptor chain. FGF1 binding to the receptor involves extensive contacts with D2, the linker region, and residues 254 to 260 of D3, creating a continuous interface that buries 1700 Å^2 of total surface area. This is consistent with biochemical evidence that narrows the minimal ligand binding region of FGFR to a fragment comprising D2, the interdomain sequence and an N-terminal sequence of D3 (Wang et al., 1995).

Interactions between FGF1 and D2 are almost entirely hydrophobic and involve residues Tyr15, Gly20, Phe22, Tyr94, Ley133 and Leu135 of the ligand and residues Lys164, Leu166, Ala168, Val169 and Pro170 of domain D2. These residues are clustered together at the ligand-receptor interface and form a hydrophobic patch that is likely responsible for the high affinity of the interaction between ligand and receptor. These observations closely resemble those made in previous biochemical and structural studies (Plotnikov et al., 1999; Stauber et al., 2000; Springer et al., 1994).

In the receptor linker sequence, the guanidinium moiety of invariant Arg251 is surrounded by three hydrophobic FGF1 residues, Leu89, Leu133 and Pro134, and is hydrogen bonded to the main chain carbonyl group of FGF1 His93 through its Nε nitrogen atom. These interactions serve to immobilise the arginine side chain in the correct position to form a crucial hydrogen bond with conserved Asn95 of FGF1. The importance of this interaction is highlighted by mutagenesis data that show a 400-fold reduction in receptor affinity when the equivalent asparagine in FGF2 is replaced by alanine (Zhu et al., 1997).

A dramatic change in the conformation of the FGFR ectodomain is observed in the structure of the ternary complex. Relative to its position in the FGF:FGFR complexes without heparin (Plotnikov et al., 1999; Stauber et al., 2000), domain D3 is swivelled around the linker region by close to 180°. The pivotal point for D3 rotation is the highly conserved Pro253, which assumes a *cis* conformation in our structure, but a *trans* conformation in FGF-FGFR complexes without heparin (Plotnikov et al., 1999; Stauber et al., 2000). The conformational change in the receptor ectodomain allows the formation of a novel set of ligand-receptor interactions. Conserved Arg255 in D3 packs against Leu89 and points towards invariant Glu87 of FGF1 in an ionic interaction that, in binary complex B, also involves hydrogen bonding. Involvement of Glu87 in a charge-charge interaction has been demonstrated by mutagenesis experiments that show a 10-fold reduction in receptor binding affinity for a glutamate to glutamine FGF2 mutant (Zhu et al., 1995). Glu87 is kept in a favourable position for interaction with Arg255 by a hydrogen bond with neighbouring Tyr97. Invariant Ile257 of D3 is partially buried by van der Waals contacts with Leu89 and the main chain atoms of Glu90 and Glu91 of FGF1, and His254 of D3.

The only spliceform-specific FGF1-D3 interaction in the ternary complex is a hydrophobic contact between Val51 and Ile350. Further FGF1-D3 interactions that only occur in one subunit are hydrogen bonds between Arg88 and the main chain carbonyl group of Leu258 in A and between Glu91 and the main chain nitrogen of Ala260 in B.

Direct interaction between receptors in the ternary complex involves a region at the top of domain D2. The loops between ß-strands B and C come in direct contact and additional contact are provided by the N-terminal tails of D2 that flank. No clear hydrophobic interface is recognisable and the total surface area buried is relatively small, 400Å^2.

5. OVERVIEW OF HEPARIN CONFORMATION AND BINDING IN THE TERNARY COMPLEX

The entire length of the protein-bound heparin decasaccharide is visible in the electron density map. Nine out of the ten hexoses of the heparin fragment in the refined structure of the ternary complex are involved in contacts with either the ligand or the receptor. The most outstanding feature of the heparin molecule in the ternary complex is the asymmetry of its association with FGF1 and FGFR2. In its path through the complex, the oligosaccharide fills the heparin binding site of one receptor chain with the disaccharide at its non-reducing end, contacts one ligand with its central section and interacts with the other ligand near the reducing end.

The total amount of surface area buried at the protein-heparin interface is large at 2240 Å^2 (divided in 617 Å^2 with FGF1 and 631 Å^2 with domain D2 in subunit A, and 992 Å^2 with FGF1 in subunit B). This is caused by the close interaction of several sulphate groups against the protein surface of both ligand and receptor, and the packing of the aliphatic portions of lysine and arginine residues against the sugar rings. Probably as a consequence of the extensive contacts deriving from the protein-heparin association, a marked distortion of the heparin molecule is present that has not been observed in previous structural studies of FGF-heparin complexes (Faham et al., 1996; DiGabriele et al., 1998). The helical axis shows a 34° kink between disaccharides 2 and 3, accompanied by a decreased helical rise between disaccharide 2 and 3 (7.10 Å) and between 3 and 4 (7.87 Å).

6. HEPARIN-PROTEIN INTERACTIONS IN THE TERNARY COMPLEX

Due to the asymmetry of the protein-heparin contacts in the ternary complex the two sets of sugar binding residues in the two ligands are only partially overlapping. We have divided the heparin decasaccharide in 5 disaccharides, numbered 1 to 5 from the reducing to the non-reducing end, where each disaccharide begins with a glucosamine followed by an iduronic acid.

FGF1 in binary complex A binds to a fraction of the heparin molecule spanning six sugars, from IdoA-1 to glucosamine in disaccharide 4 (GlcN-4). Heparin binding is mediated by Asn18 and a set of mainly basic residues located in the loop between β-strands 10 and 11, namely Lys112, Lys113, Asn114, Lys118, Arg119, Arg122, Gln127 and Lys128. These residues were previously identified as responsible for heparin binding in structural and mutagenesis experiments (Faham et al., 1996; DiGabriele et al., 1998; Thompson et al., 1994). The heparin binding region of FGFR2 observed in the structure of the ternary complex comprises residues in the highly conserved sequence 161-KMEKRLHAVPAANTVKFR-178 located in ß-strands A' and B of domain D2, thus confirming biochemical data that had previously identified this FGFR sequence as essential for heparin binding (Kan et al., 1993). All basic residues in this sequence, with the exception of Arg165, are involved in ionic and hydrogen bonding interactions with the sulphate groups of GlcN-5 and the carboxylate group of IdoA-5. Both sugars in disaccharide 5 come in close contact with the protein surface. The 6-O-sulphate group of GlcN-5 is involved in a particularly large range of interactions, that include hydrogen bonds to the backbone nitrogen of Lys176 and the side chain of His167, a salt bridge with Lys164 and vdW contacts with the protein backbone of Thr174 and Val175. An essential role for 6-O-sulphate groups of heparin in receptor binding has been proposed (Guimond et al., 1993; Rusnati et al., 1994). This interaction might be crucial for recognition of the correct heparan sulphate binding site by a 1:1 FGF-FGFR complex. The role of Lys164 in neutralising the 6-O-sulphate group of GlcN-5 appears particularly important as the sulphate group is only partially solvent exposed. Binary complex B only contacts the heparin through the ligand, while the receptor heparin-binding site remains unoccupied. FGF1 in B interacts with five monosaccharide units, from GlcN-1 to GlcN-3. Residues in the sugar binding loop that contact heparin are Lys112, Arg119 and Arg122.

In total, a heparin length of 7 sugar residues is involved in FGF dimerisation. In addition to the set of interactions between FGF1 and heparin described in previous structural studies (DiGabriele et al., 1998), we note two novel contacts between heparin and the protein surface of both ligands. The first involve FGF1 in binary complex A and the 2-N-sulphate group of GlcN-2 that packs against the peptide bond joining Lys118 and Arg119 and fits in a pocket formed by the side chains of Lys112, Lys118, Arg119 and Arg122. In the second interaction the 2-N-sulphate group of GlcN-1 is hydrogen bonded to Trp107 of binary complex B and comes in close contact with Pro121 and Lys105 that are stacked above and below the tryptophan side chain.

7. IMPLICATIONS FOR HEPARIN BINDING

The structure of the ternary complex reveals that heparin links two FGF1 molecules into a dimer that lacks a protein-protein interface, as observed in the crystal structure of a FGF1-heparin complex (DiGabriele et al., 1998), but crucial differences in the relative position of heparin and FGF1 are observed here. Superposition of FGF1 protamer B of the heparin-linked dimer (DiGabriele et al., 1998) (PDB ID: 2axm) with FGF1 of subunit A reveals that, while a very good agreement is obtained between the two heparin molecules, the FGF1 molecule in B is rotated by about 120° degrees around an axis approximately aligned with the pseudo three-fold axis of the ß-trefoil. This rotation allows FGF1, while remaining in contact with heparin through the heparin-binding loop, to make an additional, novel set of contacts with heparin. Central to this new heparin binding site is Trp107. An FGF2 mutant that had the conserved tryptophan residue mutated to alanine showed a drastic reduction in mitogenic potency, and it was originally proposed that the amino acid sequence of the loop containing the tryptophan residue constituted a secondary receptor binding site (Springer et al., 1994). Our structure shows that Trp107 is involved in a specific contact with heparin that is likely to be important for the correct positioning of FGF1 relative to heparin and thus for the recruitment of the second binary complex in an active heteropentamer.

The structure of the ternary complex shows that a heparin octasaccharide would probably still support simultaneous binding to two ligands and one receptor chain, through interactions of Lys161 with the 2-O-sulphate group of IdoA-4 and of Lys164 with the carboxylate function of IdoA-4. Finally, our structure predicts that maximisation of the heparin-protein interactions in the ternary complex would be achieved with a heparin fragment spanning twelve sugars. A dodecasaccharide would in fact be able to make an additional set of contacts with Lys208 and Arg210 on strand D of domain D2. Nothing would be gained by longer oligosaccharides as no additional interactions would be available.

8. COMPARATIVE ANALYSIS OF CRYSTAL PACKING INTERACTIONS IN FGF-FGFR COMPLEXES WITH AND WITHOUT HEPARIN

Comparative analysis of protein-protein interactions in crystals of FGF-FGFR complexes with and without heparin shows that they are to some degree conserved. We observe that heteropentamers related by the 3_1 screw axis are packed in the crystal in such a way as to bring two D2 domains together, one from each complex, in a fashion reminiscent of the canyon-forming interaction observed in crystals of FGF-FGFR complexes without heparin. This deep cleft was proposed to be the heparin binding site in the ternary complex (Plotnikov et al., 1999; Stauber et al., 2000). In our structure the non-reducing end of heparin reaches the entrance of such a cleft, leaving it otherwise empty. The steric hindrance of this interaction induces a significant movement away from the heparin in one of the D2 domains, that significantly deforms the cleft, and causes the loss of the secondary interaction site between ligand and receptor belonging to different binary complexes.

This observation leads us to conclude that the canyon-like cleft formed by dimerisation in the crystal of two 1:1 FGF:FGFR complexes is unlikely to be the physiologically relevant binding site for heparan sulphate. Indeed, it could be a crystallisation artefact induced by the concentrated sulphate solutions used in the crystallisation buffers, that would drive the basic face of two D2 domains together, creating a canyon-like cleft. This study provides evidence for a different mechanism of heparin-induced receptor dimerisation.

9. IMPLICATIONS FOR FGFR ACTIVATION MECHANISM

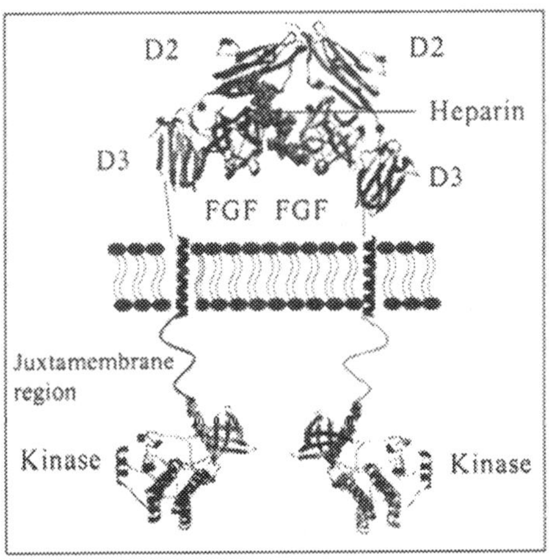

Figure 3. Schematic diagram of receptor membrane and kinase

We have determined the structure of an FGF-FGFR-heparin ternary complex. We propose that the structure presented in this study represent the active dimeric state of FGFR when bound to the ligand and heparin. The highly asymmetric binding mode of the heparin in the ternary complex argues in favour of a stepwise assembly of an active heteropentamer. In this model, one 1:1 FGF-FGFR complex would first become heparin-bound, and receptor dimerisation would follow the recruitment of a second binary complex to the heparin. Heparin would play an essential role in the assembly, by crosslinking ligand and receptor of the first binary complex, and by mediating binding of a second binary complex through interactions with the ligand. A specific contact of heparin with a set of FGF residues that were proposed to form a second receptor binding site (Springer et al., 1994) is probably crucial to the second binding event.

The receptor conformation observed in the absence of heparin, which differ in the relative position of domain D3, might also have physiological relevance. FGF and FGFR would initially associate with the receptor ectodomain in the conformation previously observed (Plotnikov et al., 1999; Stauber et al., 2000) (Figure 3). Binding of heparin to the 1:1 FGF-FGFR complex would cause the D3 domain to disengage from FGF and rotate in the conformation observed in our structure. A *cis-trans* proline switch would ensure that the receptor only assumes its active conformation when tightly bound to the correct site on the heparan sulphate glycosaminoglycan. The complex would be stabilised in the active conformation by a set of interactions involving invariant Ile257, Glu87 and conserved Arg255. The biological relevance of this hypothesis is supported by the high or absolute conservation of a group of residues in FGFR2 (Arg255, Ile257, Asp283) and FGF1 (Glu87) which appears difficult to explain based on one receptor conformation alone.

The proposed model for FGFR dimerisation, lacking a significant protein-protein interface, would easily accommodate for heterodimerisation between different ligand and receptor types. We further note that as a consequence of the receptor conformation observed in our structure a hydrophobic surface of domain D3 becomes fully exposed, and possibly poised to drive higher order receptor oligomerisation. In the crystal two heteropentamers related by the 6_2 screw axis are linked to one another by apposition of the hydrophobic surfaces of two D3 domains, that buries 1200 $Å^2$ of total surface area. Different oligomerisation states of a receptor tyrosine kinase could be used by the cell as a means to regulate the intensity of the signal and activate different signalling pathways (Marshall et al., 1995). Evidence of FGFR clustering within the focal adhesion complex has been reported (Pellegrini et al. 2000).

10. REFERENCES

Bernfield M. *et al.* "Functions of cell surface heparan sulfate proteglycans." *Annu. Rev. Biochem.* 68 (1999): 729-777.

Burke D.F., Wilkes D., Blundell T.L. and Malcolm S. "Fibroblast growth factor receptors: lessons from the genes." *Trends Biochem. Sci.* 23 (1998): 59-62.

Cappellen D. *et al.* "Frequent activating mutations of FGFR3 in human bladder and cervix carcinomas." *Nat. Genet.* 23 (1999): 18-20.

Chesi M. *et al.* "Frequent translocation t(4;14)(p16.3;q32.3) in multiple myeloma is associated with increased expression and activating mutations of fibroblast growth factor receptor 3." *Nat. Genet.* 16 (1997): 260-264.

DiGabriele A. *et al.* "Structure of a heparin-linked biologically active dimer of fibroblast growth factor." *Nature* 393 (1998): 812-817.

Faham S., Hileman R.E., Fromm J.R., Linhardt R.J. and Rees D.C. "Heparin structure and interactions with basic fibroblast growth factor." *Science* 271 (1996): 1116-1120.

Faham S., Linhardt R.J. and Rees D.C. "Diversity does make a difference: fibroblast growth factor-heparin interactions." *Curr. Opin. Struct. Biol.* 8 (1998): 578-586.

Galzie Z., Kinsella A.R. and Smith J.A. "Fibroblast growth factors and their receptors." *Biochem. Cell. Biol.* 75 (1997): 669-685.

Guimond S., Maccarana M., Olwin B.B., Lindahl U. and Rapraeger A.C. "Activating and inhibitory heparin sequences for FGF-2 (basic FGF). Distinct requirements for FGF-1, FGF-2, and FGF-4." *J. Biol. Chem.* 268 (1993): 23906-23914.

Harpaz Y. and Chothia C. "Many of the immunoglobulin superfamily domains in cell adhesion molecules and surface receptors belong to a new structural set which is close to that containing variable domains." *J. Mol. Biol.* 238 (1994): 528-539.

Huhtala M.T., Pentikainen O. and Johnson M.S. "A dimeric ternary complex of FGFR1, heparin and FGF-1 leads to an 'electrostatic sandwich' model for heparin binding." *Structure* 7 (1999): 699-709.

Johnson D.E. and Williams L.T. "Structural and functional diversity in the FGF receptor multigene family." *Adv. Cancer Res.* 60 (1993): 1-41.

Kan M. *et al.* "An essential heparin-binding domain in the fibroblast growth factor receptor kinase." *Science* 259 (1993): 1918-1921.

Kan M. *et al.* "Divalent Cations and Heparin/Heparan Sulfate Cooperate to Comtrol Assembly and Activity of the Fibroblast Growth Factor Receptor Complex." *J. Biol. Chem.* 271 (1996): 26143-26148.

Klein S., Roghani M. and Rifkin, D.B. "Fibroblast growth factors as angiogenesis factors: new insights into their mechanism of action." *Exs* 79 (1997): 159-192.

Lin X., Buff E.M., Perrimon N. and Michelson A.M. "Heparan sulfate proteoglycans are essential for FGF receptor signaling during Drosophila embryonic development." *Development* 126 (1999): 3715-3723.

Marshall C.J. "Specificity of receptor tyrosine kinase signaling: transient versus sustained extracellular signal-regulated kinase activation." *Cell* 80 (1995): 179-185.

Naski M.C. and Ornitz, D.M. "FGF signaling in skeletal development." *Front. Biosci.* 3 (1998): D781-794.

Ornitz D.M. "FGFs, heparan sulfate and FGFRs: complex interactions essential for development." *Bioessays* 22 (2000): 108-112.

Pantoliano M.W. *et al.* "Multivalent ligand-receptor binding interactions in the fibroblast growth factor system produce a cooperative growth factor and heparin mechanism for receptor dimerization." *Biochemistry* 33 (1994): 10229-10248.

Pellegrini L, Burke D.F., von Delft F., Mulloy B. and Blundell T.L.. "Crystal Structure of fibroblast growth factor receptor ectodomain bound to ligand and heparin." *Nature* 407 (2000):1029-1034.

Pellegrini L. *et al.* "The role of heparin in the complex formation between fibroblast growth factor 2 and its high affinity receptor: comparative modelling and biochemical studies." *Biochem. Soc. Trans.* 26 (1998): 545-549.

Plopper G.E., McNamee H.P., Dike L.E., Bojanowski K. and Ingber D.E. "Convergence of integrin and growth factor receptor signaling pathways within the focal adhesion complex." *Mol. Biol. Cell.* 6 (1995): 1349-1365.

Plotnikov A.N., Schlessinger J., Hubbard S.R. and Mohammadi M. "Structural Basis for FGF Receptor Dimerization and Activation." *Cell* 98 (1999): 641-650.

Rapraeger A.C., Krufka A. and Olwin B.B. "Requirement of heparan sulfate for bFGF-mediated fibroblast growth and myoblast differentiation." *Science* 252 (1991): 1705-1708.

Rusnati M. *et al.* "Distinct role of 2-O-, N-, and 6-O-sulfate groups of heparin in the formation of the ternary complex with basic fibroblast growth factor and soluble FGF receptor-1." *Biochem. Biophys. Res. Commun.* 203 (1994): 450-458.

Schumacher B., Pecher P., von Specht B.U. and Stegmann T. "Induction of neoangiogenesis in ischemic myocardium by human growth factors: first clinical results of a new treatment of coronary heart disease." *Circulation* 97 (1998): 645-650.

Spivak-Kroizman T. *et al.* "Heparin-Induced Oligomerization of FGF Molecules Is Responsible for FGF Receptor Dimerization, Activation and Cell Proliferation." *Cell* 79 (1994): 1015-1024.

Springer B.A. *et al.* "Identification and concerted function of two receptor binding surfaces on basic fibroblast growth factor required for mitogenesis." *J. Biol. Chem.* 269 (1994): 26879-26884.

Stauber D., DiGabriele A. and Hendrickson W. "Structural Interactions of fibroblast growth factor receptor with its ligands." *Proc. Natl Acad. Sci. USA* 97 (2000): 49-54.

Thompson L.D., Pantoliano M.W. and Springer B.A. "Energetic Characterization of the Basic Fibroblast Growth Factor-Heparin Interaction: Identification of the Heparin Binding Domain." *Biochemistry* 33 (1994): 3831-3840.

Wang F., Kan M., Xu J., Yan G. and McKeehan W.L. "Ligand-specific Structural Domains in the Fibrobast Growth Factor Receptor." *J. Biol. Chem.* 270 (1995): 10222-10230.

Wang F., Kan M., Yan G., Xu J. and McKeehan W.L. "Alternately spliced NH2-terminal immunoglobulin-like Loop I in the ectodomain of the fibroblast growth factor (FGF) receptor 1 lowers affinity for both heparin and FGF-1." *J. Biol. Chem.* 270 (1995): 10231-10235.

Zhu H. *et al.* "Analysis of high-affinity binding determinants in the receptor binding epitope of basic fibroblast growth factor." *Protein Eng.* 10 (1997): 417-421.

Zhu H. *et al.* "Glu-96 of basic fibroblast growth factor is essential for high affinity receptor binding. Identification by structure-based site-directed mutagenesis." *J. Biol. Chem.* 270 (1995): 21869-21874.

Zhu X. *et al.* "Three-dimensional structure of Acidic and Basic Fibroblast Growth Factor." *Science* 251 (1991): 90-93.

11. ACKNOWLEDGEMENTS

We would like to thank Paresh Gadhavi for discussions and early work on FGFR overexpression; Dimitri Chirgadze, Franck von Delft, Emilio Parisini and Joe Patel for help with X-ray data collection and MAD experiments; Ruslan Sanishvili for excellent user support at the SBC beamline; Mladen Vinkovic for help with the use of SHELX; Marko Hyvonen for advice on protein refolding and help with the figures and Martin Symmons for advice on production of selenomethionyl protein. We are grateful to Barbara Mulloy for providing the heparin. We are grateful to Sue Malcolm for insightful comments on the phenotype of FGFR mutations in Apert syndrome and for the FGFR cDNA. We are grateful to Dave Fernig for the FGF cDNA. Heparin lyase I was a kind gift of Dr. K. Johansen, Leo Pharmaceuticals, Denmark. This research was supported by grants from the MRC and the Wellcome Trust. Use of the Argonne National Laboratory Structural Biology Center beamlines at the Advanced Photon Source was supported by the U.S. Department of Energy, Basic Energy Sciences, Office of Science.

Coordinates are deposited in the Protein Data Bank with accession number 1e0o.

J. GRÖTZINGER

IL-6 TYPE CYTOKINE RECEPTOR COMPLEXES

Abstract. The typical protein fold of most cytokines is a bundle of four antiparallel helices. This 'four-helix bundle' seems to be unique for cytokines and has not been found in other proteins. The interleukin-6 (IL-6) type cytokine family comprises six members. IL-6 type cytokines interact with three receptor subunits instead of the usual two subunits. A tetramer would be the simplest model to describe such a receptor complex, but present orthodoxy describes the active complexes of IL-6 and ciliary neurotrophic factor (CNTF) as hexamers. Here, we summarize the structural and biochemical information on IL-6 type cytokines and discuss interactions between these cytokines and their individual receptor subunits. Contradictory results regarding the stoichiometry and assembly of signalling receptor complexes are rationalized by a new, unique model. The model stipulates that a ligand induced transition from an active tetrameric to an inactive hexameric complex serves as a molecular switch that turns off cytokine signals in the presence of supraoptimal cytokine concentrations.

1. INTRODUCTION

Ten years ago, Fernando Bazan published two seminal papers which were based on theoretical considerations, in which he postulated that most of the cytokines known at the time shared a typical protein fold, the so-called four-helical bundle (Bazan, 1990; Bazan, 1991). This four-helical bundle consists of four antiparallel α-helices connected by three non-helical loops (Figure1A). Today the four-helical cytokines comprise IL-2 (Interleukin-2), IL-3, IL-4, IL-5, IL-6, IL-7, IL-9, IL-10, IL-11, the p35 subunit of IL-12, IL-13, IL-15, G-CSF (granunocyte colony stimulating factor), GM-CSF (granunocyte macrophage colony stimulating factor), EPO (crythropoictin), GH (growth hormone), PRL (prolactin), CNTF (ciliary neurotrophic factor), CT-1 (cardiotrophin-1), LIF (leukemia inhibitory factor), OSM (oncostatin M), IFN α (interferon α) and INFγ (Rubinstein et al., 1998). These cytokines play a pivotal role in the orchestration of the mammalian immune response, but also in the development of different organ systems, most prominently the nervous system (Rubinstein et al., 1998).

In an extension of his original work Bazan further differentiated the four-helical cytokines into short chain (IL-3, IL-4, IL-5, GM-CSF) and long chain cytokines (the remainder) (Sprang and Bazan, 1993). Where available, the crystal or NMR structures of these cytokines have indeed confirmed Bazan's landmark prediction. Moreover, this fold has so far not been identified in any non-cytokine protein and may therefore be regarded as the typical cytokine fold. Bazan also suggested that receptors of four-helical cytokines possess a typical cytokine receptor homology region (CHR) consisting of two fibronectin type III-like domains with four characteristic cysteine residues in the N-terminal and a conserved WSXWS motiv in

M. Federwisch, M. Leyck Dieken and P. De Meyts (eds.), Insulin & Related Proteins, 201–212.

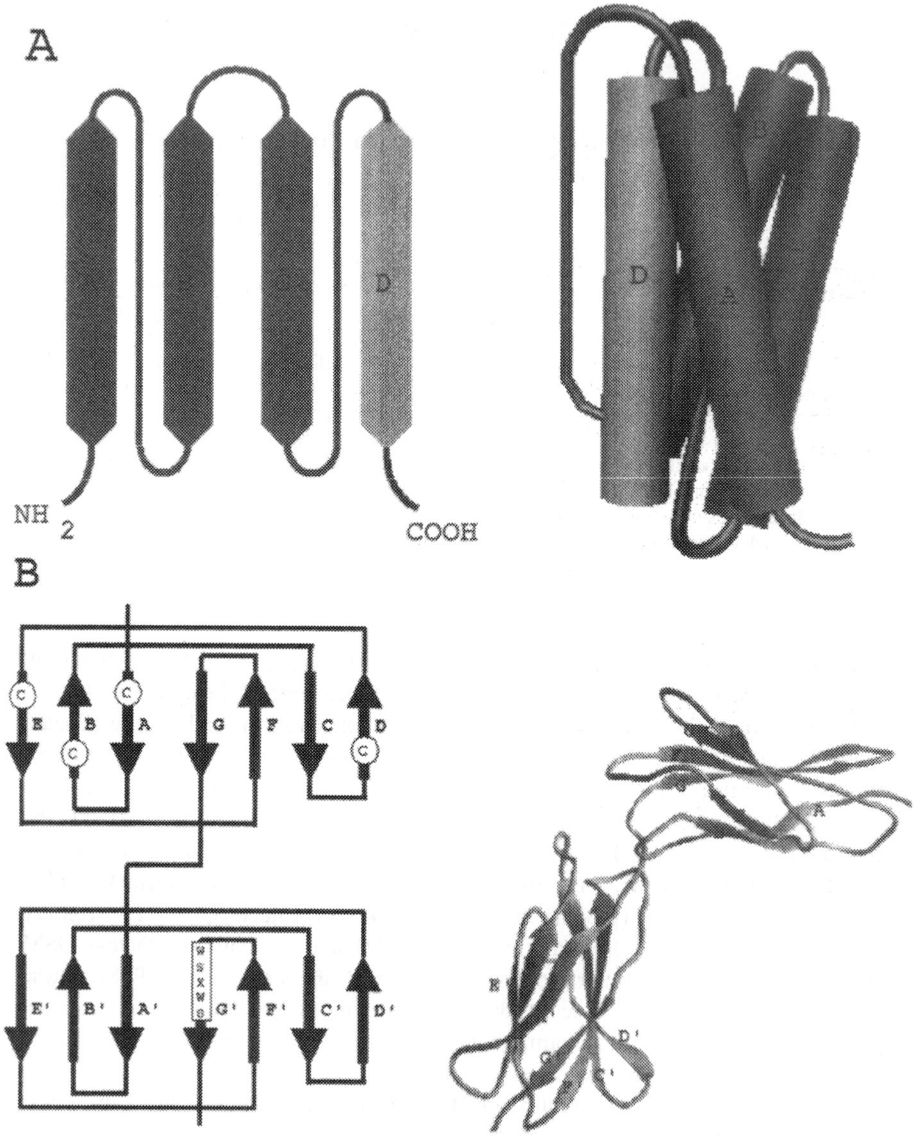

Figure 1. Cytokine and cytokine receptor proteins.
(A) Schematic representation of the 'four helix bundle' cytokines with a up-up-down-down fold. (B) Schematic representation of a cytokine-receptor-homology region consisting of two fibronectin type III-like domains which contain four conserved cysteins (C) and the WSXWS sequence motiv (left panel) and the x-ray structure of the corresponding region of gp130 (right panel) (Bravo et al., 1998).

the C-terminal domain (Bazan, 1990; Sprang and Bazan, 1993) (Figure 1B). Remarkably, the OSMRß could be cloned with a strategy based on this prediction (Mosley et al., 1996).

Four-helical cytokines induce multi-subunit receptor complexes. Families of four-helical bundle cytokines have been defined based on the observation that certain signal transducing receptor subunits are shared between several cytokines. Thus IL-6 type cytokines (IL-6, IL-11, CNTF, CT-1, LIF, OSM) use gp130 as part of their signal transduction complex, whereas IL-3, IL-5 and GM-CSF share the ß-subunit of the IL-3 receptor. IL-2, IL-4, IL-7, and IL-15 induce receptor complexes which contain the common γ-chain of the IL-2 receptor complex (Rubinstein et al., 1998; Taga and Kishimoto, 1997). In contrast to receptors with known intrinsic tyrosine kinase activity such as the EGFR (epidermal growth factor receptor) and PDGFR (platelet derived growth factor receptor), however, cytoplasmic parts of receptors of four-helical cytokine receptors characteristically lack kinase activity. Instead, Janus-Tyrosine Kinases (JAKs) are constitutively bound to the membrane proximal cytoplasmic domains of the receptors and are activated upon ligand-induced dimerization of the receptor subunits responsible for signal transduction (Heinrich et al., 1998; Watowich et al., 1999).

The growth hormone receptor complex was the first four-helical cytokine receptor to be crystallized in a ligand bound form (De Vos et al., 1992) and has therefore become paradigmatic for the understanding of four-helical cytokine receptor complexes: one growth hormone (GH) molecule binds to two receptor molecules via two contact epitopes designated site I and site II. Remarkably, the two receptors use identical amino-acids residues to bind the two different epitopes of the cytokine (Clackson and Wells, 1995; De Vos et al., 1992; Wells, 1996); however, the two interaction sites differ in their free energy of binding. One paradigmatic conclusion derived from these studies was that generally cytokines are recognized by their cognate receptors at sites equivalent to site I and site II of GH (De Vos et al., 1992; Wells, 1996). In the following we will explain that this paradigm does not hold true for IL-6 type cytokines. Based on new experimental data available for the cytokines and receptors of this family we developed a new model for the IL-6 receptor complex.

2. THE IL-6 FAMILY OF CYTOKINES

IL-6 type cytokines are key players in an unusually large spectrum of diseases (Kallen et al., 1997) and are therefore subject to intense investigation in a search for antagonists and hyperagonists of this signaling system. IL-6 and IL-11 signal via a homodimer of gp130, whereas CNTF, LIF, OSM and CT-1 use a heterodimer of gp130 and LIFR. OSM can also assemble a heterodimer of gp130 and OSMR (Taga and Kishimoto, 1997). In contrast to LIF and OSM that can directly induce heterodimerization of gp130 and LIFR (and OSMR, respectively), IL-6, IL-11,

CNTF and possibly CT-1 (Pennica et al., 1996; Robledo et al., 1997) must first bind to specific, non-signalling α-receptors, before they can subsequently induce dimerization of signal-transducing receptor subunits (also termed ß-receptors) (Taga and Kishimoto, 1997). Interestingly, in the case of IL-6, IL-11 and CNTF the α-receptors exist in membrane bound as well as in soluble form (März et al., 1999; Rose-John and Heinrich, 1994). The question arises immediately as to which binding sites on the cytokines are occupied by which receptors?

3. INTERACTIONS OF IL-6 TYPE CYTOKINES WITH THEIR RECEPTORS

The existence of three distinct receptor binding epitopes has been clearly demonstrated for IL-6, IL-11 and CNTF. In contrast to GH, site I (end of AB-loop, C-terminal D-helix) is the interaction site for the cognate α-receptors, i.e. it is not occupied by a signal-transducing ß-receptor. Site II (A/C-helix) is contacted by gp130 in all these molecules, whereas a third ß-receptor binding epitope not present on GH, site III, is occupied by a second gp130 molecule (IL-6, IL-11) or serves as a specific LIFR binding site on CNTF (Ehlers et al., 1994; Kallen et al., 1999; Savino et al., 1994; Savino et al., 1994). Site III consists of the C-terminal A-helix, the N-terminal AB-loop, the BC-loop with adjacent amino-acid residues, the C-terminal CD-loop and the N-terminal D-helix (de Hon et al., 1995; Ehlers et al., 1994; Kallen et al., 1997). For LIF it has been shown that site II is occupied by gp130 and site III by the LIFR, but it is somewhat unclear whether and how site I interacts with any of the two CHR of the LIFR (Hudson et al., 1996; Robinson et al., 1994; Vernalis et al., 1997). This situation somewhat resembles that in G-CSF which signals via a homodimer of G-CSFR contacting site II and site III of G-CSF. It has been shown that there is no α-receptor binding site (site I) on G-CSF (Reidhaar-Olson et al., 1996; Young et al., 1997). Only sparse data are available on the interaction sites of OSM and CT-1 with their receptors (Staunton et al., 1998).

Thus, the paradigm of the GH/ $(GHR)_2$ complex may not be valid for all cytokines and in particular not for IL-6 type cytokines.

The specific α-receptors of IL-6, IL-11 and CNTF belong to the hematopoietic receptor superfamily (Figure 1B). They consist of two membrane proximal domains that fulfill the criteria of a cytokine receptor homology region (Bazan, 1990; Sprang and Bazan, 1993; Wells, 1996): two conserved cystein bridges in the N-terminal and a conserved WSXWS motiv in the C-terminal domain. A third N-terminal Ig-like domain is dispenable for the assembly of a functional receptor complex (Vollmer et al., 1996; Yawata et al., 1993). However, the Ig-like domain stabilizes the receptor during intracellular trafficking through the secretory pathway (Vollmer et al., 1999).

The IL-6 α-receptor binds to site I of IL-6 with a K_D of around 10^{-9} M (Yamasaki et al., 1988; Zohlnhöfer et al., 1992). Although it is not involved in signaling, the mechanism of interaction with site I closely follows that described for the interaction of GH with GHR site I (Barton et al., 1999; De Vos et al., 1992;

Grötzinger et al., 1997). In fact, amino acid residues of IL-6 and IL-6R participating in the interaction at site I were identified by site directed mutagenesis guided by a structural model of the IL-6/IL-6R complex derived from the GH/ $(GHR)_2$ complex (Grötzinger et al., 1997; Horsten et al., 1997; Kalai et al., 1997; Kalai et al., 1997; Kalai et al., 1996; Savino et al., 1994; Savino et al., 1994; Wells, 1996). Whereas both domains of the CHR of the GHR contribute equally to the site I affinity, it appears that the C-terminal CHR domain of the IL-6R accounts for more then 90% of the binding energy to the cytokine (Özbek et al., 1998). Molecular-modeling studies suggest that this might even be more pronounced with regard to the IL-11/IL-11R complex (K. Schleinkofer, unpublished results).

In contrast to the GHR, all ß-receptors of the IL-6 type family (gp130, LIFR, OSMR) have three membrane proximal fibronectin-type-III (FN-III) like domains (Figure 2). In gp130 these domains carry one CHR and one Ig-like domain, whereas LIFR and OSMR possess a second CHR or the C-terminal domain of a CHR, respectively, on top of the Ig-like domain (see Figure 2).

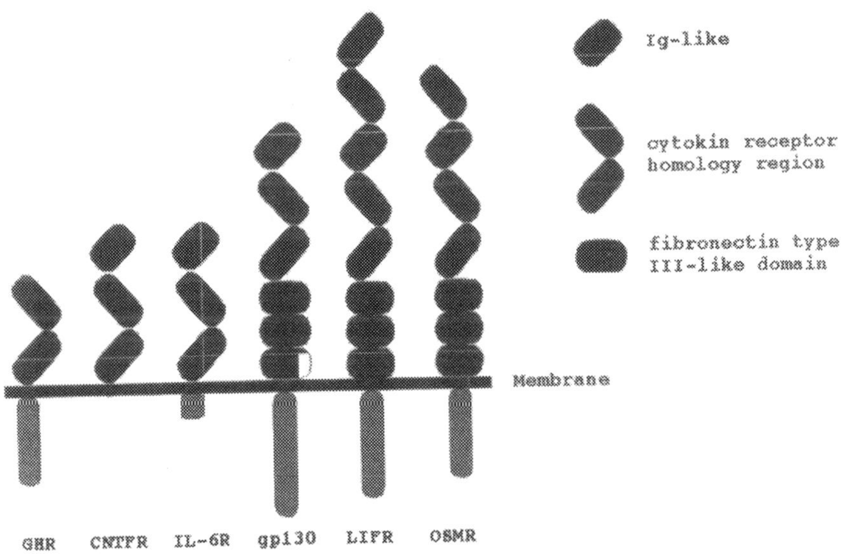

Figure 2. Schematic representation of the domain structure of GHR, CNTFR, IL-6R, gp130, LIFR and OSMR.
The cytoplasmatic portions of the receptors are shown in grey. The extracellular parts are coded as indicated in the Figure.

gp130 mutants with deleted FN-III domains are not capable of signal transduction, but can still bind to the IL-6/IL-6R via the three N-terminal domains (D_1-D_3) (Horsten et al., 1997; Horsten et al., 1995). Apparently, the N-terminal domains contain two distinct cytokine-binding epitopes (Hammacher et al., 1998; Kurth et al., 1999). Molecular modeling guided mutagenesis revealed that analogously to the GHR, the CHR of gp130 is one of these epitopes contacting site II of the cytokine (Horsten et al., 1997; Kurth et al., 1999). Two amino-acid residues in the CHR of gp130 (Phe192; D_2 and Val252; D_3) are essential for binding to the complex of IL-6/IL-6R and IL-11/ IL-11R (Horsten et al., 1997; Kurth et al., 1999). The X-ray and NMR structures now available for the complete gp130 CHR (D_2–D_3) and its isolated C-terminal domain (D_3), respectively, revealed that the side chains of these amino-acid residues are solvent exposed (Bravo et al., 1998; Kernebeck et al., 1999). Val252 is located in the flexible loop between the ß-strands B and C of the C-terminal domain (Kernebeck et al., 1999) (see Figure 1B) which allows its side chain to contact that of residue Phe191 located in the E-F loop of the N-terminal domain (D_2) of the gp130 CHR (Bravo et al., 1998). This contact might be released when gp130 binds to the IL-6/IL-6R complex.

The function of the Ig-like domain of gp130 was clarified by an Ig-deletion mutant of gp130 that showed largely impaired signaling and binding to the IL-6/IL-6R complex (Hammacher et al., 1998; Kurth et al., 1999). Therefore, the gp130 Ig-like domain which is not present in the GHR constitute a second cytokine binding epitope on gp130. Epitope swapping experiments with IL-6 and CNTF (see above) suggested that site III of IL-6 type cytokines interacts with the Ig-like domain of gp130 and LIFR, respectively (Kallen et al., 1999). This hypothesis might explain the observation that the species specificity between human and murine LIF is mediated by the Ig-like domain of the LIFR (Owczarek et al., 1997). Otherwise, it is unknown at present which of the OSMR or LIFR domains establish the contact to the different epitopes on the cytokines.

4. HEXAMER OR TETRAMER – OR BOTH ?

The existence of three different receptor binding regions on IL-6 and CNTF suggests that two ß- and one α-receptor subunit must bind to these epitops in order to assemble a functional receptor complex. The simplest model of these receptor complexes would therefore be a tetramer. However, soluble forms of the IL-6 receptor components have been expressed as recombinant proteins and were used to assess the stoichiometry of the ligand-receptor complex by gel filtration and analytical ultracentrifugation (Ward et al., 1994). This approach indicated the existence of a hexameric IL-6 receptor complex consisting of two IL-6 molecules, two soluble α-units (IL-6R) and two soluble ß-subunits (gp130) (Ward et al., 1994) (Figure 3). An alternative stoichiometric analysis of the IL-6R complex by a series of elegant immunoprecipitation experiments performed with recombinant soluble proteins also assumed a hexameric IL-6 recepor complex (Paonessa et al., 1995). It should be noted, though, that both types of experimental approaches were performed at micromolar concentrations of soluble proteins (Paonessa et al., 1995;

Ward et al., 1994). Also using immunoprecipitation experiments it was concluded that in solution the CNTF receptor complex exists as a hexamer consisting of two CNTF, two soluble CNTF-R molecules and one soluble gp130 and LIFR molecule each (De Serio et al., 1995).

However, the hexamer model cannot account for the results of recent experiments with a chimeric molecule in which site III had been shown to be a LIFR specific binding site. Furthermore, site III had been transferred from CNTF to IL-6 (Kallen et al., 1999). This transfer of a complete LIFR binding epitope resulted in an IL-6/CNTF chimera which still bound to the IL-6R, but required a heterodimer of gp130 and LIFR for signaling instead of a gp130 homodimer. The specific biological activity of the chimeric cytokine on responsive cells was identical to that of human LIF (Kallen et al., 1999). As Figure 3 shows, the hexameric receptor model of CNTF predicts that the CNTF site II and site III can bind to both, gp130 and LIFR. Therefore, the existence of a specific LIFR binding epitope on CNTF strongly argues against a hexameric composition of the active membrane bound CNTF receptor complex. However, the above results can easily be explained in the framework of a tetrameric model of the IL-6R complex recently suggested by us (Grötzinger et al., 1997).

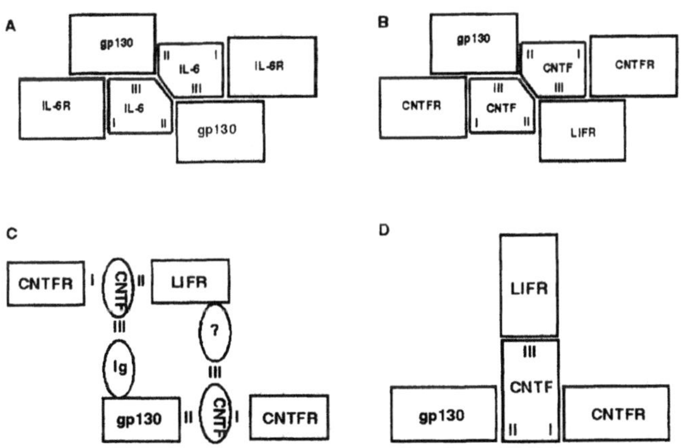

Figure 3. Models of the IL-6-receptor and CNTF-receptor complexes. (A) Paonessa's model of the IL-6-receptor complex (Paonessa et al., 1995) (B) De Serio's model of the CNTF-receptor complex (De Serio et al., 1995). (C) Simpson's model of the IL-6-receptor complex adapted to CNTF (Kallen et al., 1999; Simpson et al., 1997). Note that there is no information available on domains of the LIFR involved in binding to cytokines. (D) Tetrameric model of the CNTF-receptor complex (Kallen et al., 1999).

Recently, the EPOR was shown to exist as a preformed dimer on the cell surface (Livnah et al., 1999; Remy et al., 1999). In fact, results published by Auguste and

Gascan (Auguste et al., 1997) clearly demonstrate that gp130 and OSMR also form an inactive heterodimer on the cell surface that is only phosphorylated following ligand binding. Also, the LIFR can be coprecipitated on appropriate cells with antibodies against gp130 even in the absence of LIF (K.-J. Kallen, unpublished observation) which suggests the existence of an inactive gp130/LIFR dimer. Interestingly, cristallization of the CHR of gp130 revealed that in the absence of ligand these CHRs contact each other. Since there were no N-terminal Ig-like domains present in the recombinant gp130 proteins used, no conclusion concerning a contact between the Ig domains can be drawn (Bravo et al., 1998). One might expect that on the cell membrane gp130 also exists as a preformed, but inactive homodimer. Starting from our tetrameric model of the IL-6 receptor complex we would predict that the two gp130 molecules would contact each other via their CHR and Ig-like domain (Figure 4) and not only via their CHRs as in the preformed EPOR homodimer (Livnah et al., 1999). The symmetry of the gp130 homodimer assembly suggested here has an important corollary.

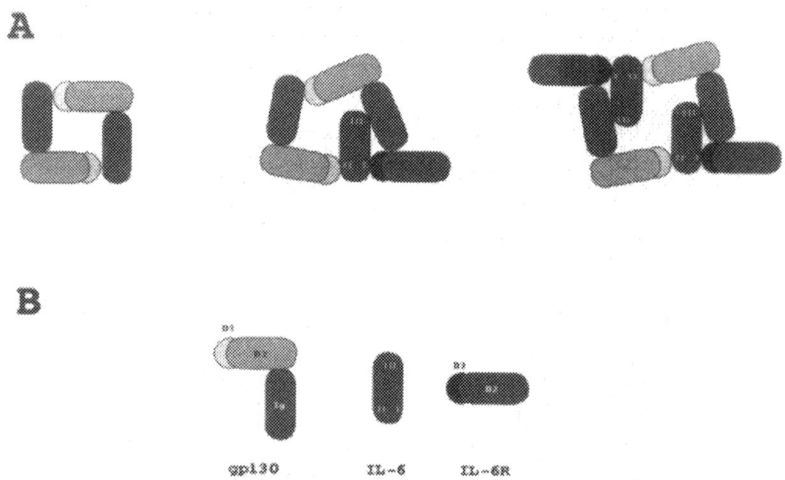

Figure 4. Tetramer-Hexamer transition model of the IL-6 receptor complex. (A) The proposed model of the IL-6/IL-6R interaction with gp130. A preformed, symmetrical gp130 dimer (left) associates with one IL-6/IL-6R complex forming the tetramer (middle) which then is able to bind another IL-6/IL-6R complex forming the hexameric state (right). The three N-terminal domains of gp130 are color coded in red (D_1, Ig-like), yellow (D_2) and orange (D_3). The two domains of the IL-6R CHR are colour coded as light grey (D_2) and black (D_3) and IL-6 is depicted in green. Receptor interaction site in IL-6 are denoted as I, II and III.

If IL-6 bound to the IL-6R partitions into a preformed gp130 homodimer, a tetrameric complex will assemble (Figure 4). However, due to the symmetry of the original gp130 homodimer the tetramer would still be able to bind another IL-6/IL-

6R resulting in a hexameric complex (Figure 4). Notably, the relative position of the two gp130 subunits in the tetrameric differs from that in a hexameric receptor complex (Figure 4). Lower concentrations of IL-6/IL-6R might therefore lead to the formation of active tetramers, whereas higher concentrations would shift the tetramer into an inactive hexamer. It should be noted that such a hexameric state relies on a symmetrical arrangement of ligand recognition sites and would therefore be impossible for cytokines which signal via heterodimeric ß-receptor subunits such as CNTF, LIF, CT-1 and OSM. Interestingly, the dose response curves of IL-6 and IL-11 which signal via a homodimer of gp130 are all bell shaped (Curtis et al., 1997; van Dam et al., 1993). Thus, supraoptimal cytokine concentrations lead to a decline of the biological response, a phenomenon predicted by the assumption of a tetramer – hexamer shift. On the other hand, CNTF does not show a decline of the biological response over an extremely wide concentration range as predicted by our model (Kallen et al., 1999). Rather than reflecting an artefact due to the high concentrations of the soluble receptor subunits, the formation of a hexamer at higher concentrations of IL-6 might be the reason for the self-antagonization of IL-6 at higher concentrations. Thus, the mode of assembly of the IL-6 receptor complex could constitute one element of the extremely tight regulation (Curtis et al., 1997; Müller-Newen et al., 1998) by which organisms try to counteract surplus IL-6. It is noteworthy that a similar self-antagonizing capacity is already known for insulin which also signals via a homodimeric receptor (De Meyts, 1994).

5. CONCLUDING REMARKS

The question whether the functional complex of IL-6, IL-6R and gp130 on the cell surface is a tetramer or a hexamer has still to be answered experimentally, although there is accumulating evidence that the signaling complex is a tetramer. Of note, even the stoichiometry of a whole cytokine/ receptor assembly in the (so far not solved) crystall structure might reflect the situation in solution, due to the high concentrations of the molecules used. Thus, further biochemical experimentation is needed to elucidate a possible transition between tetrameric or hexameric receptor complex states.

6. REFERENCES

Auguste P., Guillet C., Fourcin M., Olivier C., Veziers J., Pouplard Barthelaix A. and Gascan H.. "Signaling of type II oncostatin M receptor." *J. Biol. Chem.* 272 (1997): 15760-15764.

Barton V.A., Hudson K.R. and Heath J.K. "Identification of three distinct receptor binding sites of murine interleukin-11." *J. Biol. Chem.* 274 (1999): 5755-5761.

Bazan J.F. "Structural design and molecular evolution of a cytokine receptor superfamily." *Proc. Natl. Acad. Sci. USA* 87 (1990): 6934-6938.

Bazan J.F. "Neuropoietic cytokines in the hematopoietic fold." *Neuron* 7 (1991): 197-208.

Bravo J., Staunton D., Heath J.K. and Jones E.Y. "Crystal structure of a cytokine-binding region of gp130." *EMBO J.* 17. (1998): 1665-1674.

Clackson T. and Wells J.A. "A hot spot of binding energy in a hormone-receptor interface." *Science* 267 (1995): 383-386.

Curtis D.J., Hilton D.J., Roberts B., Murray L., Nicola N. and Begley C.G. "Recombinant soluble interleukin-11 (IL-11) receptor alpha-chain can act as an IL-11 antagonist." *Blood* 90 (1997): 4403-4412.

de Hon F.D., ten Boekel E., Herrman J., Clement C., Ehlers M., Taga T., Yasukawa K., Ohsugi Y., Kishimoto T., Rose-John S., Wijdenes J., Kastelein R., Aarden L.A. and Brakenhoff J.P.J. "Functional distinction of two regions of human interleukin 6 important for signal transduction via gp130." *Cytokine* 7 (1995): 398-407.

De Meyts P. "The structural basis of insulin and insulin-like growth factor-I receptor binding and negative co-operativity, and its relevance to mitogenic versus metabolic signalling." *Diabetologia* 37 (1994): 135-148.

De Serio A., Graziani R., Laufer R., Ciliberto G. and Paonessa G. "In vitro binding of ciliary neurotrophic factor to its receptors: evidence for the formation of an IL-6-type hexameric complex." *J. Mol. Biol.* 254 (1995): 795-800.

De Vos A.M., Ultsch M. and Kossiakoff A.A.. "Human growth hormone and extracellular domain of its receptor: crystall structure and of the complex." *Science* 255 (1992): 306-312.

Ehlers M., Grötzinger J., deHon F.D., Müllberg J., Brakenhoff J.P., Liu J., Wollmer A. and Rose-John S. "Identification of two novel regions of human IL-6 responsible for receptor binding and signal transduction." *J. Immunol.* 153 (1994): 1744-1753.

Grötzinger J., Kurapkat G., Wollmer A., Kalai M. and Rose-John S. "The family of the IL-6-type cytokines: Specificity and Promiscuity of the Receptor Complexes." *Proteins: Structure, Function, and Genetics* 27 (1997): 96-109.

Hammacher A., Richardson R.T., Layton J.E., Smith D.K., Angus L.J., Hilton D.J., Nicola N.A., Wijdenes J. and Simpson R.J. "The immunoglobulin-like module of gp130 is required for signaling by interleukin-6, but not by leukemia inhibitory factor." *J. Biol. Chem.* 273 (1998): 22701-22707.

Heinrich P.C., Behrmann I., Müller-Newen G., Schaper F. and Graeve L. "Interleukin-6-type cytokine signalling through the gp130/Jak/STAT pathway." *Biochem. J.* 334 (1998): 297-314.

Horsten U., Müller-Newen G., Gerhartz C., Wollmer A., Wijdenes J., Heinrich P.C. and Grötzinger J. "Molecular modeling-guided mutagenesis of the extracellular part of gp130 leads to the identification of contact sites in the interleukin-6 (IL-6).IL-6 receptor.gp130 complex." *J. Biol. Chem.* 272 (1997): 23748-23757.

Horsten U., Schmitz-Van de Leur H., Müllberg J., Heinrich P.C. and Rose-John S. "The membrane distal half of gp130 is responsible for the formation of a ternary complex with IL-6 and the IL-6 receptor." *FEBS Lett.* 360 (1995): 43-46.

Hudson K.R., Vernallis A.B. and Heath J.K. "Characterization of the receptor binding sites of human leukemia inhibitory factor and creation of antagonists." *J. Biol. Chem.* 271 (1996): 11971-11978.

Kalai M., Montero Julian F.A., Brakenhoff J.P., Fontaine V., De Wit L., Wollmer A., Brailly H., Content J. and Grötzinger J. "Analysis of the mechanism of action of anti-human interleukin-6 and anti-human interleukin-6 receptor-neutralising monoclonal antibodies." *Eur. J. Biochem.* 249 (1997): 690-700.

Kalai M., Montero Julian F.A., Grötzinger J., Fontaine V., Vandenbussche P., Deschuyteneer R., Wollmer A., Brailly H. and Content J. "Analysis of the human interleukin-6/human interleukin-6 receptor binding interface at the amino acid level: proposed mechanism of interaction." *Blood* 89 (1997): 1319-1333.

Kalai M., Montero-Julian F.A., Grötzinger J., Wollmer A., Morelle D., Brochier J., Rose-John S., Heinrich P.C., Brailly H. and Content J. "Participation of two Ser-Ser-Phe-Tyr repeats in interleukin-6 (IL-6)- binding sites of the human IL-6 receptor." *Eur. J. Biochem.* 238 (1996): 714-723.

Kallen K.-J., Grötzinger J., Lelièvre E., Vollmer P., Aasland D., Renné C., Müllberg J., Meyer zum Büschenfelde K.-H., Gascan H. and Rose-John S. "Receptor recognition sites of cytokines are organized as exchangeable moduls: transfer of the LIFR binding site from CNTF to IL-6." *J. Biol. Chem.* 274 (1999): 11859-11867.

Kallen K.-J., Meyer zum Büschenfelde K.H. and Rose-John S. "The therapeutic potential of interleukin-6 hyperagonists and antagonists." *Exp. Opin. Invest. Drugs* 6 (1997): 237-266.

Kernebeck T., Pflanz S., Müller-Newen G., Kurapkat G., Scheek R.M., Dijkstra K., Heinrich P.C., Wollmer A., Grzesiek S. and Grötzinger J. "The signal transducer gp130: solution structure of the carboxy-terminal domain of the cytokine receptor homology region." *Protein Sci.* 8 (1999): 5-12.

Kurth I., Horsten U., Pflanz S., Dahmen H., Küster A., Grötzinger J., Heinrich P.C. and Müller-Newen G. "Activation of the signal transducer glycoprotein 130 by both IL-6 and IL-11 requires two distinct binding epitopes." *J. Immunol.* 162 (1999): 1480-1487.

Livnah O., Stura E.A., Middleton S.A., Johnson D.L., Jolliffe L.K. and Wilson I.A. "Crystallographic evidence for preformed dimers of erythropoietin receptor before ligand activation." *Science* 283 (1999): 987-990.

März P., Otten U. and Rose-John S. "Neuronal Activities of IL-6 Type Cytokines often Depend on Soluble Cytokine Receptors." *Eur. J. Neurosci.* (1999), in press.

Mosley B., De Imus C., Friend D., Boiani N., Thoma B., Park L.S. and Cosman D. "Dual oncostatin M (OSM) receptors. Cloning and characterization of an alternative signaling subunit conferring OSM-specific receptor activation." *J. Biol. Chem.* 271 (1996): 32635-32643.

Müller-Newen G., Küster A., Hemmann U., Keul R., Horsten U., Martens A., Graeve L., Wijdenes J. and Heinrich P.C. "Soluble IL-6 receptor potentiates the antagonistic activity of soluble gp130 on IL-6 responses." *J. Immunol.* 161 (1998): 6347-6355.

Owczarek C.M., Zhang Y., Layton M.J., Metcalf D., Roberts B. and Nicola N.A.. "The unusual species cross-reactivity of the leukemia inhibitory factor receptor alpha-chain is determined primarily by the immunoglobulin-like domain." *J. Biol. Chem.* 272 (1997): 23976-23985.

Özbek S., Grötzinger J, Krebs B., Fischer M., Wollmer A., Jostock T., Müllberg J. and Rose-John S. "The Membrane Proximal Cytokine Receptor Domain of the Human Interleukin-6-Receptor is Sufficient for Ligand Binding but not for gp130 Association." *J. Biol. Chem.* 273 (1998): 21374-21379.

Paonessa G., Graziani R., De Serio A., Savino R., Ciapponi L., Lahm A., Salvati A.L., Toniatti C. and Ciliberto G. "Two distinct and independent sites on IL-6 trigger gp 130 dimer formation and signalling." *EMBO J.* 14 (1995): 1942-1951.

Pennica D., Arce V., Swanson T.A., Vejsada R., Pollock R.A., Armanini M., Dudley K., Phillips H.S., Rosenthal A., Kato A.C. and Henderson C.E. "Cardiotrophin-1, a cytokine present in embryonic muscle, supports long-term survival of spinal motoneurons." *Neuron* 17 (1996): 63-74.

Reidhaar-Olson J.F., De Souza-Hart J.A. and Selick H.E. "Identification of residues critical to the activity of human granulocyte colony-stimulating factor." *Biochemistry* 35 (1996): 9034-9041.

Remy I., Wilson I.A. and Michnick S.W. "Erythropoietin receptor activation by a ligand-induced conformation change." *Science* 283 (1999): 990-993.

Robinson R.C., Grey L.M., Staunton D., Vankelecom H., Vernallis A.B., Moreau J.F., Stuart D.I., Heath J.K. and Jones E.Y. "The crystal structure and biological function of leukemia inhibitory factor: implications for receptor binding." *Cell* 77 (1994): 1101-1116.

Robledo O., Fourcin M., Chevalier S., Guillet C., Auguste P., Pouplard Barthelaix A., Pennica D. and Gascan H. "Signaling of the cardiotrophin-1 receptor. Evidence for a third receptor component." *J. Biol. Chem.* 272 (1997): 4855-4863.

Rose-John S. and Heinrich P.C. "Soluble receptors for cytokines and growth factors: generation and biological function." *Biochem. J.* 300 (1994): 281-290.

Rubinstein M., Dinarello C.A., Oppenheim J.J. and Hertzog P. "Recent advances in cytokines, cytokine receptors and signal transduction." *Cytokine Growth Factor Rev.* 9 (1998): 175-181.

Savino R., Ciapponi L., Lahm A., Demartis A., Cabibbo A., Toniatti C., Delmastro P., Altamura S. and Ciliberto G. "Rational design of a receptor super-antagonist of human interleukin-6." *EMBO J.* 13 (1994): 5863-5870.

Savino R., Lahm A., Salvati A.L., Ciapponi L., Sporeno E., Altamura S., Paonessa G., Toniatti, C. and Ciliberto G. "Generation of interleukin-6 receptor antagonists by molecular-modeling guided mutagenesis of residues important for gp130 activation." *EMBO J.* 13 (1994): 1357-1367.

Sprang S.R., and Bazan J.F. "Cytokine structural taxonomy and mechanism of receptor engagement." *Curr. Opin. Struct. Biol.* 3 (1993): 815-827.

Staunton D., Hudson K.R. and Heath J.K. "The interactions of the cytokine-binding homology region and immunoglobulin-like domains of gp130 with oncostatin M: implications for receptor complex formation." *Protein Eng.* 11 (1998): 1093-1102.

Taga T. and Kishimoto T. "gp130 and the Interleukin-6 Family of Cytokines." *Annu. Rev. Immunol.* 15 (1997): 797-819.

van Dam M., Müllberg J., Schooltink H., Stoyan T., Brakenhoff J.P., Graeve L., Heinrich P.C. and Rose-John S. "Structure-function analysis of interleukin-6 utilizing human/murine chimeric molecules. Involvement of two separate domains in receptor binding." *J. Biol. Chem.* 268 (1993): 15285-15290.

Vernalis A.B., Hudson K.R. and Heath J.K. "An antagonist for the leukemia inhibitory factor receptor inhibits leukemia inhibitory factor, cardiotrophin-1, ciliary neurotrophic factor, and oncostatin M." *J. Biol. Chem.* 272 (1997): 26947-26952.

Vollmer P., Oppmann B., Voltz N. and Rose-John S. "The Immunoglobulin-like domain of the human IL-6 Receptor is important for protein Stability and Processing." *Eur. J. Biochem.* (1999), in press.

Vollmer P., Peters M., Ehlers M., Yagame H., Matsuba T., Kondo M., Yasukawa K., Büschenfelde K.H. and Rose-John S. "Yeast expression of the cytokine receptor domain of the soluble interleukin-6 receptor." *J. Immunol. Methods* 199 (1996): 47-54.

Ward L.D., Howlett G.J., Discolo G., Yasukawa K., Hammacher A., Moritz R.L. and Simpson R.J. "High affinity interleukin-6 receptor is a hexameric complex consisting of two molecules each of interleukin-6, interleukin-6 receptor, and gp-130." *J. Biol. Chem.* 269 (1994): 23286-23289.

Watowich S.S., Liu K.D., Xie X., Lai S.Y., Mikami A., Longmore G.D. and Goldsmith M.A. "Oligomerization and scaffolding functions of the erythropoietin receptor cytoplasmic tail." *J. Biol. Chem.* 274 (1999): 5415-5421.

Wells J.A. "Binding in the growth hormone receptor complex." *Proc. Natl. Acad. Sci. USA* 93 (1996): 1-6.

Wells J.A. "Hematopoietic Receptor Complexes." *Annu. Rev. Biochem.* 65 (1996): 609-634.

Yamasaki K., Taga T., Hirata Y., Yawata H., Kawanishi Y., Seed B., Taniguchi T., Hirano T. and Kishimoto T. "Cloning and expression of the human interleukin-6 (BSF-2/IFN beta 2) receptor." *Science* 241 (1988): 825-828.

Yawata H., Yasukawa K., Natsuka S., Murakami M., Yamasaki K., Hibi M., Taga T. and Kishimoto T. "Structure-function analysis of human IL-6 receptor: dissociation of amino acid residues required for IL-6-binding and for IL-6 signal transduction through gp130." *EMBO J.* 12 (1993): 1705-1712.

Young D.C., Zhan H., Cheng Q.L., Hou J. and Matthews D.J. "Characterization of the receptor binding determinants of granulocyte colony stimulating factor." *Protein Sci.* 6 (1997): 1228-1236.

Zohlnhöfer D., Graeve L., Rose-John S., Schooltink H., Dittrich E. and Heinrich P.C. "The hepatic interleukin-6 receptor. Down-regulation of the interleukin-6 binding subunit (gp80) by its ligand." *FEBS Lett.* 306(1992): 219-222.

F. BOST, Y.-M. YANG, C. LIU, W. CHARBONO, N. DEAN,
R. MCKAY, X.-P. LU, O. POTAPOVA, M. PFAHL AND
D. MERCOLA

INHIBITION OF GROWTH FACTOR STIMULATED PATHWAYS FOR THE TREATMENT OF PROSTATE CANCER

Abstract. Recent studies have revealed that the alternate Map Kinase pathway, termed the Jun Kinase pathway, is commonly activated by the Epidermal Growth Factor and Serum in a variety of human tumor lines, especially prostate carcinoma. This pathway activates the transcription factor c-Jun, a factor required for transformation by a number of cellular oncogenes such as activated c-Ras. In order to evaluate the role of Jun Kinase in prostate cancer, high affinity antisense oligonucleotides complementary to the two major isoforms of JNK were prepared and antisense JNK-2 was shown to block prostate cell line growth in vitro. Application of antisense to mice bearing PC3 prostate carcinoma cell xenografts by daily systemic I.P. treatment for 27 days led to tumor growth inhibition and subsequently complete regression of 14/15 of the solid tumors. Inhibition of JNK activity in treated tumors was confirmed. These studies reveal an important function of JNK, especially JNK-2, in prostate carcinoma and suggest a new treatment modality.

1. INTRODUCTION

1.1. Meeting Professor Wollmer

Axel Wollmer visited Oxford with Dietrich Brandenburg in 1971 when I (D.M.) was a postdoctoral fellow in Dorothy Hodgkin's laboratory. The synthesis of insulin and many insightful chemical studies aimed at understanding the mechanism of action of insulin had been carried out in Aachen over a number of years as delightfully reviewed by Professor Zahn in this volume. With this visit started a new wave of studies that would wed the insights obtained from the three dimensional structure determined in Oxford in 1969 (Adams et al., 1969) with the chemical prowess of the Deutschen Wollforschungsinstitut and optical studies. The promise of the approach was well foreshadowed by the chemical crosslinking studies of Professor Zahn in which he showed that the N-terminus of the A-chain of monomeric insulin in solution was likely very close to the C-terminus of the B-chain (Zahn and Meienhofer, 1958a; 1958b) and, indeed, this was borne out in the crystal structure (Blundell et al., 1971; 1972). Added to this chemical approach, Axel Wollmer was introducing new spectroscopic methods of circular dichroism and fluorescence spectroscopy with the promise of following events as they occur in physiological-like solution setting. Moreover he was providing his uncompromising interest in

M. Federwisch, M. Leyck Dieken and P. De Meyts (eds.), Insulin & Related Proteins, 213–231.

precision. It was clear that the three-way approach would be fruitful for years to come because the parties involved in the insulin crystal structure, the chemistry, and the biochemistry were devoted to their subjects and eager to work together. In addition, I had developed a spectroscopic method based on circular dichroism measurements for directly observing the distances between chromophores and, for the FITC covalently attached to the N-terminus of the A-chain and the C-terminus of the B-chain of insulin, observed the proximity that Professor Zahn had detected many years earlier and showed that these observations in solution were consistent with the crystal structure (Mercola et al., 1972). This gave me confidence in the methods and enthusiasm for Axel Wollmer's initiative to bring optical methods to stage center in the cooperative analysis of insulin. Axel Wollmer went on to study numerous chemically modified forms of insulin many of which had been designed by Dietrich Brandenburg and his colleagues specifically to test predictions based on the crystal structure. These included analysis of monomer interaction theories, the development of spectroscopic methods for following the dissociation of insulin and, therefore, the equilibrium constants, over several orders of magnitude, and the direct observation of the so-called "2-zinc-to-4-zinc" (T→R) transition in solution. With Wolfgang Straßburger, the activity of the tyrosyl side chains was calculated using rudimentary "monopole approximation" theory applied to the atomic coordinates of the refined crystal structure. These and many other studies are documented in over 50 publications on insulin. Hallmarks of all these studies are exceptional precision and care that provide the reader with assurance. I wonder how Axel can be prompted or cajoled into continuing these studies.

1.2. Introduction to Jun Kinase

Our interest is the mechanisms of growth factors and the use of this information in devising cancer treatment strategies. Insulin, of course, is a universally required growth factor and, for example, is an essential component of nearly all defined media that support mammalian cell growth in culture without serum. Interestingly, in contrast to all other well-studied serum and growth factors including the insulin-like growth factors (Smith and Gunnell, 2000), there is almost no indication that the insulin signal transduction pathway can be subverted as a means of enhanced growth or malignancy. However, related signal transduction pathways such as those involved in mediating the epidermal growth factor (EGF) are believed to be involved in a number of human cancers. The receptors for EGF are closely related to and activated by binding to the HER2 receptors that are known to enhance growth of many human breast carcinomas and other cancers. The EGF receptor gene is itself commonly amplified in human glioblastoma multiforme, lung cancer, and others cancers. In these cases, EGF promotes tumor growth. Indeed, lung cancers commonly produce EGF itself thereby establishing an "autocrine" loop that drives growth through the over-expressed EGF receptors.

How does the EGF receptor promote growth? It were studies on the growth promoting properties of insulin by Sturgill and co-workers that first provided insight to the role of the MAP Kinase pathway (Ray et al., 1988; Rossamondo et al., 1989) (Figure 1). Plasma membrane bound receptor tyrosine kinases (RTKs) such as those for binding of platelet-derived growth factor (PDGF), nerve growth factor (NGF), and EGF dimerize and become "activated' upon binding their corresponding ligand growth factors. This activation is an autocatalytic tyrosine phosphorylation activity. The phosphotyrosine residues reside on the cytosolic side of the transmembrane RTK structure. These phosphotyrosine groups and neighbouring amino acids define binding sites for adapter and docking proteins that lead to a membrane-bound complex including c-ras, itself a target of tumor progression in many cases. The activation of c-ras either by mutation or by constitutively activated up-stream growth factor receptors such as HER2 or autocrine driven and amplified PDGF and/or EGF receptors, implies activation of transcription of c-fos and other growth-mediating transcription factors such as c-myc that are "down-stream" in the MAPK pathway (Figure 1). The c-myc factor promotes the entry of cells into the S-phase of the cell and commitment to division. The mechanism of action of EGF has been reviewed recently (Schlessinger, 2000).

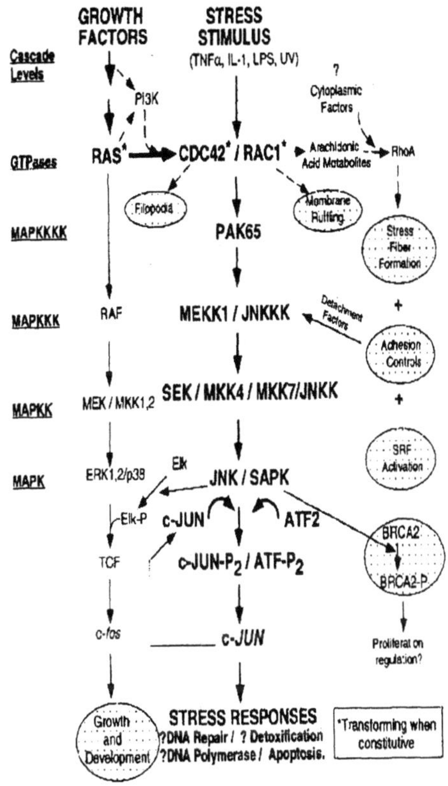

Figure 1. Outline of the signal-transduction pathway from a typical growth factor to the MAPK pathway (center) together with the generic designation of MAP kinase cascade (left). One potential mechanism for activation of the Jun Kinase (right) pathway is via activation of the Ras GTPase family members CDC42 and Pak1. Asterisk indicates kinases that are transforming when constitutively active.

1.3. Jun Kinase Mediates Transformation

The Jun Kinase or c-Jun N-terminal Kinase pathway is an alternate MAPK pathway with homologous components at each level of the signal transduction pathway

(Figure 1) (reviewed in Davis, 2000). The function of the Jun Kinase is most commonly associated with mediating apoptosis of stress such as that caused by ultraviolet irradiation or withdrawal of a trophic factor. However, apoptosis is commonly disrupted in human tumors by inactivation of the p53/Mdm2/ARF pathway (reviewed in Sherr, 2000), which occurs in over 50% of human tumors or by increased activity of the antiapopotsis factor BCL2 or by other means. In addition, there are numerous indications that the Jun Kinase pathway can promote growth. First, Jun Kinase phosphorylates the transcription factor c-Jun at two serine residues 63 and 73 in the N-terminal activation domain (Smeal et al., 1991; Binétruy et al., 1991; Smeal et al., 1992; Kyriakis et al., 1994), which greatly enhances the transactivation potential of c-Jun (Pulverer et al., 1991; Minden et al., 1994). The activation of c-Jun by N-terminal phosphorylation is an essential event for the transformation of primary fibroblasts and other model cell systems by activated oncogenes such as src, Ha-ras, c-raf, and v-sis (Smeal et al., 1992; Franklin et al., 1992; Minden et al., 1995; Clark et al., 1997). There is considerable evidence that JNK may be an important mediator of transformation not only by activated Ras (Minden et al., 1995; Clark et al., 1997) but also by activated forms of several GTPases such as Cdc42 (Wu et al., 1998, Tolkacheva et al., 1997; Glaven et al., 1999), Rac (Minden et al., 1995; Wu et al., 1998, Tolkacheva et al., 1997; Glaven et al., 1999), and C3G (Tanaka and Hanafusa, 1998; Moriguchi et al., 1997; Tanaka et al., 1997) as well as the oncogenic GTPases Vav and Dbl (Crespo et al., 1996; Crespo et al., 1997; del Peso et al., 1997), the oncogenic proteins v-Fps/Fes (Li and Smithgall, 1998), v-Crk (Tanaka and Hanafusa, 1998; Moriguchi et al., 1997; Tanaka et al., 1997), and the Met oncogene (Rodrigues et al., 1997).

1.4. Jun Kinase Is Required for the Proliferation of Certain Human Cancer Cell Lines

Although less information is available, there are indications that Jun Kinase may play an essential role in certain human cancers (Clark et al., 1997; Dickens et al., 1997; Burgess et al., 1998; Potapova et al., 1997; Bost et al., 1997; Logan et al., 1997; Clarke et al., 1998; Antonyak et al., 1997; Bost et al., 1999; Prada et al., 1993). The Bcr-abl leukemia oncogene activates JNK and requires c-Jun for transformation implicating the JNK pathway in transformation by a human leukemia oncogene (Dickens et al., 1997; Burgess et al., 1998). JNK is constitutively active in a variety of solid human tumor cells (Potapova et al., 1997; Bost et al., 1997). Moreover, EGF appears to activate JNK in a phosphoinositol 3-kinase-dependent manner (Logan et al., 1997; Clarke et al., 1998). Similarly, expression of the EGF receptor variant type III, a constitutively active naturally occurring mutation that is found in many types of human tumors, leads to transformation of NIH-3T3 cells, constitutive activation of JNK and down-regulation of the MAPK/ERK pathway (Antonyak et al., 1997). Inhibitors of phosphoinositol-3-kinase reverse these effects (Antonyak et al., 1997). In the case of A549 NSCLC cells, EGF causes preferential activation of JNK and a 2-fold enhancement of four characteristics of transformation: proliferation (Bost et al., 1997; 1999), tumor-take rate (Bost et al., 1999), growth of xenograft (Bost et al., 1999), and colony formation in soft agar (Bost et al., 1999). The *in vitro* effects are completely inhibited by expression of a nonphosphorylatable derivative of c-Jun, c-Jun(S63A,S73A), or by application of specific antisense Jun kinase-2 oligonucleotides but not by antisense JNK1 oligonucleotides indicating an essential role of JNK2 in transformation (Bost et al., 1997; 1999). Thus, at least in the lung carcinoma cell line, an EGF-based autocrine system appear to require isoform 2 of the JNK group in order to promote growth *in vitro* and *in vivo*.

Here we present *in vitro* and *in vivo* evidence that JNK is commonly required for the growth of a class of human tumor cells not previously implicated in the JNK-dependent growth mechanism, human prostate carcinoma cells. Whole serum is utilized as the growth factor source in this system.

2. MATERIALS AND METHODS

2.1. Cells

Cell lines DU145, LNCaP, and PC3 were obtained from ATCC. ALVA-31, JCA-1, PPC-1 and TSU-Pr1 were a kind gift of John C. Reed. The 267B1 line is a SV40 immortalized neonatal prostate epithelial line while Kiras-267B1 is a Kirsten-Ras transformed derivative and both were kindly provided by J. Rhim. (Prada et al., 1993). The cells were maintained in RPMI (LNCaP) or DMEM in an atmosphere of 5% or 10% CO_2 respectively both with 5% fetal calf serum (Irvine Scientific™ , Santa Ana, CA). High serum growth curves were determined in similar medium at 20% FCS.

2.2. Antisense

Antisense and scrambled sequence control oligonucleotides were developed and extensively characterized as described (Bost et al., 1997; 1999; 2000). For *in vivo* studies a single-control phosphorothioate oligonucleotide consisting of a scrambled 20 nt sequence previously used as a control in the analysis of antisense PKC was chosen: 5' –TCGCATCGACCCGCCCACTA-3' (Dean and McKay, 1994). Both antisense JNK sequences contain a single CpG sequence thought to have possible immuno-effector cell activating properties (Klinman et al., 1996). The control oligonucleotide closely approximating the average composition of JNK1AS and JNK2AS but contains three CpG dinucleotides sequences and, therefore, also serves as a control for any anti-tumor immuno-effector activity in athymic mice. The transfection of cells was performed as described (Bost et al., 1997; 2000). Briefly, oligonucleotide at the concentration indicated in the text was added to DMEM containing 10 μl/ml of Lipofectin™ (Gibco BRL® Life Technologies Inc., Rockville, MD) at 1 mg/ml original concentration. This preparation was added to 50 to 80%

confluent cells seeded the previous day. After 4.5 h, the transfection medium was generally replaced with DMEM with 0.5% or 20% FBS except for TSU-Pr1 and PPC-1 cells which were exposed to the lipofection mixture for 9 h (Dean and McKay, 1994).

2.3. *In vitro Growth Assay*

Growth curves were determined as described (Potapova et al., 1997; Bost et al., 1997). Briefly, cells are seeded into multiwell tissue culture plates at 30,000 cells/cm^2 and counted (Coulter counter) on alternate days from day three until completion of log-phase growth as defined by the observation of a plateau. The resulting curves are integrated and integrated value observed in high serum (Gs) is expressed relative to basal growth in low serum (G_B) as G_S/G_B with standard error of $G_S = \sqrt{(\Sigma\sigma^2)}$ where the σ are the standard errors estimated for the daily triplicate cell counts.

2.4. *In vivo Growth and Kinase Assays*

Athymic Harlan Sprague mice were inoculated subcutaneously with 2.0 x 10^6 freshly harvested PC3 cells (Figure 3) or PC3 cells that stably express c-Jun(S63A,S73A) termed PC3mJun cells (Potapova et al., 1997; Potapova et al., 1996), or PC3 cells that bear the empty vector pLHCX termed PC3LHCX cells (Figure 5B) (Potapova et al., 1996). Unused cells were returned to culture as a check on sterility and viability. Upon the development of visible palpable tumors treatment was started (Figure 4A, first arrow). Two groups of 10-11 animals each were treated 6 days/week with 25 mg/Kg I.P. in 0.2 mls PBS of the control oligonucleotide or vehicle alone and three groups of 15 animals each were treated with equal amounts of JNK1AS, JNK2AS or half-concentrations of both antisense oligonucleotides. Tumor volumes (Figures 4 and 5) were estimated by measurement of the greatest dimension, l, and the perpendicular dimension, w, and evaluated using the formula $V=(\pi/6)(lw^2)$ (Boven et al., 1992). All growth curves were statistically compared to each other using all tumor volume estimates measured during the treatment period (Figure 4, arrows) but excluding measurements of any tumors with visible defects by ANOVA as implemented using Systat™ (Evanston, IL) software and include corrections for multiple comparisons (Bonferroni correction). Jun Kinase activity of the tumors was determined exactly as previously described (Bost et al., 1997; 1999; 2000).

3. RESULTS

3.1. *Serum Stimulation*

We examined 9 prostate cell lines. When compared to their growth in low serum, all lines exhibited serum-inducible growth but to varying extents from 2 to over 12-fold as judged by comparison of integrated proliferation curves (see Material and Methods) (Figure 2A). JNK activities in the same two media also were examined. Several but not all lines exhibited readily detectable JNK activity even in low, 0.5% serum, suggestive of constitutive activity (Figure 2A). As for growth, when the cell lines were treated with media containing 20% serum, the cell lines exhibited a range of serum-stimulated JNK activities from 1.2 to over 9-fold (Figure 3A, top panel). Moreover, the serum-stimulated growth of all of the carcinoma cell lines appears to be directly proportional to the serum-inducible JNK activity (Figure 2A). In contrast, one normal cell line of embryonic prostate epithelium, 267B1 (Antonyak et al., 1997), exhibits the maximum observed serum-inducible JNK activity which was not linked with proportional growth whereas the Ki-ras-transformed 267B1 cells behave similarly to the human prostate carcinoma-derived cell lines (Figure 2A). Thus, a consistent relationship is observed indicating that proliferation of prostate carcinoma cells in serum is strictly proportional to JNK activity and fits a linear relationship with a Pearson correlation coefficient of 0.91 ($p = 0.04$, Bartlett Chi-square) (legend, Figure 2A).

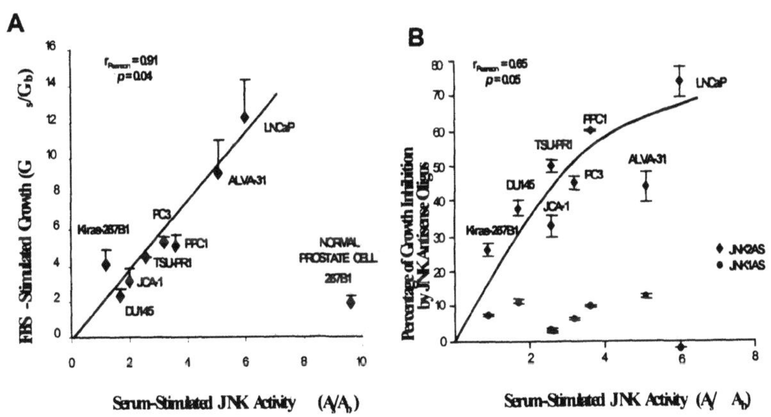

Figure 2. ***A***, *The proliferation of prostate carcinoma cells is directly proportional to the serum-inducible JNK activity. FBS-stimulated JNK activity for each of the indicated cell lines (Materials and Methods) is the ratio (A_s/A_b) of total JNK activity in complete growth medium of DME with 20% FBS (A_s) divided by the total JNK activity observed in the same medium at 0.5% FBS (A_b) for 48 h and is calculated directly from the activities shown in Figure 2A.* ***B***, *All cells were transfected with antisense oligonucleotides or mock-transfected as described (Bost et al., 1997; Clarke et al., 1998; and Materials and Methods) and proliferation redetermined as in A and expressed as percent inhibition: $100 \times (1 -([G_{S,M} - G_{S,Oligo}]/G_{S,M}))$ where G is integrated growth curve in cells x days, S indicates 20% FBS, M indicates mock transfection with lipid and subscript 'oligo' indicates the growth observed following treatment with antisense or control oligonucleotide.*

3.2. Specific Inhibition of JNK

In order to determine whether a causal link exists between growth and JNK activity, we utilized previously characterized (Bost et al., 1997; Bost et al., 1999; Bost et al., 2000) antisense oligonucleotides that are complementary to sequences common to all four alternate splice variants that make up both the JNK1 or JNK2 isoform families termed JNK1AS and JNK2AS respectively. When oligonucleotides bearing a sequence that is complementary to a chosen mRNA sequence are suitably introduced into cells, a mRNA-antisense oligonucleotide complex may form and such RNA-DNA complexes are substrates for endogenous Ribonuclease-H which then cuts the mRNA thereby functionally inactivating expression of the corresponding gene (reviewed in Bost et al., 2000). Here cells plated 24 h previously were treated with JNK1AS, JNK2AS, both JNK1AS and JNK2AS, a control mixture composed of equal amounts of the scrambled sequences of JNK1AS and JNK2AS, the lipofection agent alone, or were not treated and proliferation (Materials and Methods) was monitored for 4-6 days leading to an additional 63 proliferation curves. Treatment with either scrambled sequence control oligonucleotide had little effect. When expressed relative to mock-transfected cells, the average inhibition of proliferation for treatment of all the prostate carcinoma cell lines with the scrambled control oligonucleotides was 3.7 ± 1.5 %. Thus, for Figure 1B, inhibition by the antisense oligonucleotides is expressed relative to their respective scrambled control oligonucleotide. As for previous studies (Bost et al., 1997; 1999), suppression of protein expression at least 80% was confirmed for the cell lines examined here (*e.g* Figure 2B). Treatment with JNK1AS (Figure 2B), produced little additional inhibitory effect on the growth of any cell line, the average inhibition is 7.0 ± 4.1 % (Figure 2B, blue). In contrast, transfection with the JNK2AS oligonucleotide leads to markedly increased inhibition of growth (Figure 1B, red). First, a clear trend is apparent. The sensitivities of the various cell types to inhibition by JNK2AS increases in proportion to the ability of serum-containing media to activate JNK (Figure 1B). Therefore, a very similar rank-order distribution of cell types along the growth inhibition axis (Figure 1 B) is observed as for the order along either the growth or the JNK activity axes of Figure 2. Thus, LNCaP cells are the most growth responsive cells and the most sensitive to inhibition by JNK2AS leading to over 75% growth inhibition of a 4-day growth curve following a single treatment on day 1. Conversely, KiRas-267B1, DU145, and JCA-1 cells are the least responsive to serum and least sensitive to inhibition of JNK2 expression by JNK2AS. Second, each cell line, including those of low responsiveness to serum, exhibit a statistically significant ($p < 0.007$, Students' t-test) increase in inhibition upon treatment with JNK2AS compared to treatment with JNK1AS (Figure 2B). Since treatment with JNK1AS and JNK2AS are equally effective at reducing respective mRNA and protein (*e.g.* Figure 2B and refs. Bost et al., 1997; Bost et al., 1999; Collins et al., 1997), these results indicate that one or more JNK2 but no JNK1 isoforms are commonly required for the proliferation of prostate carcinoma cells *in vitro*.

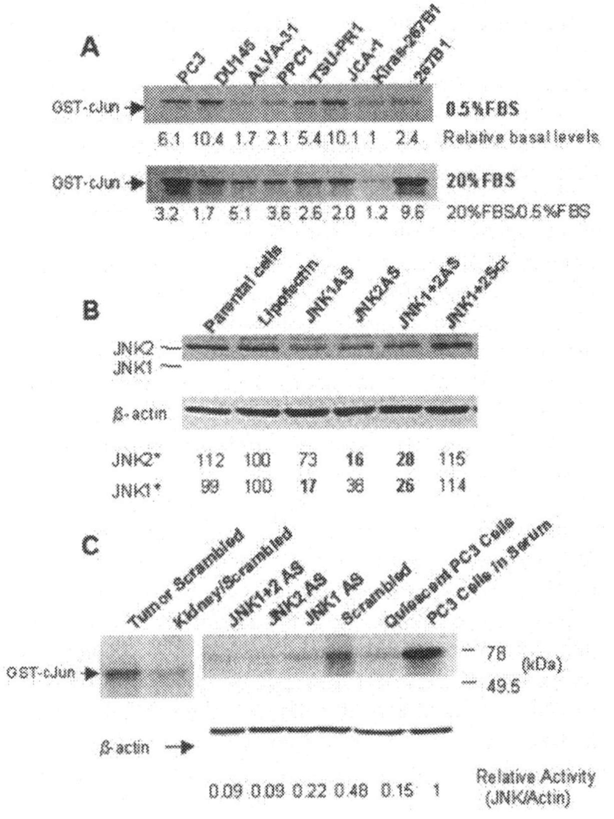

Figure 3. Representative JNK activities and JNK expression in human prostate carcinoma cells.

A. Total JNK activity was measured as the ability of cells to phosphorylate GST-c-Jun fusion protein substrate 3 h after changing to a medium containing 20% FBS (lower panel) following 0.5% FBS which was maintained for 48 h following plating (upper panel).

B. Example of reduction of JNK1or JNK2 steady state protein levels at 60 h post transfection with either 0.2 μM JNK1AS or JNK2AS for 4 h as described (Bost et al., 1997; Clarke et al., 1998).

C. JNK activity was measured for equal amounts of tumor extract of treated mice based on protein determinations and confirmed by β-actin measurements and further normalized to the activity for cultured PC3 cells. The effects of scrambled control oligonucleotide on the steady state activity of mouse kidney (lane 2) were determined in a separate experiment in parallel with a replicate control oligonucleotide-treated tumor sample (lane 1).

3.3. Inhibition and Regression of Established Tumors

To determine whether the results are indicative of the roles of JNK1 and JNK2 in tumor growth, in vivo tumor growth studies were undertaken (Materials and Methods). Human prostate carcinoma PC3 cells, which exhibit typical serum-inducible JNK activity and proliferation values (cf. Figures1A and 1B), were taken as representative and allowed to establish as readily palpable and visible xenografts before beginning systemic treatment by daily intraperitoneal injections of JNK1AS, JNK2AS, JNK1AS + JNK2AS, a single scrambled sequence oligonucleotide (Materials and Methods) or vehicle alone for 23 days (Figure 3A). The tumors of vehicle alone or scrambled oligonucleotide-treated mice readily grew to large sizes (Figure 3A) whereas those of antisense JNK-treated animals were growth inhibited. We confirmed that systemic treatment with antisense JNK but not the scrambled sequence control oligonucleotide leads to inhibition of tumor JNK activity (Figure 2C). Based on the integrated growth curve values, treatment with both JNK1AS and JNK2AS lead to substantial inhibition of tumor growth of 57% and 80% respectively with probabilities <0.0002 (ANOVA). The increased inhibition of JNK2AS compared to JNK1AS is also significant, $p = 0.012$. Combined antisense treatment lead to an inhibition of 78% compared to the average of JNK1AS alone and JNK2AS alone of 70%. Thus, these results support a role for both isoform(s) of JNK1 and JNK2 in tumor growth with JNK2 playing a dominant role. These results support the *in vitro* growth studies showing that JNK, in particular JNK2, commonly plays a role in the growth of human prostate carcinoma cells.

If the tumor growth arrest effects observed here are due to cytostasis that is promoted by the antisense treatment, it would be expected that the cytostatic tumor masses might exhibit renewed growth following the cessation of treatment. Figure 3B shows that, following a lag period of approximately one week, the tumors of one group exhibited a burst in tumor growth leading to a final growth rate (slope) equal to that of the control groups suggesting that viable but arrested tumors retained the ability to grow upon cessation of treatment.

For the remaining treatment groups, however, after day 38 of the experiment, there were few remaining intact tumors for valid estimates of tumor volume. Starting approximately half-way through the treatment period, increasing numbers of tumors (Figure 3D) developed small darkened spots invariably at the most distal site of the tumor that rapidly ulcerated leading to progressive and massive cavitation of the tumors (*e.g.* Figure 3C). The regression process was followed by measuring the diameter of the ulcer cavities and by measuring the frequency of the development of ulceration for each experimental group (Figure 3D). As the average ulcer sizes approached maximum values, visible tumor mass retreated to a circumferential rim and typically completely vanished (Figure 3D, inset). In some cases tumor regression proceeded by undermining the skin around a small ulcer mouth leaving the epithelial edges in approximation (Figure 3C) and in one case the ulcer edges reunited to form a healed site (Figure 3C, circle).

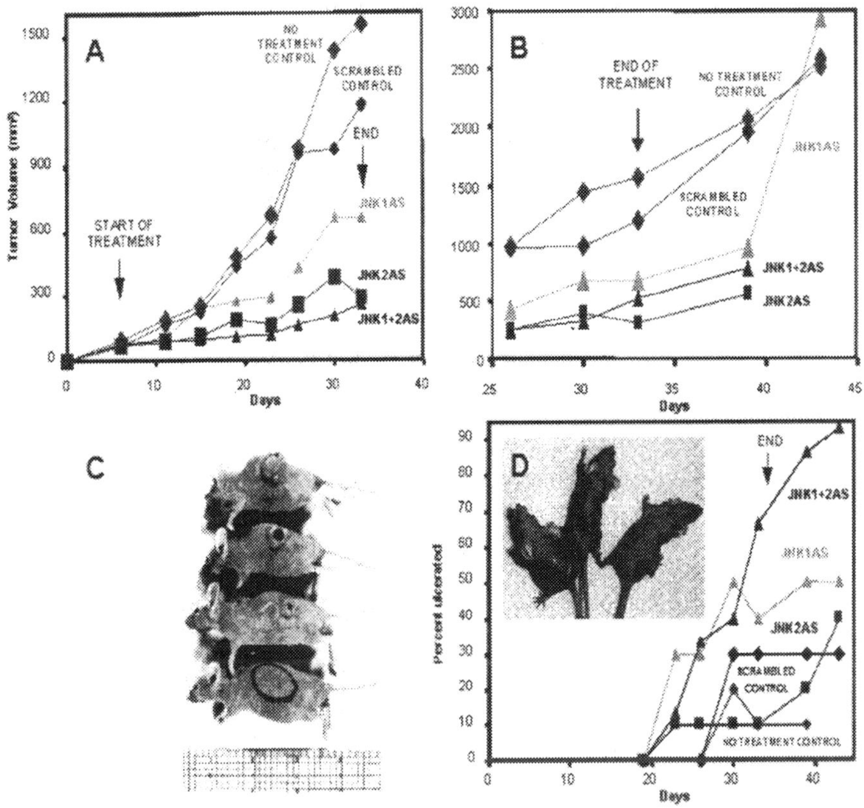

Figure 4. PC3 xenograft growth arrest by antisense JNK. ***A.*** *The average tumor volume for each treatment is plotted as a function of time. During the treatment period (downward arrows) 3 animals died, 2 in the control groups and one in the antisense groups.* ***B.*** *Tumor regrowth following cessation of treatment ("End" arrow).* ***C.*** *Examples of combined JNK1AS and JNK2AS-treated animals during the course of regression of PC3 xenografts. The circle locates a scar at the site of a former tumor for one case of apparent complete regression with healing (see text).* ***D.*** *The progression of tumor regression was followed by plotting the cumulative percent of animals in each experimental group of 11 or 15 animals that developed ulcers during the course of treatment. During the course of treatment one animal was sacrificed whereas all remaining animals (14/15 = 93%) developed progressing ulcers. D. Inset, examples of nearly completely regressed tumors at the end of the treatment period, here combined antisense JNK treated animals, d=34.*

Histology sections of tumors at the various stages of regression show that distal ulceration is preceded by subepithelial micro abscess formation (Figure 4A) together with an infiltrate of acute inflammatory polymorphorphic neutrophils in the surrounding dermis suggesting that devitalization stimulates an acute inflammatory response. Focal infiltrates of these cells are evident around the base of the expanding ulcer (not shown) indicating that these cells are responsible for the progressive ulceration. The time course of regression (Figure 3D) also indicates that the process, once started, typically extended well beyond the end of the antisense treatment period (arrow, Figure 3D) suggesting an inflammatory mechanism of continued erosion. Therefore, it is the frequency of initiation of ulceration, *i.e.* Figure 3D, rather than the actual erosion that is taken as the more relevant variable in relating regression to the treatment regimen. Remarkably, all 14 extant animals of the combined JNK1AS + JNK2AS-treated group developed distal ulcers which rapidly lead to complete regression of tumors characterized by flat encrusted sores with an average diameter of 1.4 cm (*e.g.* inset, Figure 3C). Chi-square analysis of the increased frequency of regression in the JNK1AS + JNK2AS- treated group compared to the next most commonly regressing group indicates that the increase is significant, $p = 0.022$.

Spontaneous ulceration in tumors of control animals also occurred occasionally (Figure 4D), however, this process may be distinct. Spontaneous ulceration occurred late, day ≥ 26, (Figure 4D), at time when these tumors were very large, > 1000 mm^3 (Figure 4A), whereas treatment-induced ulceration occurred in tumors commonly $\geq$ 200 mm^3 (*cf.* Figures 4A and 4D). The spontaneous ulceration of large tumors involved fewer animals with a maximum incidence of $< 30\%$ (Scrambled sequence control Figure 4D). These results indicate that, unlike cytostasis (Figure 4A), both antisense JNK1 and JNK2 treatment are required for efficient tumor devitalization and regression.

3.4. *Inhibition of Tumorigenesis by a Nonphosphorylatable c-Jun*

Previous studies indicate that N-terminal phosphorylation of the c-Jun by JNK may be a crucial event in mediating transformation by a variety of oncogenes upstream in the JNK pathway (Smeal et al., 1991; Binétruy et al., 1991; Pulverer et al., 1991; Angel and Karin, 1991; Smeal et al., 1992; Franklin et al., 1992; Kyriakis et al., 1994; Klinman et al., 1994; Minden et al., 1995; Crespo et al., 1996; 1997; Karin et al., 1997; Clark et al., 1997; Tolkacheva et al., 1997; Moriguchi et al., 1997; Tanaka et al., 1997; del Peso et al., 1997; Rodrigues et al, 1997; Dickens et al., 1997; Wu et al., 1998; Tanaka and Hanafusa, 1998; Li and Smithgall, 1998; Burgess et al., 1998; Glaven et al., 1999). Thus, inhibition of the N-terminal phosphorylation of c-Jun by over-expression of the nonphosphorylatable derivative, c-Jun(S63A,S73A) (Smeal et al ., 1991; Binétruy et al., 1991; Smeal et al., 1992; Kyriakis et al., 1994; Bost et al., 1997; Clarke et al., 1998), would be expected to prevent tumor formation. In preliminary studies of PC3 cells that stably over-express c-Jun(S63A,S73A) (Crespo et al., 1997) it was observed that these cells were entirely unable to develop tumors in female athymic mice. Here a repeated study using a different clone of c-Jun(S63A,S73A)-expressing PC3 cells (Logan et al., 1997) and using both female and male recipient athymic mice (Figure 4B). As before, empty-vector and control cells rapidly formed tumors in athymic mice whereas c-Jun(S63A,S73A)-expressing cells were entirely unable to develop tumors. This result further supports the existence of a critical role of JNK and N-terminal phosphorylation of c-Jun in xenograft formation and growth.

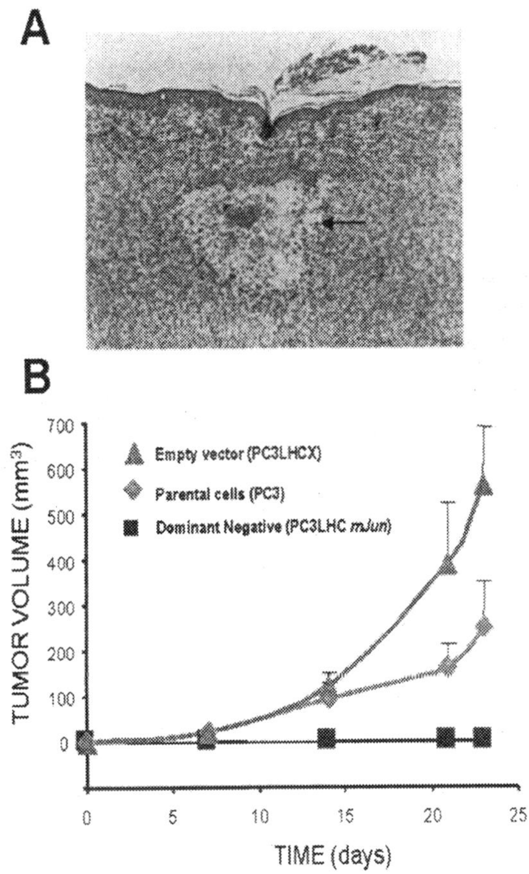

Figure 5. ***A.*** *Histology of impending ulceration. A representative PC3 xenograft of an antisense treated animal was examined. Note the subepithelial microabscess (arrow) at the distal portion of the tumor with squamous debris in lumen suggesting communication of the abscess with the surface has already occurred (hematoxyin-eosin, 10x; bar = 50 μ).* ***B.*** *Inhibition of tumorigenesis by dominant negative c-Jun(S63A,S73A). Groups of 8 athymic mice (4 males and 4 females) were inoculated and tumor volumes quantified as described (Materials and Methods). There was no significant difference between the average sizes of tumors of male and female recipients and the data were pooled. The results for the PC3LHCmJun cells are significantly different (ANOVA) from either control at time points ≥ 14 days, $p < 0.01$.*

4. DISCUSSION

4.1. Activation of JNK by Growth Factors

Numerous growth factors such as androgens, EGF, TGF-α, TGF-ß1, IGF-I, PDGF, and others have been implicated in promoting prostate epithelial cell growth by autocrine or paracrine mechanisms thereby facilitating transformation of prostate basal or epithelial cells (for reviews see Barrack 1997; Russell et al., 1998). Several of these factors such as EGF, TGF- ß1, PDGF, and TGF-ß1 are known to be strong activators of JNK (Smeal et al., 1992; Kanatani et al., 1996; Mauviel et al., 1996; Bost et al., 1997; Logan et al., 1997; Antonyak et al., 1997; Clarke et al., Mettouchi et al., 1997; 1998; Bost et al., 1999; Wong et al., 1999; Bagowski et al., 1999) and, in the case of lung carcinoma cell lines, JNK is required for transforming effects of EGF (Bost et al., 1997; 1999). Conversely, functional IGF-I receptors which are thought to be necessary for differentiation and for regulated growth of normal prostate epithelium (Connolly and Rose, 1994), are commonly down-regulated in prostate carcinoma cell lines (Bae et al., 1994; Tennant et al., 1996; Plymate et al., 1996; 1997) and in tumors in proportion to grade and stage (reviewed in Bae et al., 1994; Tennant et al., 1996; Plymate et al., 1997). Indeed, the IGF signal transduction pathway has been found to strongly inhibit the JNK pathway in several cell systems (Cheng and Feldman, 1998; Okubo et al., 1998). Thus, there is a two-way correlation between many of the conditions thought to be important in the regulation of growth of prostate tumors and the regulation of JNK activity suggesting the possibility that this alternate map kinase pathway may play an important role.

4.2. JNK2 Activity is Essential for the Growth of Many but not All Prostate Carcinoma Cell Lines

We observed that the serum-supported proliferation rate of the human prostate carcinoma cell lines is highly correlated with JNK activity ($r_{Pearson} = 0.91, p < 0.04$). In order to determine whether JNK is causally linked to the proliferation response, the expression of JNK1 and JNK2 were specifically inhibited by the use of antisense oligonucleotides. These compounds have been characterized previously (Bost et al., 2000) and used in a variety of studies (Bost et al., 1997; 1999). Following a single treatment with a lipid-antisense JNK complex, JNK mRNA and protein rapidly declines by more than 90% which persists for 48 – 72 h (Dean and McKay, 1994). Inhibition is specific in that application of JNK1AS does not alter JNK2 or housekeeping gene product levels and, conversely, JNK2AS does not alter JNK1 or housekeeping gene product levels (Bost et al., 1999; Dean and McKay, 1994). Following recovery of JNK expression starting approximately 72 h after a single treatment DNA synthesis and growth return to the normal level showing that a significant period is available for the observation of the consequences of suppressed expression (Logan et al., 1997). The recovery of JNK expression also indicates that treatment with the antisense reagents does not permanently harm the cells (Logan et al., 1997).

Applying these compounds to the nine human prostate carcinoma cell lines, it is observed that serum-supported growth as judged by direct proliferation measurements is specifically inhibited by JNK2AS but is nearly unaffected by two different scrambled sequence control oligonucleotides and only modestly effected by JNK1AS. Maximum inhibition of proliferation was approximately 75%. This may reflect the fact that for a single treatment, the decrease in target mRNA and protein levels is limited to about 80 – 90% basal levels (*e.g.* Figure 3B). In addition, proliferation is measured over the course of 6 days whereas the effects of the antisense treatment begin to reverse 3 days following a single treatment (Bost et al., 2000). Since treatment is initiated one day after the start of the proliferation study (Materials and Methods), some recovery of growth after day 4 is expected and these effects may account for the lack of inhibition above about 75%.

4.3. Antisense to JNK2 but not Antisense to JNK1 Blocks Growth of Established Tumors

In order to determine whether these growth properties are indicative of tumor growth, a cell line with typical serum-stimulated JNK activity and proliferation, PC3 cells (*cf.* Figure 1), was examined *in vivo*. Systemic treatment of tumor-bearing animals showed that cytostasis was promoted by treatment with JNK2AS to a much greater extent ($p = 0.012$) than treatment with JNK1AS consistent with the results of the *in vitro* studies. Nevertheless significant growth inhibitory activity was observed upon treatment with JNK1AS *in vivo* whereas JNK1AS was without effect on any of the cultured cell lines indicating that *in vitro* cell growth and *in vivo* tumor growth may have distinct mechanisms. Moreover, stimulation of the regression of the growth-arrested tumors with high frequency (14 of 15 animals) required both JNK1AS and JNK2AS. *In vitro*, the combination of JNK1AS and JNK2AS did not inhibit the growth of any of the cell lines beyond that of JNK2AS alone (data not shown). These results indicate that the induction of regression of established tumors is likely distinct from the mechanism of cytostasis which was promoted efficiently by JNK2AS alone. Since inflammatory cells are present even before the first visible signs of tumor breakdown and regression, we speculate that combined JNK1AS plus JNK2AS treatment promotes a cytocidal event and that the resulting dead cells stimulate and inflammatory response leading to a regression process that is due to the inflammatory cells. Indeed, recent studies indicate that JNK1 isoforms play a survival role by inhibiting apoptosis and that elimination of JNK1 leads to an increase in apoptosis (Butterfield et al., 1997; Cioffi and Monia, 2000). According, the regression process itself is not specific whereas the high frequency with which the process is *initiated* is specific to the combined use of JNK1AS and JNK2AS.

4.4. Conversely, Prostate Tumor Cells that Express a JNK Pathway Antagonist Can not Develop Tumors

Stable expression of a nonphosphorylatable form of c-Jun, c-Jun(S63A, S73A), blocked the ability of PC3 cells to initiate tumor growth further supporting the role of JNK in tumor growth. The observation also focuses attention on a phosphorylation-dependent function of c-Jun as required for the initiation of tumor growth. This is consistent with previous studies of primary fibroblasts derived from "knockout" mice with a homozygous deletion of c-Jun which show that c-Jun is required for transformation by activated ras (Johnson et al., 1996). The presence of a crucial phosphorylation-dependent function is supported by observations that the transactivation potential of c-Jun that is promoted by phosphorylation of serine residues 63 and 73 is essential for transformation of primary murine fibroblasts by activated ras and several of oncogenes (Smeal et al., 1991; Binétruy et al., 1991; Smeal et al., 1992). Thus, the results observed here with c-Jun(S63A,S73A)-expressing PC3 cells indicate that activation of the transcriptional potential of c-Jun by N-terminal phosphorylation may be important for the initiation and growth of the xenografts, *i.e.* tumorigenesis. A similar role has been attributed to c-Jun in human breast carcinoma MCF-7 cells where stable over-expression of c-Jun causes tumorigenicity in the absence of estrogens (Smith et al., 1999).

4.5. Significance of the c-Jun Substrate

C-Jun appears to be a preferential substrate of isoforms of the JNK2 family compared to JNK1 (Karin et al,. 1997; Gupta et al., 1996). This is consistent with roles of c-Jun observed here and the observation that the JNK2 family of isoforms plays a predominant role in the growth of the human prostate carcinoma cells *in vitro* and PC3 cells *in vivo*. These studies do not, however, rule out an important phosphorylation-dependent function of other substrates of JNK. The sum of results, therefore, strongly favor a model wherein the activation of JNK, in particular one or more isoforms of JNK2, leads to N-terminal phosphorylation and activation of c-Jun as one important event in mediation of oncogenic stimuli of certain human prostate carcinoma cells.

4.6. Correlation with Other Studies of Prostate Carcinoma

Recently increased expression and activity of JNK was found to correlate with grade and stage in a large series of human prostate carcinoma specimens (Magi-Galluzzi et al., 1998). Thus, for suitably screened cases of prostate carcinoma, the inhibition of JNK may be a potentially effective treatment strategy.

5. REFERENCES

Adams M.J., Blundell T.L., Dodson E.J., Dodson G.G., Vijayan M., Baker E.N., Harding M.M., Hodgkin D.C., Rimmer B. and Sheat, S. *Nature* 224 (1969): 491.

Angel P. and Karin M. *Biochim. Biophys. Acta* 1072 (1991): 129-157.

Antonyak M. A., Moscatello D. K. and Wong, A. J. *J. Biol. Chem.* 273 (1997): 2817-2822.

Bae V., Jackson-Cook C., Brothman A., Maygarden S. and Ware J. *Intl. J. Cancer* 58 (1994): 721-729.

Bagowski C.P., Stein-Gerlach M., Ghoidas A. and Ullrich, A. *The EMBO J.* 18 (1999): 5567-5576.

Barrack E.R. *Prostate* 31 (1997): 61-70.

Binétruy B., Smeal T.B. and Karin M. *Nature* 351 (1991): 122-127.

Blundell T. L., Dodson G. G., Hodgkin D. C. and Mercola, D. A. "Insulin: The Structure in the Crystal and its reflection in Chemistry and Biology." *Advances in Protein Chemistry* 26 (1972): 279-402

Blundell T.L., Cutfield J.F., Cutfield S.M., Dodson E.J., Dodson G.G., Hodgkin D.C., Mercola D.A. and Vijayan M. "Atomic Positions in Rhombohedral 2-zinc Insulin Crystals" *Nature* 231 (1971): 506-511

Bost F., Dean N., McKay R. and Mercola D.A. *J. Biol. Chem.* 272 (1997): 33422-33429.

Bost F., McKay R., Bost M., Potapova O., Dean N.M. and Mercola D. *Mol. Cell. Biol.* 19 (1999): 1938-1949.

Bost F., McKay R., Dean N., Potapova O. and Mercola, D. *Methods Enzymol.* 314 (2000): 342-362.

Boswell H.S. *Blood* 92(7) (1998): 2450-2460.

Boven E., Winograd B., Berger D.P., Dumont M.P., Braakhuis B.J., Fodstad O., Langdon S. and Fiebig H.H. *Cancer Res.* 52 (1992): 5940-5947.

Burgess G.S., Williamson E.A, Cripe L.D., Litz-Jackson S., Bhatt J.A., Stanley K., Stewart M.J., Kraft A.S., Nakshatri H., Butterfield L, Storey B., Maas L. and Heasley L. E. *J. Biol. Chem.* 272 (1997): 18261-18266.

Cheng H.L. and Feldman E.L. *J. Biol. Chem.* 273 (1998): 14560-14565.

Cioffi C. L. and Monia, B. *Methods Enzymol.* 314 (2000): 363-378.

Clark G.J., Westwick J.K. Der C.J. (1997) J. Biol. Chem. 272, 1677-1681.

Clarke N., Arenzana N., Hai T., Minden A., Prywes R. *Mol. Cell. Biol.* 18 (1998): 1065-1073.

Collins K., Jacks T. and Pavletich, N.P. (1997) Proc Natl Acad Sci U S A. 94, 2776-2788.

Connolly J. and Rose D. *Prostate* 24 (1994):167-175.

Crespo P., Bustelo X.R., Aaronson D.S., Coso O.A., Lopez-Barahona M., Barbacid M. and Gutkind J.S. *Oncogene* 13(1996): 455-460.

Crespo P., Scheugbel K.K., Ostrom A.A., Gutkindm J.S. and Butelo S.R. *Nature* 285 (1997): 169-172.

Davis R.J. *Cell* 103 (2000): 239-252.

Dean N. and McKay R. *Proc. Natl. Acad. Sci. U S A* 91 (1994): 11762-11766.

del Peso L., Hernandez-Alcoceba R., Embade N., Carnero A., Esteve P., Paje C. and Lacal J.C. *Oncogene* 15 (1997): 3047-3057.

Dickens M., Rogers J.S., Cavanagh J., Raitano A., Xia Z., Halpern J.R., Greenberg, M.E, Sawyers C.L. and Davis R.J. Science 277(5326) (1997): 693-696.

Franklin C.C., Sanchez V., Wagner G. and Woodgett J.R. *Proc. Natl. Acad. Sci. U S A* 89 (1992): 7247-7251.

Glaven J.A., Whitehead I., Bagrodia S., Kay R. and Cerione R.A. *J. Biol. Chem.* 274 (1999): 2279-2285.

Gupta S., Barrett T., Whitmarsch A.J., Cavanagh J., Sluss H.K., Dérijard B. Davis, R.J. EMBO J. 15 (1996): 2760-2770.

Johnson R., Spiegelman B., Hanahan D. and Wisdom R. *Mol. Cell. Biol.* 16 (1996): 50-11.

Kanatani Y., Kasukabe T., Okabe-Kado J., Hayashi S., Yamamoto-Yamaguchi Y., Motoyoshi K., Nagata N. and Honma Y. *Cell Growth Differ.* 7 (1996):187-196.

Karin M., Liu Z. and Zandi E. *Curr. Opin. Cell Biol.* 9 (1997): 240-246.

Klinman D.M., Yi A.K., Beaucage S.L., Conover J. and Krieg A.M. *Proc. Natl. Acad. Sci. U S A* 7 (1996): 2879-2883.

Kyriakis J.M., Banerjee P., Nikolakaki E., Dai T., Rubie E.A., Ahmad M.F., Avruch J. and Woodgett J.R. *Nature* 369 (1994): 156-160.

Li J. and Smithgall T.E. *J. Biol. Chem.* 273 (1998): 13828-13834.

Logan S.K., Falasca M., Hu P. and Sclessinger J. *Mol. Cell. Biol.* 17 (1997): 5784-5790.

Magi-Galluzzi C., Montironi R., Cangi M.G., Wishnow K. and Loda M. *Virchows Arch.* 5 (1998): 407-413.

Mauviel A., Chung K.Y., Agarwal A., Tamai K. and Uitto J. *J. Biol. Chem.* 271 (1996): 10917-10923.

Mercola D. A. and Wollmer A. "The Crystal Structure of Insulin and Solution Phenomen. Use of the Structure for the Calculation of Tyrosyl Circular Dichroism." *Structural Studies of Molecules of Biological Interest.* Dodson G.; Glusker J. P. and Sayre D. (eds.), pp. 547-582, Clarendon Press: Oxford, England 1981.

Mercola D.A.; Morris J.W.S. and Arquilla E. *Biochemistry* 11 (1972): 3860-3874.

Mettouchi, Cabon F., Montreau N., Dejong V., Vernier P., Gherzi R., MercierG. and Binétruy B. *Mol. Cell. Biol.* 17 (1997): 3202-3229.

Minden A., Lin F.X., Claret A. Karin, M. *Cell* 81 (1995): 1147-1157.

Minden A., Link A., Smeal T., Dérijard B., Cobb M., Davis R. and Karin M. *Mol. Cell. Biol.* 14 (1994): 6683-6688.

Moriguchi T., Toyoshima T., Masuyama N., Hanafusa H., Gotoh Y. and Nishida E. *EMBO J.* 16 (1997): 7045-7053.

Okubo Y., Blakesley V.A, Stannard B., Gutkind S. and LeRoith D. *J. Biol. Chem.* 273 (1998): 25961-25966.

Plymate S.R., Tennant M., Birnbaum R.S., Thrasher J.B., Chatta G. and Ware J.L. *J. Clin. Endocrinol. Metab.* 81 (1996): 3709-3716.

Plymate S.S., Bae V.L., Maddison L., Quinn L.S. and Ware J.L. *Endocrine* 7 (1997): 119-124.

Potapova O., Fakhrai H. and Mercola D.A. *Int J Cancer* 66 (1996): 669-677.

Potapova O., Haghighi A., Bost F., Liu C., Birrer M.J., Gjerset R. and Mercola D. *J. Biol. Chem.* 271 (1997): 14041-14044.

Prada D.S., Thraves P.J., Kuettel M.R., Lee M.S., Arnstein P., Kaighn M.E., Rhim J.S. and Dritschilo A. *Prostate* 23 (1993): 91-98.

Pulverer B.J., Kyriakis J.M., Avruch J., Nikolakaki E. and Woodgett J.R. *Nature* 353 (1991): 670-674.

Ray L.B. and Sturgill T.W. *Proc Natl Acad Sci U S A.* (Jun;1985) 11 (1988): 3753-3757.

Rodrigues G.A., Park M. and Schlessinger J. *EMBO J.* 16 (1997): 2634-2645.

Rossomando A.J., Payne D.M., Weber M.J. and Sturgill T.W. *Proc Natl Acad Sci U S A.* 18 (1989): 6940-6943.

Russell J., Bennett S. and Stricker P. *Clin Chem.* 44 (1998): 705-723.

Schlessinger J. *Cell* 103 (2000): 211-226.

Sherr C. *Canc. Res.* 60 (2000): 3689-3695.

Smeal T.B., Binétruy B., Mercola D., Birrer M. and Karin, M. *Nature* 354 (1991): 494-496.

Smeal T.B., Binétruy D., Mercola D., Grover-Bardwick A., Heidecker G., Rapp U.R. and Karin M. *Mol. Cell. Biol.* 12 (1992): 3507-3513.

Smith G. D. and Gunnell D. *Br. J. Med.* 321 (2000): 847-848.

Smith L.M., Wise S.C., Hendricks D.T., Sabichi A.L., Bo T., Reddy P., Brown P.H. and Birrer M.J. *Oncogene* 18 (1999): 6063-6070.

Strassburger W., Glatter U., Wollmer A., Fleischhauer J., Mercola D.A., Blundell T., Grover I., Pitts J.E., and Wood S. *Febs Lett.* 139 (1982): 295-299.

Tanaka S. and Hanafusa H. *J. Biol. Chem.* 273 (1998): 1281-1284.

Tanaka S., Ouchi T. and Hanafusa H. *Proc. Natl. Acad. Sci. U S A* 94 (1997): 2356-2361.

Tennant M.K., Thrasher J.B., Twomey P.A., Drivdahl R.H., Birnbaum R.S. and Plymate S.R. *J. Clin. Endocrinol. Metab.* 10 (1996): 3774-3782.

Tolkacheva T., Feuer B., Lorenzi M.B., Saez R. and Chean A.M. *Oncogene* 15 (1997): 727-735.

Wong C., Rougier-Chapman E.M., Frederick J.P., Datto M.B., Liberati N.T., Li J.M. and Wang X.F. *Mol. Cell. Biol.* 19 (1999): 1821-1830.

Wu W.J., Lin R., Cerione R.A. and Manor D. *J. Biol. Chem.* 273 (1998): 16655-16658.

Zahn H. and Meinhofer J. *Makromolekulare Chem.* 26 (1958a): 126-152.

Zahn H. and Meinhofer J. *Ibid.* (1958b): 153-178.

6. ACKNOWLEDGEMENTS

We thank J. Reed for providing ALVA-31, JCA-1 and TSU-Pr1; J. Rhim for providing 267B1 and Kiras267B1; and E. Adamson, T. Hunter, and C. Stein for commenting on preliminary manuscripts. Supported by USPH grants CA56834 and CA76173 (D.A.M), the Ligue Nationale Contre le Cancer (to DM and FB), the Conseil Régional de Haute Normandie (F.B.); the Ray and Estelle Spehar Fellowship of the American Cancer Society to F.B; and the California Breast Cancer Research Program (3CB-0246; D.A.M.). We thank Carrie Edgin and Mark Bickerstaff for preparation of this camera-ready manuscript.

R.W. WOODY, C. KIEFL, N. SREERAMA, Y. LU, Y. QIU AND J. A. SHELNUTT

MOLECULAR DYNAMICS SIMULATIONS OF CARBONMONOXY MYOGLOBIN AND CALCULATIONS OF HEME CIRCULAR DICHROISM

Abstract. The Soret circular dichroism (CD) spectrum of carbonmonoxy myoglobin differs strikingly for the two heme isomers: $\Delta\varepsilon_{421} = +90\ M^{-1}cm^{-1}$ for isomer A, $\Delta\varepsilon_{421} = -7\ M^{-1}cm^{-1}$ for isomer B (Aojula et al., 1988). This observation implies significant differences in the protein conformation and/or distortions of the heme from planarity between the two isomers. Molecular dynamics simulations of the two isomers have been performed, using both neutron diffraction (ND) and NMR structures for starting geometries. The geometry for isomer B was generated from the isomer A geometry by rotation of the heme by 180° about the α–γ methine carbon axis. Four ND-based trajectories for isomer A, each of 600 ps duration, gave average CD spectra that reproduced the strong positive Soret CD inferred for isomer from experiment. Four such trajectories for isomer B gave results similar to those for isomer A, and therefore disagreed with the weak negative Soret CD band inferred from experiment for isomer B. The two NMR-based trajectories for each isomer gave poor agreement with experiment. We attribute the failure of the calculations for isomer B to a poor starting structure for this isomer, and we are conducting further studies to overcome this problem. In the successful calculations for isomer A, heme-aromatic side chain coupling accounts for ~40% of the total Soret rotational strength, whereas heme-peptide coupling and intrinsic heme CD account for ~30% each. Heme nonplanarity has been analyzed by normal coordinate structural decomposition (Jentzen et al., (1995). Of the out-of-plane normal modes, ruffling correlates well with the rotational strengths of the two Soret components, which usually are opposite in sign. Correlation of ruffling with the net Soret rotational strength is relatively weak. The out-of-plane deformations of the heme are similar in isomers A and B, consistent with the similarity in the calculated Soret CD spectra.of the heme are similar in isomers A and B, consistent with the similarity in the calculated Soret CD spectra.

1. INTRODUCTION

Axel Wollmer and I first became acquainted through the literature. Axel noted the communication that Ming-Chu Hsu and I published in 1969 (Hsu and Woody, 1969) in which we provided an explanation of the circular dichroism of heme absorption bands in heme protein. Ming-Chu and I found that the sign and approximate magnitude of all four observable heme $\pi\pi^*$ bands in Mb and Hb (Hsu and Woody, 1971) could be accounted for by coupled-oscillator interactions between the heme and aromatic side chains in the protein. Axel collaborated with Jörg Fleischhauer to apply this approach to *Chironomus* hemoglobin, the monomeric heme protein that Axel was studying intensively at that time. This hemoglobin was of special interest because its Soret CD is opposite in sign (Formanek and Engel, 1968; Wollmer et al., 1970) to those of mammalian and other vertebrate Mb and Hb. Jörg and Axel (Fleischhauer and Wollmer, 1972) showed that the coupled-oscillator mechanism

M. Federwisch, M. Leyck Dieken and P. De Meyts (eds.), Insulin & Related Proteins, 233–248.

accounted for the observed Soret CD band in *Chironomus* Hb, including the sign reversal, thus greatly strengthening the case for the generality of this mechanism.

I finally met Axel and Jörg in 1982 on a visit to Aachen that had been arranged through Wolfgang Strassburger, whom I had met at a British Biophysical Society meeting held in York - another Alcuin connection! Since then I have enjoyed Axel's hospitality in Aachen many times, including a most enjoyable and productive sabbatical year with Axel and Jörg in 1995-96. So far, I have only been able to reciprocate on one occasion, when Axel and Bärbel visited us for several days in Fort Collins. I hope that we will have more opportunities now that Axel will have a more relaxed schedule.

2. HEME ISOMERISM

In this paper, we present some recent results we have obtained in our effort to understand the implications of heme isomerism for the CD of hemoproteins. Heme isomerism refers to the existence of two isomers of hemoproteins containing noncovalently bound protoheme IX, such as Mb, Hb and the cytochromes b. Three groups – those of Gerd LaMar and Klaus Gersonde (LaMar et al., 1978 a,b), Kurt Wüthrich (Keller et al., 1976) and Hans Rüterjans (Ribbing et al., 1978) – observed a doubling of the resonances of the heme, with each heme methyl or methine group giving rise to two peaks of unequal intensity. Two explanations for this doubling were proposed: protein isomerism resulting from cis-trans isomerism in a critical proline residue or from sequence variants, or heme isomerism resulting from the pseudo two-fold axis of protoheme IX (Figure 1). LaMar and co-workers (LaMar et al., 1978c) conclusively demonstrated that heme isomerism was the source of the methyl resonance doubling, using isotopically substituted protohemes. As shown in Figure 1, the α-γ axis is a pseudo two-fold axis relating the two equivalent propionate side chains of the porphyrin, which must be on the outside of the protein. Rotation about this axis also interchanges methyl and vinyl substituents on the opposite edge of the porphyrin ring. The protein discriminates between these methyls and vinyls imperfectly, with isomeric ratios for the native protein varying from ca. 10:1 for sperm whale Mb (LaMar et al., 1984) to ca. 1:1.5 for yellow fin tuna Mb (Levy et al., 1985). LaMar and coworkers have also shown (Jue et al., 1983) that the apoprotein cannot discriminate between these two kinds of substituents at all, so Mb freshly reconstituted from the apoprotein and protohemin IX has a 1:1 isomeric ratio.

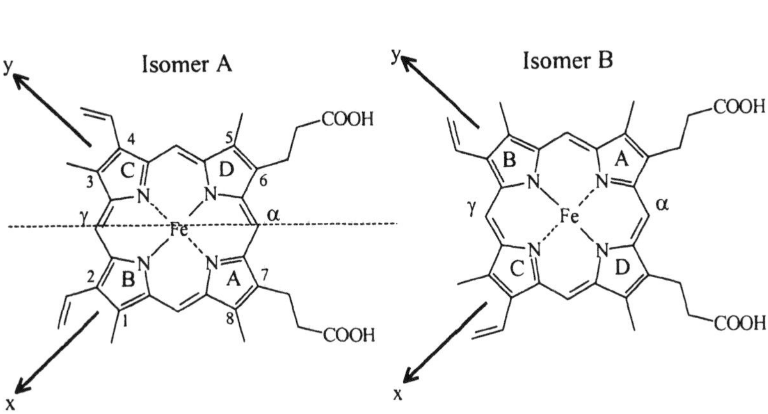

Figure 1. Orientation of protoheme IX in isomers A and B, differing by a 180° rotation of the heme about the α–γ methine carbon axis. This isomerization exchanges the position of the 2-vinyl with the 3-methyl and the 4-vinyl with the 1-methyl. The view is from the proximal side of the heme. In sperm-whale Mb, isomer A is the major form (~90%) and isomer B is the minor form (~10%).

Does heme isomerism have any physiological significance? Originally, LaMar's group (Livingston et al., 1984) reported that the two heme isomers of sperm-whale myoglobin have O_2 affinities that differ by an order of magnitude, with isomer B binding O_2 more strongly. However, two groups (Aojula et al., 1986; Light et al., 1987) have re-examined this question and found no difference in the O_2 affinity for the two isomers, beyond experimental error. Differences of possible physiological relevance have been reported for two other heme proteins, however. A difference of 27 mV in the reduction potential of the two isomers of cytochrome b_5 has been found (Walker et al., 1988), and the Bohr effect of *Chironomus thummi thummi* hemoglobin has been shown to be different for the two isomers (Gersonde et al., 1986).

In their study of the putative differences in O_2 affinity of the two sperm-whale Mb isomers, both Aojula et al. (1986) and Light et al. (1987) made a very interesting observation concerning the CD spectra of the heme isomers. They found that the Soret CD of the freshly reconstituted protein was only about half as intense as that of the native protein. Since freshly reconstituted Mb is a 1:1 mixture of the two isomers and the native protein is a 10:1 mixture, this implies that isomer B has a very weak Soret CD band, relative to that of isomer A. In a later study, Aojula et al. (1988) investigated this question quantitatively, measuring both the NMR spectrum and the CD spectrum of reconstituted Mb during the course of equilibration. A plot of the measured Soret CD for MbCO as a function of the heme isomer ratio based upon NMR showed a linear relationship from which the Soret CD of the pure A and B forms can readily be obtained by extrapolation. The A isomer has a strong positive Soret CD (+90 M^{-1} cm^{-1}), whereas the B isomer has $\Delta\varepsilon = -7$ M^{-1} cm^{-1} at the Soret maximum. Thus the two isomers give Soret CD bands of opposite sign.

Can the coupled-oscillator model of heme protein CD (Hsu and Woody, 1969, 1971) account for this large difference? If we consider the heme to be planar and neglect any changes in the protein conformation upon rotating the heme about the α-γ axis, the answer is No! First, it must be recalled that the Soret band of metalloporphyrins is doubly degenerate (or nearly so) (Platt, 1956) with the two Soret transitions polarized in the heme plane and perpendicular to each other. Since Ming-Chu and I had no *a priori* knowledge of the direction of these transition moments, we were concerned about how the absolute directions might affect our results. We showed (Hsu and Woody, 1971) that rotation of the transition moments in the plane has no effect on the net Soret rotational strengths resulting from the two components, although it does affect the distribution of the rotational strengths between the two Soret components. A simple rotation by 180° about the α-γ axis is equivalent to rotation of the Soret transition moments and thus should not affect the net Soret rotational strength, although it might affect the band shape.

Therefore, the large difference in the Soret CD for the two heme isomers requires that either non-planar distortions of the heme make significant contributions and differ for the two forms, or that there are significant differences in the heme-aromatic side chain interactions, probably due to changes in the relative position and orientation of the side chains. One hypothetical model (Moench, 1985) that could qualitatively account for the observations without requiring substantial rearrangements of the protein is as follows: contributions of heme non-planarity are comparable in magnitude to those of heme-aromatic coupling, and the heme adopts approximately enantiomeric conformations in the two isomers, whereas the protein conformation is largely unchanged. In isomer A, the non-planar heme contributions are positive, like those of the heme-aromatic interactions and give rise to the observed strong positive CD for this isomer. On the other hand, isomer B has negative contributions from the non-planar heme that slightly outweigh positive heme-aromatic coupling contributions and yield a weak negative Soret CD band!

3. METHODS

To test this and other possible explanations of the dramatic CD spectral differences between isomers A and B, we have undertaken CD calculations for the two heme isomers of sperm whale MbCO. For isomer A, high-resolution structures are available from x-ray diffraction (Yang and Phillips, 1996), neutron diffraction (Cheng and Schoenborn, 1991) and NMR (Ösapay et al., 1994). For isomer B, however, no structures are available since one can obtain at most a 50:50 mixture of isomer B with isomer A. We therefore have performed molecular dynamics simulations for both isomers A and B. In addition to providing a structure (actually many structures!) for isomer B, simulations on isomer A will provide an assessment of the effects of dynamics on the predicted CD of isomer A. This is an area in which

Axel Wollmer, Jörg Fleischhauer and I have long shared an interest and in which we have actively collaborated (Fleischhauer et al., 1994; Koslowski et al., 1996).

To construct a starting structure for dynamics on isomer B, we took the neutron diffraction structure of isomer A (PDB entry 2mb5, Cheng and Schoenborn, 1991), removed the heme, rotated it by 180° about the α-γ axis, then reinserted it in the heme pocket. The starting structures for isomers A and B were then subjected to energy minimization using the GROMOS96 force field (van Gunsteren et al., 1996), after immersion in a box of water molecules with NH_4^+, SO_4^{2-}, and Cl^- ions to assure electroneutrality. For each isomer, we carried our four simulations at 300K, each of 600 ps duration, with different sets of initial velocities. We also used two out of twelve NMR structures (Ösapay et al., 1994) as starting structures for a 600 ps simulation of isomer A and, after rotation of the heme, of isomer B.

For each of the structures along the trajectories (12,000 structures per trajectory), we calculated the intrinsic heme CD, using π-electron theory and treating the heme as a divinyl porphyrin dianion, i.e., neglecting the iron and the alkyl substituents but considering the vinyl groups as part of the π-electron system. The "standard" parameters of Weiss et al. (1965) were used and the 32 lowest energy excited configurations were used in the configuration interaction. The transition parameters for the heme were also used to calculate the coupling of the Soret transitions with the aromatic side chains and the peptide groups of the protein. The CD spectrum in the Soret region was then calculated for each structure along the trajectory, and these spectra were then averaged for the trajectory to calculate the average CD.

4. RESULTS AND DISCUSSION

4.1. Calculated Soret CD Spectra

Figure 2 compares the calculated spectra for the six trajectories (1-4 based on neutron diffraction, 5 and 6 on NMR structures) with one another and with experiment (thick line). First let's look at the spectra for isomer A. It can be seen that the four ND-based trajectories give predicted spectra that are similar in amplitude to one another and to experiment. The wavelengths for trajectories 1, 2 and 4 are also very similar, but that for trajectory 3 is somewhat red-shifted. The calculated curves are blue-shifted by about 30 nm relative to the experimental curve, but this is a discrepancy to be expected because of the neglect of the iron atom in the π-electron calculations. Finally, note that the two NMR structures give calculated curves that differ significantly in amplitude from the ND-based curves and from experiment. In the case of trajectory 6, differences in shape are predicted.

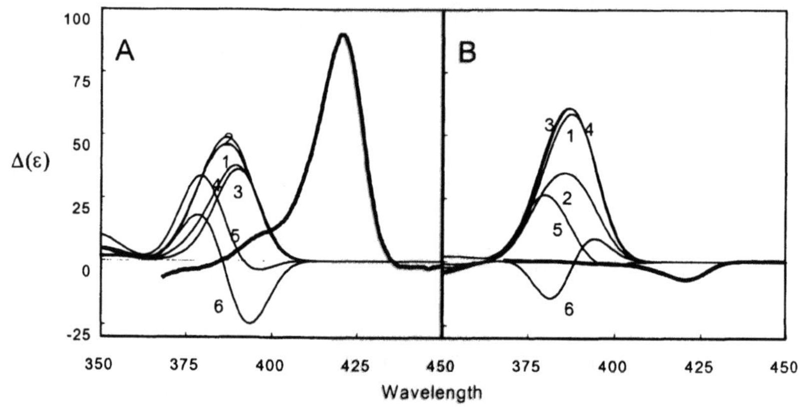

Figure 2. Soret CD spectra of MbCO, calculated from MD trajectories (thin curves), compared with CD spectra for isomers A and B based on experiment (thick curves). The curves numbered 1-4 for each isomer are based on the neutron diffraction structure (Cheng and Schoenborn, 1991). Curves 5 and 6 are based on two NMR-derived structures (Ösapay et al., 1994). The experimentally based spectrum for isomer A is calculated from the CD spectrum of the equilibrium mixture of MbCO isomers, multiplied by a factor such that $\Delta\varepsilon_{max}=+90\ M^{-1}\ cm^{-1}$, as reported for isomer A (Aojula et al., 1988). For isomer B, a factor such that $\Delta\varepsilon_{max}= -7\ M^{-1}\ cm^{-1}$ was used.

Turning to the results for isomer B, we find that the calculated CD curves for the ND-based trajectories, 1-4, are similar in amplitude to those for the ND-based trajectories of isomer A but not to the experimentally observed weak negative band. Again, as with isomer A, the NMR -based trajectories give a weaker Soret CD band and, in the case of trajectory 6, a positive couplet that contrasts with the negative couplet for trajectory 6 of isomer A.

Table 1. Soret Rotational Strength Contributions in CO-Myoglobin[a]

	Isomer A		Isomer B	
	Rotational Strength	Standard Deviation	Rotational Strength	Standard Deviation
Total	0.4973	0.0896	0.6780	0.1173
Heme	0.1464	0.0627	0.2506	0.0921
Aromatic	0.1984	0.0102	0.2984	0.0464
Peptide	0.1550	0.0357	0.1291	0.0202

[a] *Results are presented as averages and standard deviations over four 600 ps simulations based on the ND structure. Rotational strengths are given in Debye-Bohr magnetons.*

The different contributions to the Soret CD are shown in Table 1, averaged over the ND-based trajectories. For isomer A, we obtain an average of 0.50 DBM with a standard deviation of 0.09. This is to be compared with the experimental value of 0.7 DBM for the equilibrium mixture of isomers. About 40% of the rotational strength (0.20 DBM) comes from the coupling with the aromatic groups, with a standard deviation of 5% of the mean. The peptides contribute about 30% of the total, with a standard deviation of about 20% of the mean. Finally, the intrinsic heme contribution averages about 30% of the total, with a larger standard deviation, about 40% of the mean value.

For isomer B, the peptide-coupling contribution is comparable to that for A, with a comparable standard deviation. The aromatic-coupling contribution is 44% of the total, with a larger standard deviation than for isomer A. The intrinsic heme contribution is larger, with a standard deviation that is 37% of the mean. The total for isomer B (0.68 DBM) is somewhat larger than that for isomer A, with a standard deviation of about 15% of the mean, largely attributable to the spread in the intrinsic heme contributions. However, the predicted of a large positive rotational strength for isomer B disagrees with the small negative rotational strength observed for isomer B.

Overall, we can draw the following conclusions from the MD simulations:

First, the calculations for isomer A give very good agreement with experiment if the ND structure is used as the starting geometry. The intrinsic heme contributions are significant but only account for about 30% of the total. The aromatic-coupling contribution is the largest single factor. This supports the earlier work of Hsu and Woody (1969, 1970) and the results of Fleischhauer and Wollmer (1972). The significant contribution of heme coupling with peptide groups is consistent with a subsequent study of the Aachen group (Strassburger et al., 1978), which concluded that contributions of heme-peptide interactions are not negligible.

Second, the NMR structures do not provide a satisfactory starting geometry for the simulations. The two structures used lead to qualitatively different results, differing from one another, from the ND-based results, and from experiment.

Third, the simulations for isomer B have not succeeded in reproducing the observed CD behavior of this isomer. We attribute this failure to an unsatisfactory starting structure for isomer B. It appears that the protein structure for isomer B lies in a different local well of the conformational landscape, one that is not sampled to a significant extent by simulation at room temperature starting with the isomer A structure, at least on a subnanosecond time scale. There are several approaches that may provide better starting structures for isomer B: (1) simulations at higher temperatures, followed by annealing (Kirkpatrick et al., 1983); (2) simulations in which the methyl groups of isomer A are slowly grown into the vinyls of isomer B and the vinyls of A are shrunk into the methyls of isomer B (Beveridge and DiCapua, 1989; (3) simulations in four dimensions (van Schaik et al., 1993). The latter two are our preferred approaches, and we hope that one of them will provide a good starting structure for isomer B.

4.2. Heme Nonplanarity

Our interest in the intrinsic heme contribution to the CD spectra of hemoproteins has led to a closer examination of the effects of heme non-planarity. First, we considered the possible contributions of the vinyl group torsion to the heme CD. Ming-Chu Hsu (Hsu and Woody, 1971) considered this factor, as it was the only type of non-planar heme distortion for which information was available from the early x-ray diffraction studies of hemoproteins. We found no significant effect from the vinyl groups in our calculations on the static structures. In the present study, we find that there is no correlation between the intrinsic heme rotational strength and the torsional angles of either the 2- or 4-vinyl group. For the 2-vinyl group, there is a preference for torsional angles near the syn conformation, whereas the 4-vinyl group is predominantly anti. However, both Soret components show a strongly scattered pattern of rotational strengths with comparable probabilities of positive and negative rotational strengths at all vinyl torsional angles.

Next, we attempted to characterize non-planarity in the heme nucleus by considering an average twist angle determined by the bonds between the β carbons of opposite pyrroles. We averaged the dihedral angles between the C_β-C_β bonds of pyrrole rings A and C with those for rings B and D. Of course, for a planar porphyrin this twist angle is zero. It is also zero for several other types of deformation, for example doming of the heme. As shown in Figure 3, there is a good correlation between the rotational strength of the two Soret components, classified according to polarization direction, and the average twist angle. The correlation is similar for isomers A and B. Note that the individual structures along the trajectory give rise to rotational strengths for individual components as large as ±2 DBM. However, such a large rotational strength for one component is invariably associated with a large rotational strength of opposite sign for the other, and the net rotational strengths of the Soret band arising from heme distortions are relatively modest and show only a weak correlation with the average twist angle. Nevertheless, there is a distinct bias toward positive rotational strengths (Table 1).

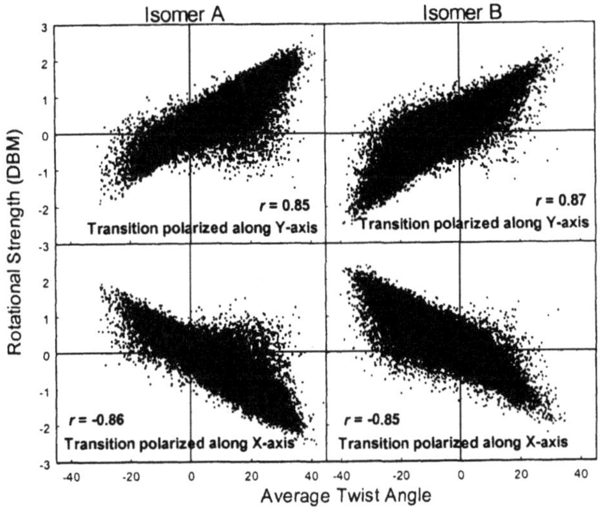

Figure 3. Correlation of the rotational strengths of the two Soret components, classified according to their polarization, with the average twist angle. The average twist angle is calculated from the virtual dihedral angles between the two sets of opposing pyrrole rings, τ_{AC} and τ_{BD}. The average twist angle is defined as $|\tau| = (|\tau_{AC}| + |\tau_{BD}|) / 2$, with the sign of τ_{AC}.

John Shelnutt and co-workers (Jentzen et al., 1995, 1996; Shelnutt et al., 1998) have developed a very attractive method for analyzing non-planar distortions in porphyrins, using the six lowest frequency normal modes for out-of-plane vibrations. The method is called normal-coordinate structural decomposition (NSD). Shelnutt and co-workers have shown that distortions along these normal modes can describe the out-of-plane deformations of the porphyrins in the several hundred hemoproteins for which high-resolution structures are available in the Protein Data Bank. This approach is illustrated in Figure 4 for the case of cytochrome c. The distorted porphyrin is shown at the top left and the total distortion (the total deviation from the mean-square plane) is just over 1Å. The six fundamental types of distortions (normal modes) are shown on the left. The distortions are called saddling, ruffling, doming, waving (x), waving (y) and propellering, in approximate order of increasing energy and decreasing amplitude for an unperturbed porphyrin. We see that the heme in cytochrome c is primarily deformed by ruffling, which by itself (solid bar) accounts for most of the observed deviations. There are also small amplitudes of saddling and waving, but these are in opposite directions and make only a small net contribution.

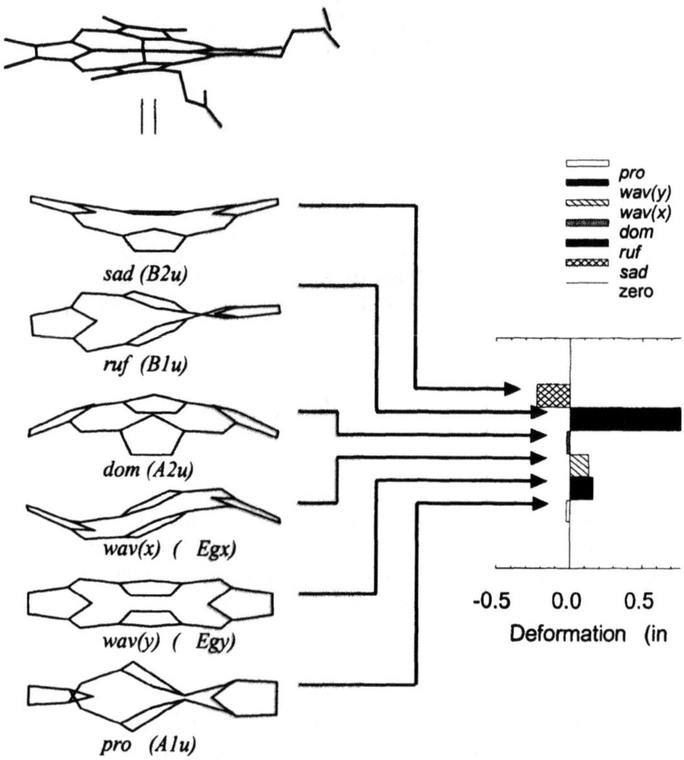

Figure 4. Types of heme distortions and their contribution to the description of the deformation of the heme in horse-heart cytochrome c (PDB file 1hrc, Bushnell et al., 1990). The heme structure is shown at the top left, with the deformation modes shown below. On the right, deformations are quantitated, with the modes indicated by shading as shown at the top right.

In the present work, the average structure over various time periods for one A and one B trajectory has been analyzed by NSD. In addition, the individual structures over a 5-ps interval have been analyzed for a trajectory of isomer A. (The code for the NSD analysis program requires a manual start for each structure studied, and so it is far too labor intensive to examine a full trajectory point by point!)

The analysis of the average structures is shown in Figure 5. At the bottom right, the 600-ps average for isomer A shows modest deformations in all six normal modes, with the largest for wav (y), ruffling and propellering. Examination of each 100-ps interval shows that larger deformations occur in the various segments, but these vary in direction over the whole trajectory, thus giving rise to the smaller global average.

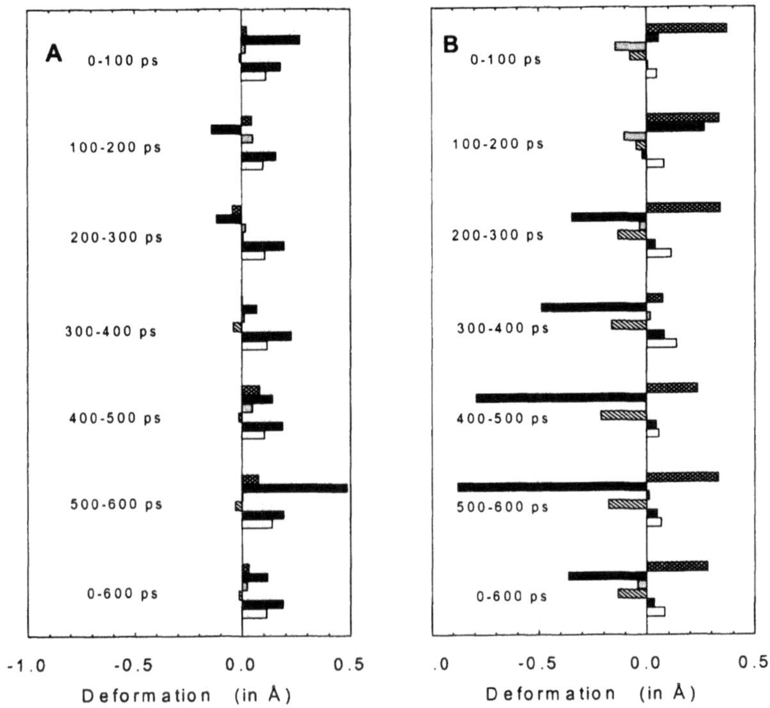

Figure 5. Out-of-plane deformations of the heme in trajectories for isomers A (left) and B (right), based upon the neutron diffraction structure (Cheng and Schoenborn, 1990). The analyses were performed for structures averaged over 100 ps intervals and, at the bottom, the entire 600 ps interval. The identities of the modes are indicated by shading, according to the key in Figure 4.

For isomer B, the deformations of the average structure over the entire trajectory are larger than for isomer A, with ruffling being dominant, followed in importance by saddling and wav (x). The 100-ps interval average structures for isomer B show some variation in time, e.g., the sign of the ruffling deformation reverses between the second and third 100-ps interval, and the doming deformation nearly vanishes in the fifth 100-ps interval, but the most striking effect is that the ruffling deformation increases strongly in the last half of the simulation.

ruf	→	-ruf
sad	→	-sad
dom	→	-dom
wav(x)	→	-wav(y)
wav(y)	→	-wav(x)
pro	→	-pro

Figure 6. Transformation properties of the out-of-plane normal modes describing heme deformations. The modes on the left transform to the modes on the right when the heme is rotated by 180 °about the α–γ methine carbon axis.

It can be seen from Figure 5 that the time-averaged structures for isomers A and B frequently have the opposite signs of deformation, especially the prominent ruffling distortion that, as will be seen, is especially important for CD. However, in comparing the deformations for isomers A and B, one must take into account the effect of the two-fold rotation about the α-γ axis that interconverts the two forms. When this is considered, the rules for transformation are given in Figure 6. We see then that, in fact, the structure averaged over the full 600 ps has qualitatively the same deformations for each normal mode. This extends to most of the structures averaged over 100-ps intervals as well. Thus, it appears that the time-average deformations imposed on the heme by the protein are qualitatively the same for isomers A and B. This is consistent with the result noted earlier, that the net intrinsic rotational strength for the heme transitions is positive for both isomers. This may be an artifact of the present calculations for isomer B, however. If we can get into the right region of conformational space for isomer B in further simulations, the time-average conformations may be different and the intrinsic heme contribution to the Soret rotational strength may have a net negative value when averaged over the trajectory.

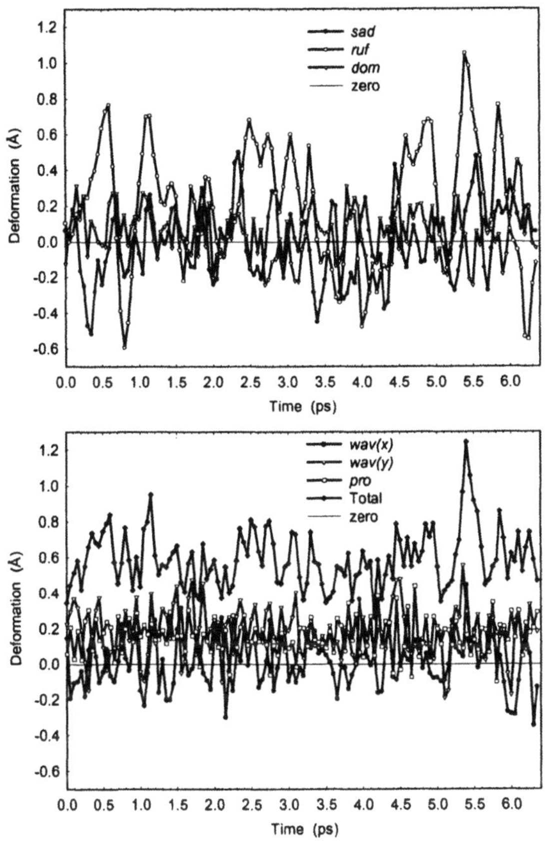

Figure 7. Time course of the individual deformation modes for the initial 6.4 ps of an MD simulation for isomer A, starting from the neutron diffraction structure. Top: sad, ruf, and dom deformations; Bottom, wav(x), wav(y), pro and total deformations.

When we examine the individual structures, we find rapid fluctuations in all of the deformations, as seen in Figure 7 for the 1-6 ps interval of a trajectory for isomer A. Each of the deformations reverses sign a number of times over the 5-ps interval. A Fourier analysis of these time series has been performed. The most conspicuous feature from this analysis is a peak in the ruffling deformation with a frequency of ca. 50 cm^{-1}, comparable to the frequency predicted from a vibrational analysis (Jentzen et al., 1997) of Cu-porphine (88 cm^{-1}). However, none of these deformations give simple vibrational patterns because of interactions with the heme pocket.

We have sought correlations between the normal coordinates and the average twist angle discussed above, and between the normal coordinates and the sign of the intrinsic rotational strengths of the Soret components. Only the ruffling component correlates with these other parameters. It is therefore not surprising that the x- and y-components of the Soret band correlate well with the ruffling distortion and none of the other types of distortion.

In summary, the NSD analysis of the MD trajectory has revealed two important conclusions:

- Ruffling distortions are the most important contributors to the inherent CD of porphyrins, and
- The distortions of the heme stabilized by the protein are qualitatively the same for isomers A and B, as emerges from the simulations thus far. However, it is very possible (perhaps even likely) that simulations using a more suitable starting structure for isomer B will reveal qualitative differences.

5. REFERENCES

Aojula H.S., Wilson M.T. and Drake, A. "Characterization of haem disorder by circular dichroism." *Biochem. J.* 237 (1986): 613-616.

Aojula H.S., Wilson M.T., Moore G.R. and Williamson D.J. "^{1}H-NMR and CD studies of haem orientational disorder in sperm-whale myoglobin and human haemoglobin." *Biochem. J.* 250 (1988): 853-858.

Beveridge D.L. and DiCapua F.M. "Free energy via molecular simulation: Application to chemical and biochemical systems." *Ann. Rev. Biophys. Biophys. Chem.* 18 (1989): 431-492.

Bushnell G.W., Louie G.V. and Brayer G.D. "High-resolution three-dimensional structure of horse heart cytochrome c." *J. Mol. Biol.* 214 (1990): 585-595.

Cheng X. and Schoenborn B.P. "Neutron diffraction study of carbonmonoxy-myoglobin." *J. Mol. Biol.* 220 (1991): 381-399.

Fleischhauer J. and Wollmer A. "Influence of aromatic amino-acid side chains on sign of Soret Cotton effect in Chironomus hemoglobin (CTTIII)." *Z. Naturforsch. B* 27 (1972): 530- 532.

Fleischhauer J., Grötzinger J., Kramer B., Krüger P., Wollmer A., Woody R.W. and Zobel E. "Calculation of the CD spectrum of cyclo (L-Tyr-L-Tyr) based on a molecular dynamics simulation." *Biophys. Chem.* 49 (1994): 141-152.

Formanek H. and Engel J. "Optical rotatory dispersion of a respiratory hemeprotein of Chironomus thummi." *Biochim. Biophys. Acta* 160 (1968): 151-158.

Gersonde K., Sick H., Overkamp M., Smith K.M. and Parish D.W. "Bohr effect in monomeric insect haemoglobins controlled by O_2 off-rate and modulated by haem-rotational disorder." *Eur. J. Biochem.* 157 (1986): 393-404.

Hsu M.-C. and Woody R.W. "Origin of the rotational strength of heme transitions in myoglobin." *J. Am. Chem. Soc.* 91 (1969): 3679-3681.

Hsu M.-C. and Woody R.W. "The origin of the heme Cotton effects in myoglobin and hemoglobin." *J. Am. Chem. Soc.* 93 (1971): 3515-3525.

Jentzen W., Simpson M.C., Hobbs J.D., Song X.-Z., Ema T., Nelson N.Y., Medforth C.J., Smith K.M., Veyrat M., Mazzanti M., Ramasseul R., Marchon J.-C., Takeuchi T., Goddard, W.A. III and Shellnut J.A. "Ruffling in a series of Ni(II) *meso*-tetrasubstituted porphyrins as a model for the conserved ruffling of the heme of cytochromes *c*." *J. Am. Chem. Soc.* 117 (1995): 11085-11097.

Jentzen W., Song X.-Z. and Shelnutt J.A. "Structural characterization of synthetic and protein-bound porphyrins in terms of the lowest-frequency normal coordinates of the macrocycle." *J. Phys. Chem. B* 101 (1997): 1684-1699.

Jue T., Krishnamoorthi R. and LaMar G.N. "Proton NMR study of the mechanism of heme-apoprotein reaction for myoglobin." *J. Am. Chem. Soc.* 105 (1983): 5701-5703.

Keller R., Groudinsky O. and Wüthrich K. "Contact-shifted resonances in the ^{1}H NMR spectra of cytochrome b_5. Resonance identification and spin density distribution on the heme group." *Biochim. Biophys. Acta* 427 (1976): 497-511.

Kirkpatrick S., Gelatt C.D., Jr. and Vecchi M.P. "Optimization by simulated annealing." *Science* 220 (1983): 671-680.

Koslowski A., Botterweck H., Fleischhauer J., Kurapkat G., Wollmer A. and Woody R.W. "Calculation of the CD spectrum of ribonuclease." *Progr. Biophys. Mol. Biol.* 65, Suppl. 1 (1996): 43.

LaMar G.N., Overkamp M., Sick H. and Gersonde K. "Proton nuclear magnetic resonance hyperfine shifts as indicators of tertiary structural changes accompanying the Bohr effect on monomeric insect hemoglobins." *Biochemistry* 17 (1978a): 352-361.

LaMar G.N., Budd D.L., Viscio D.B., Smith K.M., and Langry K.C. "Proton nuclear magnetic-resonance characterization of heme disorder in hemoproteins" *Proc. Natl. Acad. Sci. USA* 75 (1978b): 5755-5759.

LaMar G.N., Davis N.L., Parish D.W. and Smith K.M. "Heme orientational disorder in reconstituted and native sperm whale myoglobin." *J. Mol. Biol.* 168 (1983): 887-896.

LaMar G.N., Toi H. and Krishnamoorthi R. "Proton NMR investigation of the rate and mechanism of heme rotation in sperm whale myoglobin: evidence for intramolecular reorientation about a heme twofold axis." *J. Am. Chem. Soc.* 106 (1984): 6395-6401.

Levy M.J., LaMar G.N., Jue T., Smith K.M., Pandey R.K., Smith W.S., Livingston D.J. and Brown W.D. "Proton NMR study of yellowfin tuna myoglobin in whole muscle and solution." *J. Biol. Chem.* 260 (1985): 13694-13698.

Light W.R., Rohlf R.J., Palmer G., and Olson J.S. "Functional effects of heme orientational disorder in sperm whale myoglobin." *J. Biol. Chem.* 262 (1987): 46-52.

Livingston D.J., Davis N.L., LaMar G.N., and Brown W.D. "Influence of heme orientation on oxygen affinity in native sperm whale myoglobin." *J. Am. Chem. Soc.* 106 (1984): 3025-3026.

Moench S.J. "CD and ^{1}H NMR studies of native hemeproteins and hemeproteins containing modified hemes." (1985) Ph.D. Thesis, Colorado State University.

Ösapay K., Theriault Y., Wright P.E. and Case D. "Solution structure of carbonmonoxy myoglobin determined from nuclear magnetic resonance distance and chemical shift constraints." *J. Mol. Biol.* 244 (1994): 183-197.

Platt J.R. "Electronic structure and excitation of polyenes and porphyrins." in *Radiation Biology*, Hollaender, A., Ed. McGraw-Hill, New York, vol. III (1956): pp. 71-123.

Ribbing W., Krümpelmann D. and Rüterjans H. "Isomeric incorporation of haem into two monomeric haemoglobins of *Chironomus thummi thummi.*" *FEBS Lett* 92 (1978): 105-108.

Shelnutt J.A., Song X.-Z., Ma J.-G., Jia S.L., Jentzen W. and Medforth C.J. "Nonplanar porphyrins and their significance in proteins." *Chem. Soc. Rev.* 27 (1998): 31-41.

Strassburger W., Wollmer A., Thiele H., Fleischhauer J., Steigemann W. and Weber E. "Calculation of the circular dichroism of Chironomus hemoglobin in the light of the quality of its X-ray structure." *Z. Naturforsch. C* 33 (1978): 908-911.

van Gunsteren W.F., Billeter S.R., Eising A.A., Hünenberger P.H., Krüger P., Mark A.E., Scott W.R.R. and Tironi I.G. *Biomolecular Simulations: The Gromos96 Manual and User Guide*, Zürich, Hochschulverlag AG an der ETH (1996).

van Schaik R.C., Berendsen H.J.C., Torda A.E. and van Gunsteren W.F. "A structural refinement method based on molecular dynamics in four spatial dimensions." *J. Mol. Biol.* 234 (1993): 751-762.

Walker F.A., Emrick D., Rivera J.E., Hanquet B.J., and Buttlaire D.H. "Effect of heme orientation on the reduction potential of cytochrome b_5." *J. Am. Chem. Soc.* 110 (1988): 6234-6240.

Weiss C., Kobayashi H., and Gouterman M. "Spectra of porphyrins. Part III. Self-consistent molecular orbital calculations of porphyrin and related ring systems." *J. Mol. Spectrosc.* 16 (1965): 415-450.

Wollmer A., Sick H., and Gersonde K. "Conformationally induced circular dichroism in hemoglobins of Chironomus thummi thummi." *Eur. J. Biochem.* 16 (1970): 303- 312.

Yang F. and Phillips G.N., Jr. "Crystal structures of CO-, deoxy-, and met-myoglobins at various pH values." *J. Mol. Biol.* 271 (1996): 762-774.

7. ACKNOWLEDGMENTS

Support of grants from NIH (GM22994) and the Pittsburgh Supercomputer Center (MCB980029P) are gratefully acknowledged. Sandia is a multiprogram laboratory operated by Sandia corporation, a Lockheed Martin Company, for the United States Department of Energy under Contract DE-AcC04-94AL85000. This study was initiated when R.W.W. was on sabbatical in the Laboratory of Physical Chemistry, University of Groningen,. R.W.W. sincerely thanks Wilfred van Gunsteren for his guidance and for many helpful discussions, and both Wilfred and Herman Berendsen for their warm hospitality during his sabbatical. Support for the sabbatical was generously provided by a Fogarty Senior International Fellowship (F06 TW01417) and Colorado State University.

Author Index

Keywords

MIX
Papier aus verantwortungsvollen Quellen
Paper from responsible sources
FSC® C105338

If you have any concerns about our products,
you can contact us on
ProductSafety@springernature.com

In case Publisher is established outside the EU,
the EU authorized representative is:
Springer Nature Customer Service Center GmbH
Europaplatz 3, 69115 Heidelberg, Germany

Printed by Libri Plureos GmbH
in Hamburg, Germany